Principles of Computer-Integrated Manufacturing

S. KANT VAJPAYEE
University of Southern Mississippi

PRENTICE HALL
Englewood Cliffs, New Jersey   •   Columbus, Ohio

Library of Congress Cataloging-in-Publication Data
Vajpayee, S. Kant.
    Principles of computer-integrated manufacturing / S. Kant Vajpayee.
      p. cm.
    Includes bibliographical references and index.
    ISBN 0-02-422241-0
    1. Computer integrated manufacturing systems. I. Title.
  TS155.63.V34   1995
  670'.285—dc20

94-20412
CIP

Cover photo: Courtesy of Mazak Corporation, Florence, Kentucky
Editor: Steven Helba
Production Editor: Sheryl Glicker Langner
Text Designer: Jill E. Bonar
Cover Designer: Cathleen Carbery
Production Buyer: Pamela D. Bennett
Electronic Text Management: Marilyn Wilson Phelps, Matthew Williams,
    Jane Lopez, Karen L. Bretz

This book was set in Dutch 801 by Prentice Hall and was printed and bound by R. R. Donnelley & Sons Co. The cover was printed by Phoenix Color Corp.

© 1995 by Prentice-Hall, Inc.
A Simon & Schuster Company
Englewood Cliffs, New Jersey 07632

Printed in the United States of America

10 9 8 7 6 5 4 3 2 1

ISBN: 0-02-422241-0

Prentice-Hall International (UK) Limited, *London*
Prentice-Hall of Australia Pty. Limited, *Sydney*
Prentice-Hall of Canada, Inc., *Toronto*
Prentice-Hall Hispanoamericana, S. A., *Mexico*
Prentice-Hall of India Private Limited, *New Delhi*
Prentice-Hall of Japan, Inc., *Tokyo*
Simon & Schuster Asia Pte. Ltd., *Singapore*
Editora Prentice-Hall do Brasil, Ltda., *Rio de Janeiro*

◻ DEDICATION

To my "mama" (grandmother) who is no more,
but whose struggle with life,
based on dignity of hard work and honesty,
has been rewarded with her great grandchildren's
enormously enhanced quality of life
in a land she might not have dreamed of.

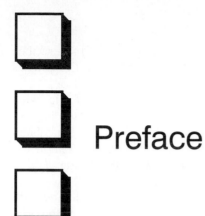

Preface

Including its indirect contribution to the economy, manufacturing accounts for more than half the wealth created in industrialized nations. Manufacturing is undergoing enormous changes, primarily due to the cost-effectiveness of computers and their recent association with communications technology. These changes are likely to render manufacturing facilities of the 21st century almost as clean as a hospital, and as organized as a theater show where things rarely go wrong. The current evolution in the entire process of making and marketing a product can be summed up by the term *computer-integrated manufacturing (CIM)*.

This book presents the basic principles of CIM and highlights the interactions among its elements. It has been developed as a text for a one-semester course on CIM for engineering technology and industrial technology students. It is equally suitable at the undergraduate or graduate level for students majoring or minoring in manufacturing, mechanical, or industrial engineering. Two-year programs in these majors should also find the text appropriate for advanced courses in CIM. Business majors specializing in manufacturing or production management will also find it useful.

The text is organized into five parts. Part I, comprising the first two chapters, explains the meaning of CIM and provides an overview of discrete manufacturing. Part II, Chapters 3, 4, and 5, discusses the principles behind the technologies of computers, communications, and databases, which I have elected to call the "brain, nerves, and heart" of CIM. Since these technologies represent the cornerstones of CIM, their fundamentals have been emphasized.

The various manufacturing technologies and related systems integrated to achieve CIM are presented in Part III (Chapters 6 through 11). The topics have been arranged in the sequence in which various activities of discrete manufacturing take place.

Success with CIM depends on a management team that is forward looking, committed, and somewhat adventurous, and on a workforce that is skilled, well-trained, and excited about what is happening at the shop floor. These two elements of CIM are covered in Part IV (Chapters 12 and 13).

Part V, an epilogue, comprises the last two chapters. Chapter 14 describes emerging technologies that are likely to influence CIM, while Chapter 15 discusses the ramifications of CIM.

Besides books, journals, and magazines, there are other resources for learning about CIM; these are presented in Appendix A. For instructors interested in reinforcing their lectures through hands-on experience, a laboratory course has been suggested in Appendix B. The experiments have been designed to highlight the interactions that come into play among various functions of manufacturing under a CIM environment.

The text has several features. Each of the five parts has a preamble and each chapter begins with introductory remarks. Wherever necessary, the text includes an historical account of developments. Boxes throughout the chapters highlight examples and describe industry experiences. Each chapter ends with a brief list of trends, a summary, key terms, exercises (including projects), and a list of suggested readings. Key terms are also explained in the glossary at the end of the text. Suggested readings have been classified into books, monographs and reports, journals and periodicals, and articles.

The text is supported by an instructor's manual that contains a suggested course outline and numerous exercises for tests and examinations. Feedback from instructors adopting the book and from readers, especially students, is welcome for improving the future editions.

ACKNOWLEDGMENTS

I appreciate the input of reviewers for this book: Marvin D. Simon, University of Dayton; Meenakshi Sundaram, Tennessee Technological University; R. Scott Zitek, Lorain County Community College; David Ings, Oregon Institute of Technology; Aaron C. Clark, Chesapeake College; William Cullins, Aims Community College; Michael C. Hall, Indiana Vocational Technical College; David Kirwin, Central Ohio Technical College; Robert E. Speckert, Miami University; M. Reza Ziai, University of Southwestern Louisiana; Mary Berreth, Rochester Community College; Phillip Foster, University of North Texas; Terry Richardson, Northern State University; and Boyd Larson, University of Wisconsin. Their comments and suggestions have been invaluable. I also appreciate the support I've been given from the staff at Prentice Hall, especially my editor, Steve Helba.

I must thank Paroo (my better half) and our children Dipti, Tripti, Archna, Nishant, and Arpna for their continual interest in my efforts toward this text.

S. Kant Vajpayee

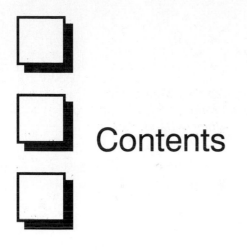

Contents

☐ PART II
Brain, Nerves, and Heart of CIM   57

3
Computer Technology and Manufacturing   59

4
Fundamentals of Communications   113

5
Database   171

❏ PART III
Technology and Systems   201

6
Product Design   203

7
Production Planning 235

8
Production 255

9
Shop-floor Control 295

10
Robotics and Material Handling   325

11
Quality   339

15
Ramifications of CIM

APPENDIX A
Other Resources

APPENDIX B
A Laboratory Course

PART I
Prologue

S ince the Industrial Revolution, mechanization and dedicated automation have influenced the practice of manufacturing. In the second half of the twentieth century, however, another technology—soft automation, based primarily on computers and microprocessors—has been having an enormous impact on manufacturing. Soft automation technology has eventually led to computer-integrated manufacturing (CIM). Part I of the text (Chapters 1 and 2) introduces CIM and provides an overview of manufacturing.

Chapter 1 explains the meaning and scope of CIM and discusses its evolution. The concept of CIM, as proposed by CASA/SME (Computer and Automated Systems Association of the Society of Manufacturing Engineers) and embodied in what has been called the CIM wheel, is presented. The difference between CIM I (what is currently practiced) and CIM II (what is required) is explained.

Chapter 2 presents an overview of manufacturing to highlight the large number of issues CIM involves. Since CIM is both challenging and interdisciplinary, it has recently been attracting professionals from fields other than manufacturing, such as computer science, electronics, communications, and business administration. Chapter 2 has been developed specifically for readers who are interested in CIM, but lack the background to appreciate the intricacies of manufacturing. This chapter stresses that (a) we live in a world full of manufactured goods, (b) manufacturing, like any other industrial activity, produces goods for the satisfaction of consumers, and (c) the ultimate aim of manufacturing is profit (in the case of a free-market economy) or the best possible product for a given cost (in the case of a socialistic economy).

Each chapter end includes the following:

- A section on *trends* describing current developments,
- A *summary* to capsulate the material covered,
- A list of *key terms* that are defined in the glossary,
- A set of *exercises* including projects, and
- A list of *suggested readings*.

CHAPTER 1
The Meaning and Scope of CIM

"Computer-integrated manufacturing is contagious."

—Joseph Harrington

"CIM is an amorphous beast. It will be different in every company."
—Leo Roth Klein, Manufacturing Control Systems, Inc.

"It has been called a strategy, a product, a direction, and a vision. It has been the subject of thousands of books, articles, speeches and conferences. Manufacturers have invested billions of dollars in it. Yet nobody can agree on what 'it' is."
—"In Search of CIM," *ASKhorizons*, fall 1989, p. 7

"The term computer-integrated manufacturing does not mean an automated factory."
—Joseph Harrington

"CIM is not applying computers to the design of the products of the company. That is computer-aided design (CAD)! It is not using them as tools for part and assembly analysis. That is computer-aided engineering (CAE)! It is not using computers to aid in the development of part programs to drive machine tools. That is computer-aided manufacturing (CAM)! It is not materials requirement planning (MRP) or just-in-time (JIT) or any other method for developing the production schedule. It is not automated identification, data collection, or data acquisition. It is not simulation or modeling of any materials handling or robots or anything else like that. Taken by themselves, they are the application of computer technology to the process of manufacturing. But taken by themselves they only create the islands of automation."
—Leo Roth Klein, Manufacturing Control Systems, Inc.

"A forum is needed to get out the horror stories that have occurred in some CIM implementations. This will allow people to realize that they are not alone and it is not their own personal failure. There is a need to recognize that we are dealing with a problem that is bigger than any individual. There is a need to document successes as well as failures."
—*CIM Integration Tools* (based on a roundtable discussion),
SME Blue Book series, p. 17

Computer-integrated manufacturing (**CIM**) is a broad term covering all technologies and **soft automation** used to manage the resources for cost-effective production of tangible goods. Chapter 1 introduces CIM by explaining its meaning and scope and presenting an historical account of its development and current potential.

1.1   INTRODUCTION TO CIM

The term CIM comprises three words—computer, integrated, and manufacturing. Though all three words are equally significant, the first two are secondary—merely adjectives modifying the last one (manufacturing). CIM is thus the application of computers in manufacturing in an integrated way. All types of computers, from **personal computers** (**PCs**) to mainframes, may be used in CIM.

The middle term, integrated, in CIM is very appropriate. It brings home the point that integration of all the resources—capital, human, technology, and equipment—is vital to success in manufacturing. Implicitly, CIM discourages any haphazard application of computers, and other technologies, that results in isolated **islands of automation**. Integration is achieved through timely and effective communication, which CIM relies on heavily. Since the computer is the basis of integration, communication within the context of CIM is strongly computer-oriented.

Although computers and computer communications have been with us since the 1950s, CIM is relatively new. It began to draw attention only in the 1980s. Why this late? For two reasons. First, until recently computers had been too expensive to be cost-effective in manufacturing. Only business functions, such as accounting and payroll, and to some extent inventory management, could justify the high costs. The low cost and improved capabilities of today's computer systems have changed that. The second reason for the delayed "birth" of CIM and its slow progress is the sheer complexity of integration, arising from the large number of tasks that interact in **discrete manufacturing** in today's sophisticated market.

Integrated manufacturing by itself is not a new concept. But CIM—which orchestrates the factors of production and its management—is. CIM is an umbrella term under which all functions of manufacturing and associated acronyms, such as **computer-aided design** and **computer-aided manufacturing** (**CAD/CAM**), **flexible manufacturing system** (**FMS**), and **computer-aided process planning** (**CAPP**) find a place.

Discrete manufacturing has always presented a challenge because of the large number of factors involved and their interaction. CIM is being projected as a panacea for this type of industry, which produces 40% of all goods. Process industries, where volume is high enough to justify **hard** or **dedicated automation**, may also benefit from CIM.

Definition of CIM

CIM means exactly what it says: computer-integrated manufacturing. It describes integrated applications of computers in manufacturing. A number of observers have attempted to refine its meaning:

One needs to think of CIM as a computer system in which the peripherals, instead of being printers, plotters, terminals, and memory disks, are robots, machine tools, and other processing equipment. It is a little noisier and a little messier, but it's basically a computer system.

—Joel Goldhar, dean, Illinois Institute of Technology

CIM is a management philosophy, not a turnkey computer product. It is a philosophy crucial to the survival of most manufacturers because it provides the levels of product design and production control and shop flexibility to compete in future domestic and international markets.

—Dan Appleton, president, DACOM, Inc.

CIM is an opportunity for realigning your two most fundamental resources: people and technology. CIM is a lot more than the integration of mechanical, electrical, and even informational systems. It's an understanding of the new way to manage.

—Charles Savage, president, Savage Associates

CIM is nothing but a data management and networking problem.

—Jack Conaway, CIM marketing manager, DEC

The preceding comments on CIM have different emphases. For example, Goldhar considers CIM a computer system, whereas both Appleton and Savage see it as a management objective. In Conway's view, CIM is data management and communications. Although these individuals view CIM differently, the underlying message is the same: orchestrated use of the various resources improves productivity and quality.

An attempt to define CIM is analogous to a group of blind persons trying to describe an elephant by touching it; each has a different description depending upon the body part touched. Nevertheless, several definitions of CIM have been attempted. The one put forward by Shrensker (1990) for the **Computer and Automated Systems Association of the Society of Manufacturing Engineers (CASA/SME)** is perhaps the most appropriate. According to him, "CIM is the integration of the total manufacturing enterprise through the use of integrated systems and data communications coupled with new managerial philosophies that improve organizational and personnel efficiency."

CIM Wheel

CASA/SME has suggested a framework, the **CIM wheel**, to elucidate the meaning of CIM. Formed by SME in 1975, CASA is an interest group of manufacturing professionals. The CIM wheel, developed by CASA/SME's Technical Council, is shown in Figure 1.1. It depicts a central core (integrated systems architecture) that handles the common manufacturing data and is concerned with information resource management and communications. The radial sectors surrounding the core (wheel hub) represent the various activities of manufacturing, such as design, material processing, and inspection. These activities have been grouped under three categories—manufacturing planning and control, product/process, and factory automation—as

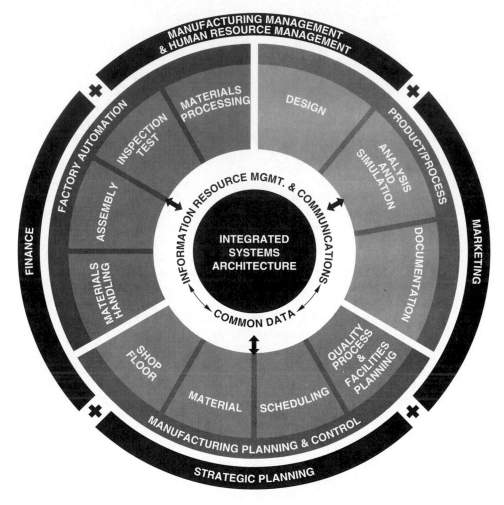

Figure 1.1

CASA/SME's CIM wheel—an embodiment of the concept of computer-integrated manufacturing

Source: Courtesy of the Society of Manufacturing Engineers. The CIM "Wheel" developed by the Technical Society of Manufacturing Engineers' Computer and Automated Systems Association (CASA/SME). Copyright © 1985 SME, second edition, revised November 5, 1985. Reproduction of this CIM Wheel is from the Society of Manufacturing Engineers (SME).

depicted in the wheel's inner rim. The outer rim represents the upper management functions, grouped into four categories: strategic planning, marketing, manufacturing and human resource management, and finance.

The CIM wheel depicted in Figure 1.1 is the expanded version of an earlier model. The outer rim was added in 1985 to emphasize the need of including both management and technology functions within the scope of CIM. As the wheel illustrates, CIM is broad enough to encompass all aspects of the manufacturing enterprise and its management, including those of personnel and finance.

Concept or Technology?

Is CIM a concept or a technology? This is a common question. CIM is a concept, an environment, an objective, a strategy. However, the aim of operating in a CIM environment cannot be achieved without modern technology. From this viewpoint, CIM can also be thought of as a technology. For upper management, CIM is a concept, a blueprint for success; for middle management and line managers, it is a technology, a physical realization of resources that are more capable and flexible. CIM can help upper management improve long-term strategic planning while it helps those who operate the plant to do so more efficiently.

"Some people view CIM as a concept, while others merely as a technology. It is actually both. A good analogy of CIM is *man*, for what we mean by the word man presupposes both the mind and the body. Similarly, CIM represents both the concept and the technology. The concept leads to the technology which, in turn, broadens the concept," according to Vajpayee and Reiden (1991).

1.2   EVOLUTION OF CIM

CIM has been evolving since the mid-1970s; however, until 1980 it was merely a concept. The 1980s, especially the second half, saw CIM expand into a technology. By now, industry has realized that CIM is a necessity rather than a luxury.

Computer-integrated manufacturing continues to evolve so that any claim that a "true" CIM plant exists is debatable. Progress in this direction has been phenomenal, however, and several full-blown CIM plants will probably be operating by the turn of the century. Today, numerous companies market an array of products that, when put together intelligently, can convert an average manufacturing facility into a CIM operation.

Primary factors that have led to the development of the CIM concept and associated technologies include the following:

1. Development of **numerical control (NC)**
2. The advent and cost-effectiveness of computers
3. Manufacturing challenges, such as global competition, high labor cost, regulations, product liability, and demand for quality products
4. The capability-to-cost attractiveness of microcomputers.

Development of Numerical Control

Numerical control is a technique for controlling machine tools and other equipment using symbols that include numbers. Its origin and development are interesting.

In 1947, the U.S. Air Force faced a machining problem concerning turbine blades. Parsons Corp. took the challenge of developing a new method for moving the tool, using numbers, to achieve the desired accuracy. Parsons worked with the Massachusetts Institute of Technology to produce the first NC milling machine in 1954. Since then, NC has developed, along with microelectronics and computers, into a sophisticated technology that drives modern machining centers, **flexible machining cells (FMCs)**, and FMSs. Chapter 8 covers numerical control in greater depth.

Computers

While NC was evolving on its own, computers were being developed independently. Of the various factors behind the evolution of CIM, the advent of computers is undoubtedly most important. A computer is an effective tool for carrying out repetitive tasks. Moreover, it can store large amounts of data and handle various types of mathematical analyses. Manufacturing involves these three tasks.

Initially, computers were introduced in manufacturing to support business functions such as payroll and accounting. These were then common applications of the computer in service industries. Next, computers were extended to inventory control. Further inroads into manufacturing had to wait until computer graphics matured as a technology. Today, with enhanced capabilities and lower costs, computers represent an attractive technology.

Early applications of computers in administrative and financial functions of manufacturing were feasible due to their cost-effectiveness. These functions involve well-defined, repetitive data-handling tasks that are suitable for computers. In the 1950s and 1960s, however, no one could foresee that computers would be integrated with production functions.

As expected, business computers accelerated the "mushrooming" of data processing departments reporting to various management functions, especially finance. Over the years, these departments became so large and specialized that the idea of computers at the shop floor seemed unlikely. Today, plant computers are a reality. In fact, in smarter companies, the plant's computing needs are affecting the choice of business computers. Chapter 3 has been devoted entirely to computers, since they represent the "brain" of CIM.

Computer-Aided Design (CAD)

A product must be designed before it can be manufactured. Many manufacturing companies may seem to be engaged in production only, with no visible design department. A closer look, however, would reveal that an outside designer is involved.

Design involves both creative and repetitive tasks. The repetitive tasks within design are very appropriate for computerization. In the early 1970s, using computers for tasks such as generating drawings and documentation proved cost-effective. Since then, computer use in design has exploded, especially with advances in computer graphics. The computer's help in design and drafting is commonly expressed by the term **computer-aided design (CAD)**. Some people use the term **CADD (computer-aided design and drafting)** to emphasize the drafting task. CAD is extremely important; with products conceived at the design stage, CAD is in fact the cornerstone of CIM. In recognition of this, a separate chapter on this topic (Chapter 6) has been included in the text.

Computer-Aided Manufacturing (CAM)

By the time computer use in design began, NC technology had matured to become cost-effective for applications in machining. An important task in NC is part programming. A part program is simply a set of statements comprehensible to the **machine control unit** (**MCU**) that oversees slide and tool movements and other auxiliary functions. In the case of components with complex geometries, part programmers had to carry out lengthy calculations, for which it was logical to use computers. This gave rise to MCUs with built-in **microprocessors**—the building blocks of computers. The use of computers in extending the applications of NC technology, especially to part programming, was earlier termed **computer-aided machining** (**CAM**), and the associated technology was called **computer numerical control** (**CNC**). Later, CAM became an acronym for computer-aided manufacturing (note that manufacturing is a broader term than machining). CAM was coined to distinguish computer use in tasks other than design. With increasing use of computers in nondesign functions other than NC, the meaning of CAM has broadened over the years. Earlier, it used to denote computer use in part programming only. Today, it means any nondesign function of manufacturing that is computer-aided. In fact, in the literal sense at least, CAM is as broad in scope as CIM, except that the latter emphasizes integration. Since lack of integration is the main stumbling block to further use of computers in manufacturing, CIM is currently more talked about than CAM.

From CAD and CAM to CAD/CAM

As the use of computers in design and machining broadened under CAD and CAM, it became evident that certain tasks were common to both. For example, both design and manufacturing require data on tolerances. Part geometries created during CAD can readily be saved in a database for later use. Why reenter such information again at the MCU? Though manufacturers knew that design and manufacturing represent a continuum, in the beginning CAD and CAM developed independently of each other. Later, when the drawbacks of CAD and CAM compartmentalization were realized, the logical move from CAD and CAM was to CAD/CAM. The slash between CAD and CAM was meant to reinforce the shared functions of design and manufacturing. By now, the two have come close enough not to require the slash; perhaps the term **CADCAM** should replace CAD/CAM.

Islands of Automation

In the 1970s, another concept—**flexible automation**—came into being. Flexible automation is based on computer systems and permits variations in the task through software modifications. The term was coined to distinguish this type of automation from **Detroit-type automation**, which is inflexible. A robot is an example of flexible automation. To denote the flexibly automated independent units of processes or workstations, a new terminology—island of automation—came into popular use. The idea was first to create islands, and then connect them. By the early 1980s, though, it became evident that the linking of the islands was an insurmountable task (see Box 1.1). The difficulties of the islands approach underscored the need for integration. Today, of the three words in CIM, the middle one is the most critical.

BOX 1.1   *Trend is Less "Islands of Automation" and Towards Automation*

The first use of the computer was to automate the processing of information within the enterprise, accounting data being the first and most obvious candidate. That was logical because in the final analysis an enterprise is essentially a series of information processing steps.

As the technology of computers and their application advanced, automation of the factory became possible. But, there was a difference. While the office could accumulate data in large quantities and then process the data for any number of purposes the factory needed information as events transpired. Automation of the factory needed real time control not batch processing of accumulated information. As we mastered that change in perspective factory automation became almost a fad. Every industry and company seemed to have some activity that was begging to be automated. What frequently happened was that solving a problem through automation in one corner of the plant simply moved the problem to another area in another form.

Little was gained if automobile bodies could be automatically welded with great speed and precision if they couldn't be painted with the same efficiency. Nothing was gained by rolling steel at higher speeds and closer tolerances if a significant percentage of the finished coils were scrapped due to material handling damage. Somehow we needed a larger view of the enterprise and how it best functions as a whole. We began to think of integration as a way to make automation really work. Then we began to realize that integration was applicable even if there was little or no automation. The emphasis changed from automation to integration.

Manufacturing integration, that is a comprehensive view of a significant functional part of an enterprise, or even the whole enterprise, with the objective of coordinating the parts into an effective whole, is relatively new. In an integrated system the specific functionality of the individual parts is largely transparent with respect to the operation of the whole. Your automobile is a good example of an integrated system.

Source: "Trend Is Less 'Islands of Automation' and Towards Automation" by J. Murray and R. Ferrari. Mid-America '90 Manufacturing Conference. SME paper # MS90-168.

The primary difference between the islands approach to automation and CIM is that the latter is global while the former is local. Computer-integrated manufacturing represents the logical evolution of the islands of automation concept of the 1970s. By now, we have seen that the "islands" are not enough by themselves without a "big picture" of the entire manufacturing operations that would determine how the islands would be interconnected.

Evolution of the CIM Concept

During the 1970s, computers had made inroads into the technical areas of manufacturing in the form of CAD, CAM, and islands of automation. Companies that led these developments continued to maintain the competitive edge provided by such technologies. These were, and are, the companies that made computers and other high-tech products with sufficient profit margins to justify expensive investments. They continued to exploit the computer-based technologies, but when attempts were made to minimize or eliminate duplication of manufacturing data generation, the islands of automation were usually difficult, if not impossible, to connect. An integrated approach to computer-based automation was the next logical step, giving rise to the concept of computer-integrated manufacturing. Although the concept of CIM had been introduced in the 1970s, CIM was not recognized as a necessary, albeit expensive, technology until the 1980s.

CIM developed primarily in response to the demands of the marketplace for more product variety, better quality, and lower prices. In the 1960s and 1970s, manufacturers coped with the latter requirement by concentrating only on one factor: direct labor cost. They invested in new equipment and processes that minimized labor cost. Sometimes, they moved operations offshore to countries with lower wages. In the 1990s, the emphasis has shifted to other cost elements, since in most cases direct labor represents today only a small portion (10–15%) of the total manufacturing cost.

The push for CIM in the 1980s came from the challenges manufacturers were facing in an ever-changing global market. This is true even in the 1990s and will remain so in the future. The various challenges involve complex interrelated issues that can be grouped as technological, social, and political. The major issues are product quality, global competition, government regulations, high wages, and product liability. These are further discussed in Chapter 2.

Computer-Aided or Computerized?

Currently, the term computer-aided is quite often used in the manufacturing field as an adjective. It is an appropriate term, since in most cases computers only aid. However, with the expected impact on CIM of emerging technologies, such as **artificial intelligence (AI)** in the form of **knowledge-based systems (KBS)**, neural-networks-based computers, and machine vision, manufacturing terms may in the future carry the adjective *computerized*, rather than computer-aided. The word computerized represents a higher level of sophistication. Computerizing processes under CIM results in *unmanned* automation.

1.3   CIM II

According to the summary of a 1986 roundtable discussion (Savage, 1987), most current CIM efforts are aimed at interfacing the existing systems. These efforts are called **CIM I**, which stands for **computer-interfaced manufacturing**. True integration is distinguished by use of the term **CIM II**, or computer-integrated manufacturing. CIM II is what CIM really means. CIM I represents the fourth generation of development of computer-integrated manufacturing, whereas CIM II represents the fifth. The various generations show parallels with developments in computer technology and management practices, as shown in Table 1.1. Note the correspondence between CIM II and the networked management environment.

CIM creates an operational environment similar to what existed in the eighteenth century. Then, when manufacturing was just a craft, the craftsman represented the ultimate in integration. He was the company, the president, the worker,

Table 1.1
The five generations of computers and their impact on manufacturing and management environments

Generation	Computer Innovations	Manufacturing Impact	Management Environment
1	Vacuum tube	Payroll/inventory	Small/entrepreneurial
2	Transistors	CAD/**MRP II**	Hierarchical/functional
3	Integrated circuits	CAD/CAM	Matrix
4	**LSI/VLSI**	CIM I	Data interfaced
5	Parallel processing	CIM II	Networked

and the salesperson. His brain did what the corporate computer does in today's CIM operation. He had to conceptualize the product, determine its market, finance the project, design the product, make it, and sell it. Since his was a one-man operation, all the elements of a full-blown CIM were interacting in his manufacturing business. This is true even today for small businesses. Later, growth and expansion along with mechanization and automation caused a gradual compartmentalization of manufacturing into separate functions, such as design, marketing, and production. This brought division of labor into play, giving rise to further specialization, under **Taylorism**, during the first half of the twentieth century.

Today, with increasing computerization, CIM brings manufacturing full circle, back to the craft-type operational environment. Harrington (1986) calls it reintegration, since CIM effectively restores the unity, the strength, and the flexibility of the craftsman while retaining the volume and variety of today's production. According to him,

> This re-integration was not apparent until electronic data processing technology came into common use. But once it was introduced, the potential, and its vast implications soon became apparent. It is this electrifying perception—this new and broadened perspective on an old art now becoming a science, this new concept of the structure of manufacture—that people have termed computer integrated manufacturing. (p. 29)

A comprehensive recent essay on CIM, appearing as Box 1.2, sums up its evolution and the current status rather well.

1.4   BENEFITS OF CIM

If CIM is so useful, what are its benefits? Similar to the variations in CIM definition, the benefits to companies depend on their experiences with CIM. Moreover, the benefits may be interrelated. For example, a reduction in inventory translates into higher profits.

In general, CIM benefits can be grouped into tangible and intangible categories, as listed in Table 1.2.

Table 1.2

Benefits of CIM
Reprinted with permission from
*Computer-Integrated
Manufacturing Handbook*
(p. 4.413) by E. Teicholz and
J. N. Orr (Eds.), 1987. New York:
McGraw-Hill.

Tangible Benefits
• Higher profits
• Less direct labor
• Increased machine use
• Reduced scrap and rework
• Increased factory capacity
• Reduced inventory
• Shortened new product development time
• Fewer missed delivery dates
• Decreased warranty costs

Intangible Benefits
• Higher employee morale
• Safer working environment
• Improved customer image
• Greater scheduling flexibility
• Greater ease in recruiting new employees
• Increased job security
• More opportunities for upgrading skills

BOX 1.2   *In Search of CIM*

It has been called a strategy, a product, a direction and a vision. It has been the subject of thousands of books, articles, speeches and conferences. Manufacturers have invested billions of dollars in it. Yet nobody can agree on what "it" is.

"It" refers to CIM—Computer Integrated Manufacturing. For nearly twenty years, the phrase has dominated the discourse of manufacturers. Arguments have ranged from what parts of a business are included in CIM to what constitutes integration. Debate has also raged over the role of humans. Traditionally, CIM has been envisioned as a mass of machines linked together by a web of electronic interconnections—the "factory of the future." But new models have placed people at the center of CIM. It has been called 20% technological and 80% cultural. No matter what

you believe, one thing is certain—CIM has been 100% confusing.

So what exactly is this mysterious thing called Computer Integrated Manufacturing?

The phrase was coined in 1973 by James Harrington in his landmark book *Computer Integrated Manufacturing*. A consultant for a major consulting firm, Harrington feared that his creation would be misapplied. "He didn't want it turned into an acronym," states Jack Conaway, Manager of CIM Strategic Programs at Digital Equipment Corporation. "He was afraid that it would lose its meaning." Unfortunately, Harrington's fears were well-founded. CIM has become a buzzphrase subject to an endless variety of interpretations.

Conaway believes that the scope of the definition depends on who offers it. "Vendors tend to define CIM in terms of what they're selling or would like to sell," he observes. "Small vendors may define CIM in terms of shop floor automation. Others define it as integrating the whole manufacturing department or the whole company or the whole enterprise."

He personally finds the term useless. "It has become devoid of meaning," he states. "It especially confuses discussions with people in process manufacturing. They think that CIM is a discrete manufacturing term."

The multiple definition problem is viewed as a historical phenomenon by John Monroe, Director of Strategic Consulting Organization, Manufacturing Application Group at Hewlett-Packard Company. "Confusion about CIM abounds because the definition is constantly evolving," he explains. "Originally CIM meant a way of getting people out of the production process. Then it evolved into making the production process more consistent. That phase included a lot of talk about robots. Today, CIM is viewed by HP and our most advanced customers as a tool to help manufacturers achieve a competitive advantage."

Mistakes Based on Past Definitions

Wrangling over the definition of CIM is more than simple academic debate. It ultimately affects implementation strategy and the bottom line. For example, many costly CIM mistakes can be traced to inappropriate objectives arising from faulty definitions.

According to Conaway, the classic mistake occurred several years ago when CIM was defined as a method for eliminating direct labor on the shop floor. "The whole effort was misdirected," he explains. "Direct labor was only 5 to 8% of the product cost. Trying to reduce indirect labor would have been a much more appropriate goal."

The other major mistake was made by manufacturers that defined CIM solely as a technical system. "They planned CIM projects as if they were some kind of deterministic, technical system," recalls Conaway. "But CIM systems are socio-technical. The people component is not deterministic. You must account for the human, organizational and change management aspects of the system."

Current Views

In order to move away from the confusion surrounding CIM, Digital Equipment Corporation now talks about the Computer Integrated Enterprise. "We don't limit our discussion of automation and integration to the manufacturing department or the manufacturing company," explains Conaway. "Our scope is broader with CIE."

He defines an enterprise as a line of business within an industry. "In addition to company functions such as sales, marketing, service, engineering and manufacturing, an enterprise includes at least one step into the supplier and distributor chains," explains Conaway. "In some lines of business it may go further than one step. It depends on the business."

As an example, he cites the food and beverage industry. "Food and beverage manufacturers rely heavily on advertising," Conaway explains. "They need up-to-date information about a product's profit per linear foot on retailers' shelves. So it's important for them to reach all the way through their distributors down to the retailer for information." He also notes that many manufacturers are now stepping further into their supplier and distributor chains for information. "They want to know the quality of materials purchased by their suppliers," states Conaway. "They're also concerned about timeliness. Basically, they want to do better planning."

At Hewlett-Packard, CIM is currently viewed as an enterprise-wide system that can help manufacturers achieve a competitive advantage. "First, you understand why customers buy your product and why you want them to buy your product," explains Monroe. "Then you put together a manufacturing system that delivers the advantage you want your customers to perceive. The real benefit of CIM is that it gives you a manufacturing capability that delivers the competitive advantage that motivates people to buy your product."

Monroe views CIM as a "third pillar" of manufacturing after TQC (**total quality control**) and JIT (**just-in-time**). "TQC focusses on understanding your processes and making them predictable," he explains. JIT is a philosophy of eliminating waste and simplifying your processes. CIM lets you manage your factory as a strategic weapon." He envisions technology that would connect a collection of machines to achieve a factory which is automatically controlled. It would also allow access to information anywhere in the production cycle and the value chain. "If you couple that with a desire to achieve a strategic result," he states, "you can achieve competitive advantage in manufacturing."

Current Applications

Today, many manufacturers apply CIM selectively in their operations. A good example is the Dana Corporation, a $5 billion manufacturer and distributor of transportation related equipment. Its Victor Products Division in Lisle, Illinois, manufactures gaskets and sealing systems using CIM systems to develop new products.

"We have a few prototype lines where we are running the full book definition of CIM," states Warren Smith, Victor's Manager of Information Systems. "We exchange CAD drawings electronically with our customers. We use the programmable controllers on the CNC punch presses. We download directly from CAD to that CNC punch press and produce the product right from there. But most of our operations don't use the textbook definition of CIM."

Smith believes that you must look carefully at when to use automation and when to use people. "We could implement a material handling vehicle to pick up tools and set them in the press," he states. "But it wouldn't make good business sense for us. Automation has to serve a business purpose where you use it."

Conaway agrees. Using its CIE model, Digital deploys automated equipment and systems where they will enhance business objectives. For example, in its Springfield, Massachusetts-based disk operation, the company makes extensive use of automation on the shop floor. Incoming parts shipments are bar-coded on the loading dock and immediately placed in an **Automated Storage and Retrieval System (AS/RS)**. The AS/RS is then accessed by automated robots and conveyors. Requests to the AS/RS are made by people working on the line or robots working in some of the cells. The AS/RS then picks the parts and sends

them down the conveyor. The system also includes workcell graphics for assembly plants. Line workers look at graphic representations of the assembly on a workstation next to a conveyor that automatically brings the appropriate parts.

"The whole system came out of a business plan," states Conaway. "In the disk business, the time for introduction of new products keeps shrinking and the reliability of products keeps improving. So the critical success factors in that particular business were time to market and quality. That's what the automation is designed to address."

CIM Futures

What's next for CIM? Conaway believes it will continue to expand in scope. "It will have to evolve from an abstract marketing concept into a technology architecture—principles, rules and guidelines for implementing these systems," he explains. "People are tired of living in the stratosphere. They want to get past defining the scope of CIM. They want to concentrate on its elements and how they fit together."

Some of the specifics Conaway wants answered include: How do heterogeneous databases communicate with each other? What kind of process controllers do you need layered over the applications? What are the application protocols for two applications to talk to each other? How is this layered into a network connecting heterogeneous computing equipment running different operating systems? Each systems architect needs to be able to custom craft an architecture that answers these questions and also satisfies the enterprise business goals.

Monroe also believes that CIM will continue to evolve. "CIM has a working definition which is moving in time," he states. "The definition will evolve as new technology becomes available and as people comprehend that effective competition involves an appropriate balance between automation and management."

Source: From "In Search of CIM" by P. Cole, *ASKhorizons*, *2*(4), pp. 7–9. Reprinted with permission of *ASKhorizons* magazine, a publication of ASK Computer Systems Inc. (Mountain View, CA), copyright 1989.

TRENDS

❏ Interest in CIM is worldwide, with many people calling it a do-or-die phenomenon. World-class manufacturers are already adopting CIM.

❏ Computer-integrated manufacturing is moving rapidly from concept to technology. The technology for CIM is available, albeit in fragments, and it is possible to tailor it to a company's specific needs.

❏ CIM solutions are usually not generic; they are customized for a particular company. There is, however, a trend to package CIM solutions in a generic way.

❏ CIM's impact on manufacturing is in essence evolutionary; but its impact on discrete manufacturing is revolutionary.

❏ CIM is expensive, but its cost is declining at a phenomenal rate. Developments in microelectronics feeding through PCs and other digital systems are rendering CIM more and more viable.

❏ According to Salas (1989), four key trends will cause an explosion of CIM applications in the 1990s:

1. Standard system architectures and other technologies simplifying integration and lowering total costs
2. CIM suppliers teaming with each other to reduce integration complexity and risk
3. Electronic information sharing both internally and externally
4. People within a company learning to work together, refocusing on common manufacturing goals.

❏ CIM is likely to do to manufacturing what mechanization did to farming, with fewer workers producing in fewer factories all the goods we will need at a price that is affordable.

SUMMARY

This chapter introduced the concept and history of computer-integrated manufacturing, or CIM. Simply stated, CIM involves the integrated application of computers in manufacturing. CIM is an umbrella term covering all manufacturing functions and activities, both technical and managerial.

The concept of integration is not new to manufacturing; it is as old as industry itself. The craftsman of the eighteenth century represented the ultimate in integration by designing, marketing, and financing—all on his own. As today's version of integration, CIM is the orchestrated use of computers in manufacturing. CIM is feasible today because of the lower cost and enhanced capabilities of computers.

Computers were first used in manufacturing for business functions, such as payroll, and in inventory control. Next, computers were applied to design under CAD, followed by CAM, and then CAD/CAM. By 1975, CIM became a concept, and the 1980s saw it expand into a technology. Besides the computer, NC technology, evolving on its own, had a significant impact on the evolution of CIM. Since both NC and computers work on coded characters (numbers, letters, and other symbols), the two have merged together as a technology. This has accelerated the progress of CIM, the most talked-about topic in manufacturing today.

KEY TERMS

Note: Key terms are defined in the glossary at the end of the book.

Artificial intelligence (AI)

Automated storage and retrieval system (AS/RS)

CIM wheel

Computer-aided design (CAD)

Computer-aided design and manufacturing (CAD/CAM, CADCAM)

Computer-aided design and drafting (CADD)

Computer-aided manufacturing (CAM)

Computer-aided process planning (CAPP)

Computer and Automated Systems Association of the Society of Manufacturing Engineers (CASA/SME)

Computer-integrated engineering (or enterprise) (CIE)

Computer-integrated manufacturing (CIM) (CIM II)

Computer-interfaced manufacturing (CIM I)

Computer numerical control (CNC)

Dedicated automation

Detroit-type automation

Discrete manufacturing

Flexible automation

Flexible machining or manufacturing cell (FMC)

Flexible machining or manufacturing system (FMS)

Hard automation

Islands of automation

Just-in-time (JIT)

Knowledge-based system (KBS)

Machine control unit (MCU)

Microprocessor

Numerical control (NC)

Personal computer (PC)

Soft automation

Taylorism

Total quality control (TQC)

EXERCISES

Note: Exercises marked * are projects.

1.1   Explain the meaning of CIM in approximately 200 words.

1.2   What is a CIM wheel? Discuss its strengths and weaknesses as an elucidator of CIM's scope.

1.3   Is CIM a concept, a technology, or both? Justify your answer.

1.4   Trace the evolution of CIM in approximately 200 words.

1.5   What is the difference between CIM I and CIM II?

1.6*   Provide your own definition of CIM. Compare it with those of your class-mates or colleagues.

1.7*   Draw a wheel model that you think better represents the CIM concept. Why do you think your wheel is better?

1.8*   Visit a local manufacturing facility, tour the plant, and talk with the management to gauge its awareness of CIM. Write a two-page report on your visit.

1.9*   Study a recent article on CIM and present it to the class as a 10-minute talk, followed by a group discussion. (With participation of the other students, this can turn into a seminar on the state-of-art in CIM.)

1.10   Circle T for true or F for false or fill in the blanks in the following.

    a.   The islands-of-automation concept of the 1970s led to the development of CIM.   T/F

    b.   All the functions of CIM are covered in CAD, CAM, FMC, and FMS.   T/F

    c.   Discrete manufacturing produces 40% of manufactured goods.   T/F

    d.   NC technology evolved because of computers.   T/F

    e.   CIM I represents the fifth generation of development in manufacturing.   T/F

    f.   CIM I stands for _____
       and CIM II stands for _____

    g.   The CIM wheel was developed by _____

    h.   The term CIM was coined by _____

SUGGESTED READINGS

Books

Bray, O. H. (1988). *Computer integrated manufacturing: The data management strategy*. Boston: Digital Press.

Chiantella, N. A. (1986). *Management Guide for CIM*. Dearborn, MI: CASA/SME.

Foston, A. L., Smith, C. L., & Au, T. (1991). *Fundamentals of computer-integrated manufacturing*. Englewood Cliffs, NJ: Prentice-Hall.

Goetsch, D. L. (1988). *Fundamentals of CIM technology*. New York: Delmar.

Harrington, J. Jr. (1985). *Computer integrated manufacturing*. Malabar, FL: Robert E. Krieger.

Mitchell, F. H. (1991). *CIM systems: An introduction to computer-integrated manufacturing*. Englewood Cliffs, NJ: Prentice-Hall.

Ranky, P. G. (1986). *Computer integrated manufacturing*. Englewood Cliffs, NJ: Prentice-Hall International.

Rembold, U., Blume, C., & Dillmann, R. (1985). *Computer-integrated manufacturing technology and systems*. New York: Marcel Dekker.

Scheer, A. W. (1988). *CIM—Computer integrated manufacturing*. New York: Springer-Verlag New York Inc. (Based on research visits to CIM development centers of leading computer manufacturers and industrial enterprises in Europe, this book addresses CIM as a total concept and approach to industrial organization. Originally in German, the English edition, containing several case studies, is aimed at North American readers.)

Teicholz, E., & Orr, J. N. (Eds.). (1987). *Computer-integrated manufacturing handbook*. New York: McGraw-Hill.

Vail, P. S. (1988). *Computer integrated manufacturing*. Boston: PWS-KENT.

Monographs and Reports

Beard, T. L. (Ed.). (1994). *Modern machine shop 1994 NC/CIM guidebook*. Cincinnati: Gardner.

CASA/SME Technical Council, SME Blue Book series. Dearborn, MI: CASA/SME.

Martinez, M. R., & Leu, M. C. (Eds.). (1983). *Computer integrated manufacturing*. PED-Vol. 8. New York: ASME.

Nazemetz, J. W., Hammer, W. E., Jr., & Sadowski, R. P. (Eds.). (1985). *Computer integrated manufacturing: Selected readings*. Norcross, GA: Industrial Engineering and Management Press.

Savage, C. M. (1987). *Fifth generation management* (pp. 3–5). SME Blue Book series. Dearborn, MI: SME.

Shrensker, Warren L. (1990). *CIM: A working definition*. SME Blue Book series. Dearborn, MI: SME.

Journals and Periodicals

COMMLINE: The Journal of Computerized Manufacturing. Wheeling: IL.

Computer-Integrated Manufacturing Systems. Guildford, Surrey, UK: Butterworth-Heinemann Ltd.

Industrial Engineering. Norcross, GA: Industrial Engineering and Management Press.

International Journal of Computer Integrated Manufacturing. New York: Taylor & Francis Ltd.

Manufacturing Engineering. Dearborn, MI: SME.

Mechanical Engineering. New York: ASME.

Modern Machine Shop. Cincinnati: Gardner.

Production. Cincinnati: Gardner.

Articles

Cole, P. (1989). In search of CIM. *ASKhorizons*, 2(4), 7–9 (appears in Chapter 1 as Box 1.2).

Harrington, J., Jr. (1986, July-August). Why computer integrated manufacturing? *COMMLINE*, 6, 29.

Klein, L. R. (1990). MRP-JIT-WCM-CIM—they are not just TLAs! Mid-America 1990 Manufacturing Conference—Management Briefing Proceedings. Dearborn, MI: SME

Murray, J., & Ferrari, R. (1990). Trend is less "islands of automation" and towards integration. Mid-America 1990 Manufacturing Conference, SME paper # MS90-168.

Salas, R. (1989, August). The future impact of CIM. *Production*, pp. 78–79.

Vajpayee, S. K., & Reiden, C. E. (1991, fall). Computer integrated manufacturing and its ramifications. *Journal of Industrial Technology*, 7(4), 31–35.

CHAPTER 2

Manufacturing: An Overview

"The starting point of any economy is manufacturing—the making of things. Money not backed up by manufacturing is nothing more than waste paper."
　　　　　—Hajime Karatsu, *Tough Words for American Industry*

"The U.S. cannot be without a manufacturing sector. It's the only thing in this country that creates wealth. You don't create wealth by cutting one another's hair or giving manicures. Publishing daily newspapers does not create any national wealth. Some people for some silly reason began to think we were heading for an all-service economy. We can't be a world power shuffling securities in New York."
　　　　　—David M. Roderick, ex-chairman, USX Corp.

"Look at manufacturing systems today. Most of them assume that the business of manufacturing takes place within the four walls of the factory. But manufacturing doesn't only occur within the four walls of a factory. In the real world, the manufacturing business extends far beyond the factory floor. It includes distribution, sales, and accounting. It involves a whole range of third parties—vendors, customers, subcontractors, dealers, distributors, resellers. And it involves operations around the world. The manufacturing business is much more complex than just the manufacturing process."
　　　　　—Sandra Kurtzig, CEO and president, ASK Computer Systems Inc.

This chapter presents an overview of manufacturing processes to elucidate the need for CIM. The role of manufacturing in our lives, and in the "life" of a nation for which it creates wealth, is discussed. The chapter deals with the basics of manufacturing, especially discrete-parts manufacturing and batch production. The two contexts in which the term manufacturing is used—one specific and the other broad—are noted. The concept of manufacturing as a **system**, with CIM as its realization, is emphasized. Also included are the manufacturing challenges that have been encouraging a move toward CIM. The chapter serves as a primer to the principles and complexities of discrete-parts manufacturing, and to the "why" of computer-integrated manufacturing. Its purpose is to help the reader comprehend the various issues CIM needs to address.

2.1   THE WORLD OF MANUFACTURING

Before reading any further, pause for a minute or two and look around you. It does not matter where you are—in your study room, in a classroom, or on a plane. Just eye the objects around you. Were these objects produced by people or by nature? My guess is that most of the objects you are glancing at are manufactured, unless you are relaxing by the beach or in a park. Even when away from home or the workplace, you would find that manufactured goods outnumber those created by nature. What does this simple, but thought-provoking, observation prove? Quite simply, that we live in a world of manufactured goods.

How long do you think it took us to create this vast world of manufactured goods? Make a guess. Isn't it amazing that we learned how to manufacture all these products during a short period of 200 years, the major ones only during the last 50 to 100 years. To fulfill our desire to live a comfortable life, we have reached a stage where we "drown" ourselves daily in manufactured goods. Our society has begun to realize that we are fast nearing the limit to growth, especially for the sake of the environment.

People are so busy enjoying the numerous manufactured products, such as cars, televisions, telephones, and coffee makers, that they hardly realize the existence of the world of manufacturing. We manufacture so many different products—from bread to bombers, rugs to robots, salt to satellites—that often we fail to perceive their vast numbers and varieties. To facilitate comprehension, the enormous array of manufactured products and the associated industries are classified according to **Standard Industrial Classification (SIC)** numbers.

Standard Industrial Classifications

In the United States, industrial activities are grouped according to SIC numbers, which appear in a directory compiled and published by the U.S. Office of Management and Budget. SICs are used to monitor the performance of establishments representing various facets of the U.S. economy. The system was created in the 1930s by a special committee appointed by the White House and the Bureau of the Budget.

Table 2.1
Standard Industrial Classification categories

SIC Code	Industry
01–09	Agriculture, Forestry, and Fishing
10–14	Mining
15–17	Construction
20–39	**Manufacturing**
40–49	Transportation, Communications, Electric, Gas, and Sanitary Services
50–59	Wholesale and Retail Trade
60–67	Finance, Insurance, and Real Estate
70–89	Services
91–97	Public Administration

Table 2.2
SIC major groups for manufactured products

Major Group	Types of Manufactured Products
20	Food and Kindred Products
21	Tobacco Manufactures
22	Textile Mill Products
23	Apparel and Other Finished Products made from Fabrics and Similar Materials
24	Lumber and Wood Products, except Furniture
25	Furniture and Fixtures
26	Paper and Allied Products
27	Printing, Publishing, and Allied Industries
28	Chemicals and Allied Products
29	Petroleum and Related Industries
30	Rubber and Miscellaneous Plastics Products
31	Leather and Leather Products
32	Stone, Clay, Glass and Concrete Products
33	Primary Metal Industries
34	Fabricated Metal Products, except Machinery and Transportation Equipment
35	Machinery, except Electrical
36	Electrical and Electronic Machinery, Equipment, and Supplies
37	Transportation Equipment
38	Measuring, Analyzing, and Controlling Instruments; Photographic, Medical, and Optical Goods; Watches and Clocks
39	Miscellaneous Manufacturing Industries

This federally mandated system provided a way to label and classify establishments engaged in productive economic activity. The SIC basis for categorizing industrial activities is appropriate to computerized assessment of the economy as a whole. Governments and businesses plan strategies based on such assessments.

Each type of industry is assigned an identifying numerical code and a logical place within the classification. Industrial products are classified into major groups, as shown in Table 2.1.

Manufacturing industries fall into the 20 to 39 range. This range covers all types of products and activities within the realm of manufacturing, as detailed in Table 2.2.

Figure 2.1 depicts a typical product in each major group. Beginning with group 20 (food and kindred products), the numbering continues counterclockwise to end with the miscellaneous group representing ordnance and associated activities (39).

A third digit further breaks down the major groups. For example, the fabricated metal products under SIC 34 include the following categories, among others:

341 Metal cans and shipping containers
342 Cutlery, hand tools, hardware
346 Forgings and stampings
349 Miscellaneous fabricated metal products

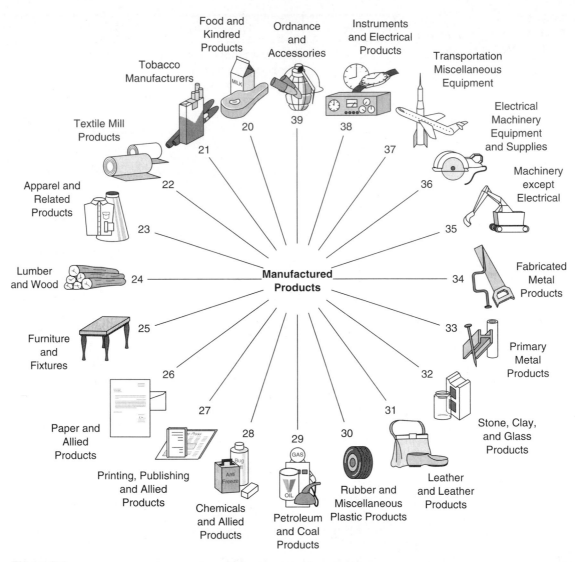

Figure 2.1
The 20 SIC classes of manufacturing businesses

Still another digit subdivides categories even further. For example, the metal cans and shipping containers under 341 are subdivided as 3411 for metal cans, 3412 for metal shipping barrels, drums, kegs, pails, and so on.

Although the SIC coding allows tracking performance for individual sectors, it should be noted that industrial activities represent a continuum. Because of that, an improvement in one sector may have ripple effects on others. For example, lower interest rates (group 60–67, finance) are known to create feverish activities in the construction industry (group 15–17).

Similarly, an upswing in manufacturing has been found to benefit transportation and other activities in group 40–49 as well as the industries in group 50–89. In the U.S., for example, every 100 new jobs in manufacturing create another 60 direct jobs in the nonmanufacturing sector. In fact, when the cumulative effect on all other sectors is accounted for, every manufacturing job creates two to three indirectly related jobs (Rowe, 1990).

As seen in Table 2.2, manufacturing industries produce a wide variety of goods. Based on demand, these industries fall into two broad categories. In the first are the industries whose products have a very large market, for example, headache pills. Such products are manufactured in large numbers and remain cost-effective even when produced by expensive dedicated automation. These products generally fall under chemical and allied industries.

In the second category are industries whose products have a limited and fluctuating demand. With a finite demand and uncertain market for their products, such industries face unique logistical problems. These industries are batch-production type, since they make products in batches to serve a variable market. CIM is useful primarily in these industries.

Methods for producing the vast array of manufactured goods vary tremendously (Lindbeck, Williams, and Wygant, 1990; Kruppa and Lindbeck, 1985), making manufacturing a complex activity. Of the various types, discrete-parts manufacturing presents the most difficult operational environment for efficiency and logistics. Discrete-parts manufacturing denotes operations in which individual components are manufactured and/or assembled into products that can be disassembled.

The world of manufacturing is both exciting and challenging. Manufacturing is the foundation of most business activities, including those in the service sector of the economy. It is the "hub" of industrial activity and the proving ground for new technologies. Manufacturing contributes immensely to the well-being of a nation and its people by creating wealth.

2.2   CREATION OF WEALTH

Prior to industrialization, **wealth creation** had only a limited base. It occurred through activities—such as farming, mining, and fishing—that harnessed the natural resources, with limited conversion or processing. Manufacturing was then considered not to produce any wealth. In the eighteenth century, economists Adam Smith and David Ricardo suggested that the value of manufactured goods should also be included in the assessment of national wealth. They argued that manufacturing satisfies the concept of **vendibility**—the characteristic of a product that makes it saleable.

During the latter part of the nineteenth century, the concept of **utility** was put forward. Utility is a subjective measure of the degree of buyer satisfaction. The utility concept considers all industrial activities—even services such as retail sales, transportation, and education—contributors to national wealth.

From the viewpoint of wealth creation, there are two differences between manufacturing and service industries. First, manufacturing involves production of material or tangible goods, whereas service industries produce intangible (nonmaterial) goods only. The utility of tangible goods lasts beyond production, whereas that of intangible goods disappears in the very act of their provision. (Are you left with anything tangible at the end of an airline flight, except perhaps a headache?) Second, creation of national wealth by the service industry depends on a thriving manufacturing sector. Service industries flourish in nations where the manufacturing economy's base is strong enough to support them. Service industries merely do what the term says: serve. Manufacturing, which provides the underpinnings for the entire economy, is in reality the "master."

The industrialization of a nation begins with manufacturing. As the manufacturing base expands, the standard of living, commonly measured by per capita **GNP**[1] (gross national product), improves. After providing for the necessities of life, consumers are left with money that they spend on products of service industries, such as vacations or movies. As development continues, the service sector becomes visibly important to the economy, resulting in a "back seat" for manufacturing. This may lead to growth in imports of tangible products and signal the beginning of the decline of manufacturing. The current situation in the United States represents this phase of industrial development. If manufacturing is reemphasized in time, it will arrest and eventually stop the decline. Otherwise, manufacturing industries will continue to weaken and lag behind the economies of the nations that emphasize manufacturing.

When manufacturing makes profit for the company shareholders, it also creates wealth for the nation. The "direct" contribution of manufacturing to the **national income**[2] of the U.S. since 1960 is shown in Table 2.3. As evident from this table, manufacturing's contribution has been declining since 1965. However, at eighteen percent it is still significant.

The word *direct* has been used to emphasize that the data in Table 2.3 do not include indirect benefits from manufacturing. Manufacturing contributes to the national wealth and employment indirectly, for example, through transportation, communications, and service activities. The indirect contribution is normally difficult to assess. In general, manufacturing contributes a total of 50% to 75% to the income of an industrialized nation.

Table 2.4 shows the contribution of various industrial sectors to the 1990 U.S. national income. Of the $807 billion in manufacturing, durable goods (those that last

---

[1]GNP is the monetary value of all the goods and services sold by a nation. The term **GDP** (gross domestic product), which excludes any income from overseas operations, is often used as an indicator of the national economy. The 1992 per capita GDP for the three strongest economies were: Germany, $24,585; Japan, $27,232; and the United States, $22,204 (*U.S. News and World Report*, 12 July 1993, pp. 26–27).

[2]GNP and national income are different. The national income is lower than the GNP, since it takes into account capital consumption allowance, indirect business tax and nontax liability, subsidies, and so forth. For example, the 1990 U.S. national income was $4,446 billion, while the GNP was $5,465 billion.

Table 2.3
Contribution of manufacturing to U.S. national income

Year	Manufacturing Income (in bilions)	National Income (in billions)	Manufacturing's Contribution (Percentage)
1960	$125	$ 429	29%
1965	172	584	29
1970	216	835	26
1975	318	1,315	24
1980	532	2,264	23
1985	672	3,198	21
1990	807	4,446	18

Source: *World Almanac and Book of Facts 1988* (p. 95). New York: Newspaper Enterprise Association Inc.; *World Almanac and Book of Facts 1992* (p. 142). New York: Pharos Books.

for at least three years) accounted for almost 57%, nondurable goods for the other 43%. Private industries generated almost six times the income of government enterprises. This ratio will, of course, be lower for socialistic or mixed economies in which public sector activities and government-sponsored monopoly inhibit the private sector.

As shown in Table 2.3, U.S. income from manufacturing increased more than six-fold between 1960 and 1990. However, the number of people directly employed in manufacturing during this period increased only 14%, from 16.8 million to 19.1 million, peaking in 1980 at 20.3 million. If one considers production workers only, the increase is even smaller (from 12.6 million in 1960 to 13.0 million in 1990, peaking at 14.2 million in 1980). Although employment in U.S. manufacturing industries has remained fairly constant, it has declined as a percentage of total employment. A similar trend is found in other developed countries as well.

An increase in manufacturing income without a corresponding increase in manufacturing employment results from the positive impact of technology and automation on **productivity**. One common criterion for measuring improvement in manufacturing performance is **labor productivity**. In the U.S., the Bureau of Labor Statistics of the Department of Labor tracks labor productivity in terms of output per employee hour.

How will CIM affect employment and the contribution of manufacturing to the national income? Also, what type of training and job skills are necessary for CIM? These issues are discussed in Chapter 13.

2.3   THE MEANING OF MANUFACTURING

The term *manufacturing* is used in two different contexts—one broad and the other narrow. In the broad sense, it pertains to the entire manufacturing enterprise and all its functions, including production management and technology as well as the management of all other resources. In the "global" context, the term covers all the activi-

Table 2.4
Distribution of U.S. national income, 1990

SIC code			Income in billions
	Private Industries		$3,756
01–09	Agriculture, Forestry, Fishing		103
10–14	Mining		42
15–17	Construction		225
20–39	**Manufacturing**		**807**
	Durable goods	462	
	Nondurable goods	345	
40–49	Transportation, Public Utilities, etc.		329
50–59	Trade		639
	Wholesale	262	
	Retail	377	
60–67	Finance, Insurance, and Real Estate		648
70–89	Services		963
91–97	Government, Government Enterprises		648
	Domestic Industries		4,404
	Rest of the World		42
	National Income		4,446

Source: *World Almanac and Book of Facts 1992* (p. 142). New York: Pharos Books.

ties included in the CIM wheel (Figure 1.1). This text and the concept of CIM view manufacturing from this broad perspective.

The word manufacturing also applies to another, narrower, context. A shop-floor professional is more familiar with this context, in which manufacturing denotes the functions between design and the actual production phase; for example, tool and die design, selection of cutting speeds and feeds, programming of NC equipment, and so on. The title *manufacturing engineer* or *technologist* generally refers to a professional with responsibilities related to this restricted meaning of manufacturing. In Europe, and to some extent in Asia, manufacturing in this narrow context is usually called production.

2.4   TYPES OF MANUFACTURING

From Figure 2.1 and the discussions in Section 2.1, it is evident that many types of goods are manufactured. Manufacturing entails so many processes and operations that comprehending them requires some type of categorization.

Manufacturing operations can be categorized in several ways depending on the purpose of grouping, for example, national versus international or product types. For most purposes, classifications reflect the following six criteria:

1. Continuous or discrete
2. Variety and volume
3. Raw material to final product
4. To order or to stock
5. Size
6. Machinery used

Continuous or Discrete Manufacturing

Manufacturing operations fall into two very broad groups: (a) continuous-flow or process type and (b) discrete-parts manufacturing (also known as discrete manufacturing). Continuous-flow operations typify the chemical and mining industries and oil refineries, which produce large amounts of bulk material. SIC groups 26, 28, and 29 (Figure 2.1) are some examples. Products in these groups are usually measured in units of volume or weight, batch size is large, and product variety is low. Since batches are large, designing and building special machines for their production make sense. Such machines are usually expensive, but their cost is distributed over a large volume, contributing only marginally to the unit cost. Since processes are specialized, they are difficult to modify or salvage, if for some reason the customer no longer requires the product.

Continuous-flow operations, used to manufacture "mature" products in large volumes, are relatively easier to control and operate, since production uses dedicated machines. These operations are usually fully automated, with operators minding the machines. From an integration point of view, the production task is simpler, since processing requirements (one sequentially following the other) are such that integration is built in at the equipment design stage itself. The need for flexibility is just not there. As technology improves, newer machines with built-in automation replace the old ones. Thus, while the term CIM may be new to process industries, integrated manufacturing based on the CIM concept certainly is not.

The term *discrete-parts manufacturing* denotes operations involving products that can be counted. The output of process-type industries is also counted eventually; for example, sugar in terms of number of sacks or tons. What distinguishes discrete manufacturing from process industries is the potential flexibility of its output. When demand falls in process industries, operations are simply phased out. Discrete-type operations, on the other hand, are cost-effective to modify for other products needed by the market.

A special feature of discrete manufacturing is that the end product, generally made of several components, can be disassembled and reassembled; an example is a bicycle. It is not essential for the end product to comprise several components. For example, a discrete manufacturing facility that machines only connecting rods of different shapes and sizes for automobile manufacturers produces a single-part end product. Whether single- or multiple-part, a product must be designed, raw materials procured, machines set up, tools sharpened, operators trained, and a host of other steps taken before actual production can begin. All this is, in essence, preparation for production. The preparation-for-production cost is normally the same whether one unit or hundreds of units are produced. Since it is independent of the number of units actually produced, this cost is fixed. Obviously, the burden of the fixed cost on each unit grows as batch size (number of units in the batch) declines. In mass production, where batch size is large, fixed cost per unit is obviously low. At the other extreme, in job shops with a batch size of one or two, the fixed cost per unit is relatively high.

Variety and Volume

Another way to look at manufacturing facilities is according to variety and volume. A low-variety, high-volume operation is easier to manage, since dedicated automation is possible. A high-variety, low-volume operation, on the other hand, is more difficult to operate and manage. Based on volume and variety, discrete manufacturing is of three types:

Mass production
Batch production
Job shop

Mass Production.   In mass production of discrete parts or assemblies—for example, bolts or ballpoint pens—the production volume is high. Therefore, special-purpose, dedicated equipment can be employed. Machines are considered dedicated when they are tailored to specific products. Examples of mass-produced goods include bicycles, washing machines, and video games. A mass-production facility is termed a **transfer line** when products are assembled while conveyor systems transfer them from one end of the plant to the other. A good example of a transfer line is an automobile-production facility.

Batch Production.   In **batch production** of parts or assemblies, the volume is lower, and the variety higher, than in mass production. Examples (see Figure 2.1) are books, apparel, and products under SIC code 25 or 35. When the end item is an assembled product, the producer may make some parts in house and buy others from vendors. Batch production is sometimes referred to as a midvolume, midvariety operation. The limited volume does not justify very specialized production machines; general-purpose machines are used instead. This does not, however, alter the shop-floor goal of keeping the machines running and the operators busy. An enormous amount of coordination among various production functions is essential to optimize use of the resources. In this type of application, CIM technologies such as cellular manufacturing or robotics hold promise to deliver the economies of mass production while still coping with variety. Batch production, and to some extent mass production, of discrete products provides all the challenges under CIM.

    In batch production, goods are manufactured in batches that may be repeated as required. As Figure 2.2 shows, manufacturing directly contributes 30% to the GNP in industrialized economies. Batch production accounts for 40% of this, or 12% to the GNP. Also note that three-quarters of batch production involves batch sizes of 50 or less. Thus, a typical manufacturing facility produces small batches.

Figure 2.2
Importance of batch production and small batch sizes to GNP

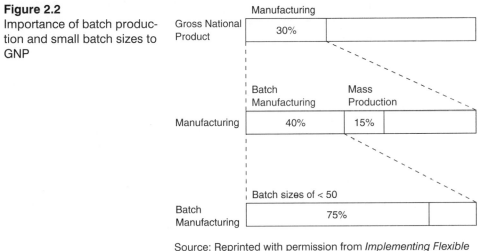

Source: Reprinted with permission from *Implementing Flexible Manufacturing Systems* by N. R. Greenwood, Copyright 1988 by John Wiley & Sons.

Job Shop Production.   The job shop represents the most versatile production facility. Within the limitations of the machines and the operators, it can manufacture almost any product. With a low production volume, sometimes as low as 1 to 10 units, the cost of product design and set up is relatively high. Production facilities for aircraft, ships, or special machine tools are examples of job shops. NC and CNC technologies can significantly improve the productivity of job shops.

Which of the three discrete-manufacturing facilities is suitable for a product depends on two factors: variety and volume. How many different products (including their models, if significantly different) are to be produced? How many of each product (i.e., of each variety) is to be produced during a given period of time? Note that the term *volume* actually means quantity—the number of units. On the basis of volume and variety, the three types of manufacturing facilities just discussed can be represented graphically as shown in Figure 2.3. The overlaps emphasize the fact that their boundaries are not rigid. The actual values on the volume and variety axes depend on the complexity of the product.

Raw Material to Final Product

On the basis of the relationship between raw material and the end product, manufacturing follows one of four different patterns: disjunctive, sequential, locational, or combinative.

Disjunctive.   In the disjunctive pattern, a single raw material is progressively processed into its various components as end products. Examples of disjunctive facilities are slaughterhouses, lumber mills, and oil refineries.

Sequential.   In sequential facilities, too, there is only one raw material as input. But, unlike disjunctive operations, which separate the raw material into compo-

Figure 2.3
Volume and variety by production type

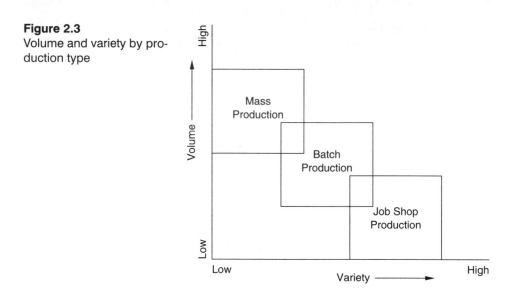

nents, it is progressively modified to become the end product. An example is a supplier's production facility that machines castings for the automobile manufacturer.

Locational.   Locational patterns involve buying, storing, and eventually distributing manufactured goods without any substantial physical modification in the product. An example is the company that buys a product in large quantities and distributes it in small packets under its own brand name. This pattern suits bulk materials,
such as sugar or rice.

Combinative.   The combinative type is basically discrete manufacturing in which
components—some produced in-house and some bought from suppliers—are
assembled, inspected, packaged, and shipped as end products. A good example is an
automobile factory.
    From a production viewpoint, the combinative pattern is the most complex.
CIM is targeted primarily at this pattern, although CIM concepts apply to the other
three as well.

To Order or to Stock

Based on the immediate destination of the end products, manufacturing may be of
two types. In the first, products are shipped directly to consumers, wholesalers, or
retailers. Such companies are said to produce "to order." Since they do not store the
end products, for finished-goods inventory is unnecessary. Capital is therefore
released and profit realized immediately following production. Job shops usually
operate in this mode. In the second type, products are stocked in finished-goods
inventory; marketing distributes them to retailers or consumers as needed. This type

of operation is said to produce "to stock." Such facilities usually produce in batch sizes that minimize the unit cost. In this type, capital is tied up until the end products can be sold.

CIM can offer significant benefits for both types of operations. To-order companies can respond rapidly to meet the needs of consumers, while to-stock companies can produce economically in smaller batch sizes, thus lowering the capital investment in finished-goods inventory.

Size

It is sometimes convenient to classify manufacturing companies on the basis of size, with criteria such as number of employees, annual sales turnover, net worth, and so forth.

Whether a company is small or large is often determined by the number of employees. While there is no standard cut-off number, the following categorization is usually practiced: small, below 100; medium, 100 to 499; large, 500 or more.

Contrary to the general perception that only large companies can afford modern facilities, the level of modernization and the sophistication of technology used are independent of the company size (see Table 3.6, Chapter 3). A 20-person operation can use CAD, **machining centers** or **co-ordinate measuring machines (CMM)**.

Machinery Used

A variety of machine tools, equipment, and processes are used in an average plant. They fall into the following functional groupings:

Metal forming
Metal cutting
Assembly
Material handling
Inspection, testing, gauging
Others, such as casting, welding, riveting, brazing, heat treatment, washing stations, plastic molding, etc.

Depending on the product and its variety and volume, discrete-manufacturing facilities use different combinations of equipment.

2.5   DISCRETE-PARTS MANUFACTURING

In discrete manufacturing, products are usually processed in batches. To keep the fixed cost low, once preparation for production (called setup) has been made, companies try to produce as many units in the batch as possible before resetting the equipment for the next batch. The process of setting up is sometimes called *retooling*, especially when it involves several tool-and-die changes as in the automotive industry.

While it saves on setup costs, producing in large volumes creates other problems. If the products cannot be sold immediately, they tie up capital. Obsolescence, spoilage, or pilferage are also potential problems.

If m represents the number of products (variety) a company manufactures, and n_j is the annual production of the jth product, then the total annual production P_a is given by

$$P_a = \sum_{j=1}^{j=m} n_j$$

A manufacturing facility designed to produce several different products (m high) in small batches is flexible to cope with frequent batch changes, and utilizes technologies that reduce the setup time.

Should the batch size be large or small? From a production viewpoint, a large batch size lowers the per-unit fixed cost. From a financial viewpoint, however, a small batch size will tie up less capital in finished-goods inventory. Marketing personnel would answer the above question with both yes and no: Yes, because with large batch sizes items are more likely to be available in the finished inventory when needed. No, since production in large batches may monopolize the facility for long periods of time for one product, creating stock-out situations for other products.

Sometimes a product is not manufactured even when a company has spent money designing it. A competitor may have launched in the meantime a similar or better product. Occasionally, a product is not manufactured even after investment in setup. Obviously, companies must recover the expense of these aborted designs and setups by marking up the prices of other products.

2.6   MANUFACTURING AS A SYSTEM

Discrete manufacturing of the midvolume, midvariety type represents a complex activity involving several interacting elements. The U.S. Air Force's **Integrated Computer-Aided Manufacturing (ICAM)** program has attempted to illustrate this complexity, as shown in Figure 2.4. Notice the various functions represented by the blocks, from "plan projects" to "support service of products," and their interactions as indicated by arrows. To facilitate analysis of such a complex activity, it is desirable to treat manufacturing as a system.

A system is defined as an integrated assembly of interacting elements designed to carry out cooperatively a predetermined function. According to this definition, a system displays four characteristics. It must (a) be an assemblage of components, (b) have interactive elements, (c) be goal-oriented, and (d) be flexible.

Assemblage.   A system is an assemblage of several distinguishable units called subsystems, elements, components, or factors. The assemblage does not need to be physical; it may just be conceptual. In a manufacturing system, the machining center,

Figure 2.4

ICAM's modeling of the manufacturing function

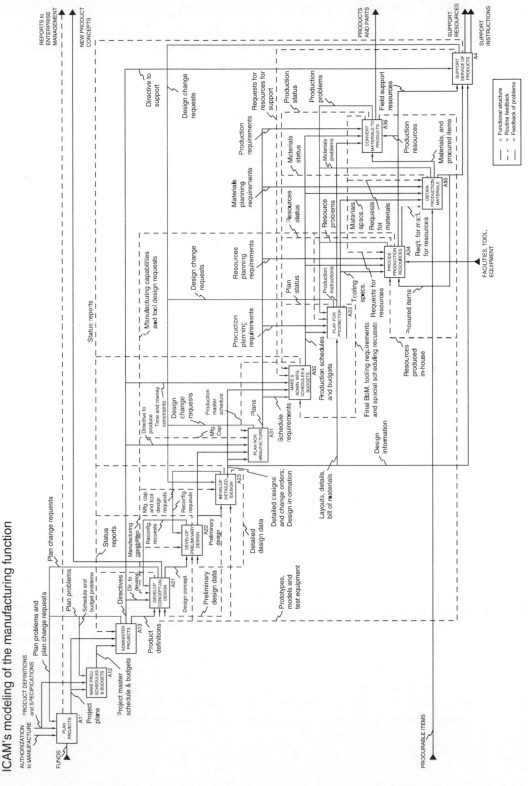

Source: Reprinted with permission from *Understanding the Manufacturing Process: Key to Successful CAD/CAM Implementation* by J. Harrington, Jr. Copyright 1984 by Marcel Dekker, Inc.

for example, represents one element, its operator another. The capital required to start and run the business is a third element. Depending on the level of refinement, a manufacturing system may comprise dozens of elements.

Interaction.   The elements of a system must interact. A mere assemblage without interaction is not a system, but a group or a set. As shown in Figure 2.4, there are a large number of interactions among the elements of a manufacturing system. Figures 2.5 and 2.6 further illustrate the interaction from two other angles. Notice in Figure 2.5 the large number of functions that a typical manufacturing facility can involve. In CIM, these functions need to be integrated using computers, wherever this is cost-effective. Figure 2.6 shows the typical organization of a manufacturing enterprise. Beginning with the stockholders, the various resources including personnel are grouped into functional or departmental categories. Grouping the tasks this way can hinder the effectiveness of CIM, if it is not managed carefully.

Goal Orientation.   A system attempts to achieve a certain set of goals. In a manufacturing system, the goal is to maximize profit if operating in a free market economy. For markets controlled by governments, the goal is to manufacture the best product (or provide the best service) for a specified cost.

Flexibility.   A system must be flexible enough to adapt to the environment in which it operates. In manufacturing, the environment includes competition, consumers, governments, the economy, and so forth. The environment can help or hurt a manufacturing system, which can respond to it effectively if flexible. In the 1990s, teenagers want red cars, an example of consumer influence. Similarly, the increase in oil prices in the 1970s, which opened the U.S. market for smaller cars, was an environmental influence on the market.

Input-Output Model

A novel way to study a complex system such as manufacturing is to treat it as an input-output model, as shown in Figure 2.7. Such a representation makes it easier to design a new system or identify problems with an existing one for seeking solutions. The interior of the block (system) in Figure 2.7 represents what the system does. In a manufacturing system, the input (I) comprises raw materials, capital, information, data, and so on. Other inputs such as wages and energy costs are used to support transformation—the activities that change the raw materials into end products. The output (O) is, of course, the manufactured products. Inside the system, the transformation (T) is responsible for adding value to the raw materials. The system operates in an environment that is conceptually external to the plant's physical boundaries.

The input-output model facilitates investigation of the system for several purposes. With the concept of input, output, and transformation, a manufacturing system can be studied to solve one or more of the following problems:

Figure 2.5
Management, production, and personnel functions of a manufacturing system

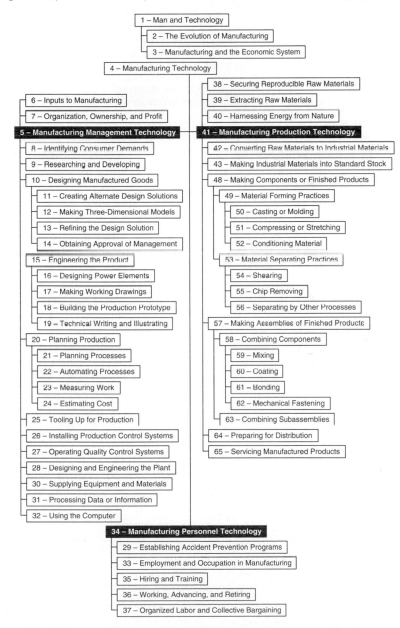

Source: Reprinted with permission from *The World of Manufacturing*, Copyright 1971 by The Ohio State University Research Foundation.

Figure 2.6
Organizational structure of a typical manufacturing enterprise

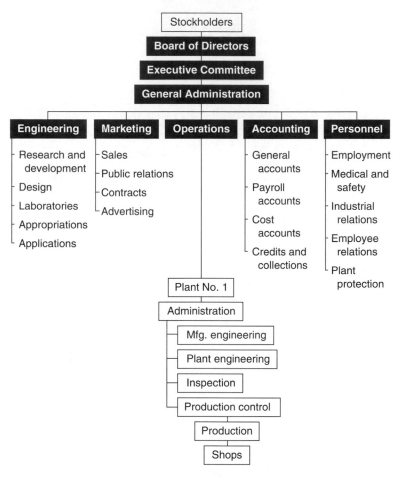

Source: Reprinted with permission from *The World of Manufacturing*, Copyright 1971 by The Ohio State University Research Foundation.

Analysis: Identify or clarify the contents of I, O, and T.

Operation: For a specific T and I, find O.

Synthesis: Determine a suitable T for given I and O.

Inversion: Find I for given O and T.

Optimization: Select the controllable parameters to operate the system at its best for a specified criterion.

    While the input-output model is a useful concept, the complexity of manufacturing is such that it allows an "exact" study only in rare cases. Nevertheless, the input-output approach is often attempted to seek a "good" solution, which may be

Figure 2.7
Input-output model of a manufacturing system

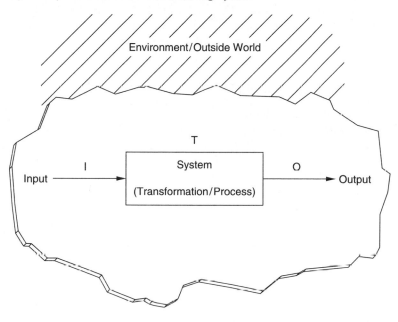

approximate due to the simplifying assumptions underlying the input, output, and transformation parameters.

Productivity

Productivity is a basic measure of how effectively the resources of an enterprise are being managed. It is defined as:

$$\text{Productivity} = \text{Output/Input} \qquad\qquad (2.1)$$

The input-output model can be used to determine the productivity of a manufacturing enterprise, or of its units or departments. It is obvious from Equation 2.1 that productivity can be improved by:

1. Increasing the output for a given input
2. Decreasing the input for a given output
3. Varying both the output and input in such a way that their ratio increases

When the input and output include all the parameters of a system, then Equation 2.1 yields its overall productivity. When only specific parameters are used, the productivity measure is partial. For example, when input includes only the labor cost, Equation 2.1 yields what is called labor productivity. In a similar way, one can assess **capital productivity** by limiting the input to capital costs only.

All technologies under CIM, including flexible automation, change the input, output, or both. Irrespective of the extent of improvement in individual parameters, determining the cost-effectiveness of a technology must reflect its effect on overall productivity.

Manufacturing Subsystems

All the tasks of a manufacturing enterprise can be represented as system elements and accordingly grouped into subsystems. To succeed, the enterprise needs to be managed properly. The management of a company represents one of the two broad subsystems of manufacturing. The other is technology, which includes processes, machinery, and so forth. These two subsystems can further be divided into their elements; for example, management may be of finance and capital, personnel, or inventories. The technology subsystem can similarly be grouped into plant maintenance, R&D, production, and so on.

In companies in which design is a major activity, a manufacturing system can be divided into three basic subsystems: business, design, and production. The business function, comprising upper management, concentrates on financial aspects and long-term planning and control. It can be considered the "software" of the manufacturing enterprise. The design and production functions, on the other hand, implement the decisions of the business function and, thus, are the "hardware" of the enterprise. In a CIM environment, these three subsystems and their various elements work together to achieve the system goal, that of maximizing profit through quality goods at a price attractive to consumers.

Despite recent advances, manufacturing remains a complex system, which makes the realization of a true CIM environment both difficult and expensive. The key requirement is that the subsystems and their elements work in an "orchestrated" way to achieve the enterprise goal.

2.7   CIM: AN IMPLEMENTATION OF THE SYSTEM CONCEPT

The idea of treating manufacturing as a system developed in the 1970s. It resulted from the success of using the system approach in other complex subjects, such as the economy of a nation. The approach facilitates mathematical analyses of system behavior in given situations. Manufacturing is so complex that analysis of the entire system as a mathematical model is extremely difficult, if not impossible. Its subsystems and their elements, however, can be, and are, analyzed to yield remarkable insight into the performance of individual departments or operations.

CIM can be thought of as a computerized implementation of the system concept. Similar to a system, CIM also requires inclusion of as many relevant manufacturing functions as are feasible. The only difference is that, while the system approach attempts to handle the problem through mathematical or quantitative models, CIM does it through computerization, with timely information controlling the operations. CIM relies on computer-based simulation and practice of the manufacturing system or its subsystems. Both CIM and the system approach begin with a "bird's-eye view" of manufacturing. An illustration of this view for a discrete manufacturing system compartmentalized into its functions is shown in Figure 2.8. Note the entry of the raw materials and the exit of the end products (at bold arrows). The CIM implementation of such an operation is objectively similar to the system approach, since both attempt to look at manufacturing in its entirety. The various tasks represented by the blocks of Figure 2.8 can be illustrated in another format, Figure 2.9, which rearranges all steps of manufacturing into four groups (subsystems): research and development (R&D), production, marketing, and management.

2.8   FINANCIAL JUSTIFICATION

As in other cases, the cost of CIM technology must be justified for an investment. Be it a robot or CNC equipment, technology for technology's sake makes little business sense; the company adopting it must benefit financially. Most often managers have several projects competing for the available capital. Recognizing that money has time value, project costs must be reduced to a common time base for comparison.

The decision to fund any CIM technology reflects the benefits expected from the investment. Methods commonly used in determining financial feasibility are (a) **payback period**, (b) rate of return, (c) present worth, and (d) uniform annual cost.

Payback Period Method.   The payback period is the time during which benefits from the investment equal the cost of investment. This method is based on the sound logic of an initial investment paying for itself in a reasonable time. The payback period (*PP*) can be expressed as

$$PP = C/(S - A) \tag{2.2}$$

where

C = required investment
S = annual savings
A = annual cost of utilizing the investment.

Data used for determining the payback period are current funds. The salvage value, which is the monetary worth of the equipment or facility being replaced (defender), is ignored. It is also assumed that investment is interest-free. The option with the shortest payback period is the obvious choice.

The payback period method has two shortcomings. First, it does not consider the time value of money. It weighs the cash flow of today equally with the value in the future. Second, it ignores the duration of the cash flow. In spite of these weaknesses, the payback period method is widely used because of its simplicity. Recently, it has been giving way to the discounted-cash-flow method, which considers time value of money.

Figure 2.8
Representation of a manufacturing system by functional blocks

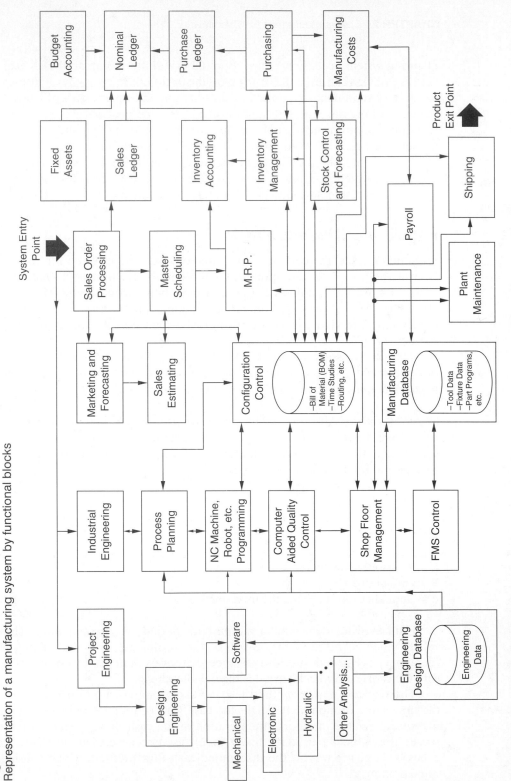

Source: Reprinted with permission from *Computer Integrated Manufacturing* by P. G. Ranky, Copyright 1986 Prentice-Hall International, UK, Ltd.

	R & D			Production				Marketing				Mgmt.
Steps in Manufacturing	Product Research	Product Design	Product Engineering	Production Engineering	Quality Control and Safety	Production		Market Research	Advertising	Distribution	Sales	(Management)

Column headers (left to right): Product Research, Product Design, Product Engineering, Production Engineering, Quality Control and Safety, Production, Market Research, Advertising, Distribution, Sales, (Management)

Steps in Manufacturing

1. Research and Marketing present findings on product ideas.
2. Management approves new product idea. Design process begins.
3. Design problem is stated.
4. Possible designs are sketched.
5. Designs are finalized.
6. Management approves design drawings.
7. Engineers advise designers.
8. Paste-up mock-up made for management approval.
9. Appearance mock-up made for advertising purposes.
10. Product design approved and given to Engineering.
11. Planning begins for Sales.
12. Engineers calculate product details.
13. Working drawings are made.
14. Production engineers advise product engineers.
15. Production planning begins. Product engineers analyze product.
16. Parts list is made.
17. Prototype is built and tested.
18. Prototype is approved.
19. Consumer research is done using prototype.
20. Advertising plans are developed.
21. Production plans are developed.
22. System of production is chosen.
23. Production flow chart is prepared.
24. Jobs are planned.
25. Equipment is chosen and ordered.
26. Parts and materials are chosen and ordered.
27. Packaging is designed and approved.
28. Operation sheets are prepared.
29. Plant layout is made.
30. Materials-handling system is planned.
31. Equipment is set up.
32. Materials arrive.
33. Production workers are trained.
34. Sales plans are developed.
35. Advertising and sales plans are approved.
36. Trial run is made.
37. Production begins.
38. Product passes final inspection.
39. Product is packaged and shipped to sales outlets.
40. Product is sold to consumer.

*Product engineers and production engineers are both manufacturing engineers. The responsibilities are different. In some cases the same person or persons may assume all responsibilities.

Figure 2.9

Various tasks of R&D, production, marketing, and management subsystems of manufacturing as the product reaches the consumer

Source: Reprinted with permission from *Manufacturing* by J. F. Fales, E. G. Sheets, G. J. Mervich, and J. F. Dinan. Copyright 1986 Glencoe Publishing Company.

While proposing CIM technologies for adoption, bear in mind that projects with payback periods of up to two years are usually approved. Those between two and five years are accepted if justified on other nonmonetary grounds, such as a company's image enhancement or environmental considerations. Projects with payback periods of more than five years are seldom financed in private industry.

In Equation 2.2, all benefits are expressed in monetary terms. Sometimes, especially in cases of CIM technologies, there may be intangible benefits that are difficult to quantify. For example, use of **statistical process control** (**SPC**) reduces rework, which can be expressed in dollars. But what about the improved image of the company as a manufacturer of high-quality products? How much is this worth? Many CIM benefits may be intangible; conventional investment analysis must be modified to account for these. Ignoring the value of intangibles may underestimate the benefits of a CIM investment.

Example 2.1

A manufacturing company is considering installation of a robotized welding station that costs $295,900. The robot is expected to yield better quality welds with a throughput (production rate) equal to that of five manual welders. Each welder costs $15 per hour, including fringe benefits, and works one shift of eight hours. A year has 250 working days. Assume that the robotized station will need supervision, requiring half the time of a semiskilled operator at a cost of $10 per hour, and that the annual cost of spare parts for the robotized facility is likely to be $500. Should the project be funded if the company plans for payback periods of less than three years?

Solution

First, determine the three cost factors: cost of investment (C), annual savings (S), and operating cost (A) of using the robotized station. It is given that

$C = \$295,900$

Savings are derived from the cost of manual operations. Given that each welder works 2,000 hours per year (250 days @ 8 hours per day), the total annual cost of five welders is

5×2000 hours $\times \$15$ per hour $= \$150,000$
$S = \$150,000$

Note that the intangible benefit of more consistent weld quality from robotized welding is not being considered here. Robotization may also offer other savings, such as less material waste and fewer rejects.

The operating cost (A) of the robot welder comprises half the cost of a semiskilled operator who would oversee it, plus the cost of spare parts. Thus,

$$A = 0.5\ (\$10\text{ per hour } \times\ 2000\text{ hours}) + \$500$$
$$= \$10,500$$

Again, note that other costs, such as maintenance and utilities, are not being taken into account.

Once the values of the three parameters are known, the payback period is calculated. In this case,

$$\text{Payback period, } PP = C/(S - A)$$
$$= 295,900/(150,000 - 10,500)$$
$$= 2.12\text{ years}$$

Since the payback period is less than three years, the project should be funded.

Rate of Return Method.   An investment involves cash flows, money going out to pay for the technology or equipment and money coming in as savings. In the rate of return—also called the **return on investment (ROI)**—method, a rate is determined by equating the cash flows of receipts with those of the disbursements. When this rate is above a cut-off level, set by management, the investment project is approved. Obviously, ROI should be higher than the prevailing bank interest rates, unless management has some other strategy.

Example 2.2
_____

If the minimum acceptable rate of return (MARR) in Example 2.1 is 15%, should the project be funded? Assume that the robotized station will have a useful life of 10 years with no salvage value. The company practices straight-line depreciation in which the value of equipment is reduced by an equal annual amount.

Solution

With 10 years of useful life and no salvage value, the yearly depreciation is $295,900/10 = \$29,590$.

$$\text{Thus, the total annual cost} = \$29,590 + \$10,500$$
$$= \$40,090$$

$$\text{The net annual savings} = \text{Labor cost saving } - \text{ annual cost}$$
$$= \$150,000 - \$40,090$$
$$= \$109,910$$

A savings of $109,910 per year will be realized for an investment of $295,900.

Thus, the return on investment (ROI) $= 109{,}910/295{,}900$
$$= 0.371$$
$$= 37.1\%$$

Since ROI is higher than the MARR of 15%, the project should be funded.

Present Worth Method.   The present worth method involves determining the present value of future money receipts and disbursements and basing the decision on these values. This method compares the project's estimated expenditures with all the revenues. If the worth of the revenues and other benefits exceeds that of the expenditures, the project is funded. When more than two alternatives compete, the one with the largest present worth is selected. A variation of the present worth method is the future worth method, which compares a project's estimated expenditures and benefits at a future time.

Uniform Annual Cost Method.   The uniform annual cost method is a variation of the present worth approach. In this method, all the cash flows are converted to an equivalent uniform series of annual payments similar to mortgage payments on real-estate properties. The project is approved if the annual worth of the revenues exceeds that of the costs.

2.9   CHALLENGES AND TRENDS

The developments that led to CIM represent the response to the challenges faced by manufacturing in the 1970s and 1980s. This section examines these developments and the resulting trends.

Manufacturing in Transition

Manufacturing is currently undergoing a transition (Naisbitt, 1984) which, though evolutionary from a technology viewpoint, is revolutionary in its impact on the economy and employment.

Manufacturing is expected to follow the path the farming sector took following mechanization. Today's manufacturing is like the farming of yesteryear—inefficient and burdened by overhead costs. In the near future, however, it is likely to achieve the efficiency of modern farming, with fewer facilities producing all the goods needed. This change will come about largely due to CIM. In which countries will these facilities be located and who will own them? Which countries will belong to the group of industrialized nations? Will there be fewer members in this group? How will CIM affect the standard of living and quality of life? These are all important questions.

Assuming that manufacturing will follow in the footsteps of farming and that CIM will make this happen, jobs lost in the manufacturing sector are expected to be replaced by new positions in **information technology**. Information technology industries trade in data, information, and knowledge, and fall in the service sector.

Profit and Performance

The objective of manufacturing is profit for the owner or shareholders of the company. Profit margins must be competitive with alternative investment opportunities. Private manufacturers always try to improve their performance with goals such as higher profit, lower unit cost, and better product quality. In government-supported operations, where profit is not the primary motive, low unit cost is usually the objective.

Companies assign managers and directors the responsibility to turn a profit. Managers make decisions that are aimed at benefiting the company. But, since decision making depends on the future, which is uncertain, managers do make mistakes. CIM helps managers make better decisions by providing timely, accurate, and reliable data and information.

Productivity and Automation

A major advantage of treating manufacturing as an input-output system (Figure 2.7) is that it directs company efforts toward increasing output, as much as possible, for a given level of input. The term commonly used to assess these efforts is productivity.

As Equation 2.2 indicates, productivity is the ratio of the output value to the cost of all the inputs. Productivity is thus an index with no dimension or unit. It is measured over a specific time and compared with its value (baseline) from a past period.

Besides overall productivity, we sometimes consider partial productivity, such as labor productivity or capital productivity. If an existing process is modified or redesigned requiring fewer person-hours (low input) to produce the same number of units (unchanged output), then obviously labor productivity has improved. But what about the cost (capital) of the redesign and modification? Could it be that the capital productivity has deteriorated? An improvement in one kind of productivity, such as labor's, may not necessarily increase the overall productivity, which really is the ultimate objective. Low labor productivity may be one of the several reasons behind poor company performance; however, careful managers do not ignore the other related issues, which may at times be difficult to assess.

Sometimes, comparisons are made among worker productivity in different countries. For example, the 1987 U.S. labor productivity was $37,800, which is the GNP divided by the number of people employed. The corresponding figures for Germany and Japan were $36,600 and $35,700, respectively. Such a comparison is valid, however, only if the countries being compared are on equal ground technologically (as they were in this case). The comparison loses its value if the countries' production facilities are vastly different in terms of technology and sophistication, for example, those in Mexico and the United States.

When is automation justified for productivity enhancement? As in the case of other resources, the cost of automation must be compared with its benefits. Introduction of computers (for that matter, any CIM technology) is in fact automation of a kind, and therefore subject to cost-benefit analysis. For example, CAD investments could be justified in the beginning when they significantly improved designers' productivity.

Pressure to improve manufacturing productivity and sustain it at a high level is not new; it is an ongoing effort. The driving force of a free market economy is always survival of the fittest, and manufacturing companies are no exceptions. CIM is the new tool, not only for survival, but also for attaining a leadership position.

Global Competition

In recent years, competition for the market of manufactured goods has turned global. With faster communications and transportation, we buy from companies and sell to customers continents away. Global competition is fundamentally healthy for advances in manufacturing, though at times it can generate political controversy. Global competition provides consumers with best-value products and a broader choice. It is a major factor in CIM developments. The competitive advantage belongs to the company that provides its customers with products of superior value—determined by the customers, not by the company's marketing division—in terms of cost, performance, and quality. This advantage is derived from rapid response to changing customer needs and demands, wherever they may be.

Gone are the days when manufacturers could relax on the strength of a few successful products. Today, they need to be alert to the moves of competitors who may be anywhere in the world. Since the marketplace has gone global, manufacturers, especially small and midsize companies, may need guidance and support from government agencies. In a recent document, *The Technological Dimensions of International Competitiveness*, the U.S. National Academy of Engineering examines the issues of technology and competitiveness. The academy reemphasizes that the three principal institutions of a nation—government, industry, and the education system—play a key role in supporting manufacturing.

Role of Marketing

The ultimate aim of manufacturers is to produce goods they can sell to users immediately at a profit. An interaction between the producer and the consumer takes place. This interaction is initiated, developed, and managed by the producer's marketing staff. Today, marketing represents a function so vital that manufacturing produces what marketing "directs" it to produce. Marketing is the primary interface between the manufacturer and the customer.

During the depression of the 1930s, consumers did not have enough money to buy goods, and no amount of advertising would have changed that. Soon after World War II, when consumers could afford to pay for more than essentials, manufactured goods were scarce. This was the perfect time to set up a manufacturing business, and substantial profits were possible with little advertising. Soon companies began to produce more than consumers demanded. As a result, companies evolved from being production-oriented to being sales-oriented. In the beginning, manufacturers adopted the "hard sell" approach, which worked well for awhile. Eventually, however, companies determined that the right approach was to predict what consumers were likely to want and to gear up for that demand accordingly—a marketing orientation. This created a whole new field to answer questions such as what to manufacture, how many to manufacture, and when to manufacture. Over the years, marketing has grown to become one of the key functions of manufacturing.

Marketing is primarily a planning task. In fact, the marketing plan guides overall business planning efforts. The business plan attempts to prevent the build-up of finished-goods inventory and to keep the plant working at a steady level. The marketing plan aims to ensure that the plant's output will sell. Besides the traditional task of dovetailing the existing production facilities with the needs of the customer, marketing also directs the production function by considering factors such as public policy, consumer purchasing habits, and innovations in technology.

Marketing identifies the need for a product, then convinces management that it will sell, which then triggers actions in design and production. Marketing determines a product need through **market research**. Market research is much more than a casual observation or a "feel" for the market. It entails careful gathering and analysis of specific data. Customer surveys are a common method of market research. Computers record and analyze the data gathered. They also help integrate marketing with other components of the manufacturing system, such as design and inventory.

By its very nature, manufacturing represents a *dynamic* system. Whenever a company finds that demand exceeds supply, it moves in. As expected, other companies follow suit. For awhile they can sell their products easily; but eventually the demand ebbs, creating a **business cycle**—a period of expansion and depression. The business cycle has a jolting effect on the economy, including manufacturing. CIM enables manufacturing companies to respond proactively to the effects of business cycles.

Research and Development (R&D)

Manufacturing industries fall into two groups, depending on whether products are designed internally or externally. In the first group, product designs are carried out in-house in conjunction with the marketing and production functions, with design launching all the technical activities. Designs reflect the perceived demands of a particular market. But designs must also consider the constraints of the existing production facilities to keep costs in line with the market. Constance (1992) claims that about 70% of the product cost is built in during the design stage. While **design for manufacturing (DFM)** has been known in the past, in recent years manufacturers have realized that a strong design-production interaction is critical to the success of the shop-floor aspects of CIM, as attested by the CAD/CAM's acceptance as a "monolithic" technology.

In the second group, product design is external to company operations; thus, the design-production link is usually weak. It is limited to ascertaining, before accepting the order, whether the existing resources will be able to produce to the client's design. External designers usually do not consider potential DFM improvements (for example, a slight modification in the specified tolerance that might significantly reduce the product cost). The manufacturer seldom bothers to find fault in the design from a DFM angle. The design-production interface that attempts to reduce product cost is replaced by an order-production interface concerned primarily with delivery dates.

In some cases, a manufacturing company produces to its own designs as well as those of clients'. Regardless of whether the design is internal or external, manufacturers continually search for cost-effective technologies to ensure up-to-date products, operations, and facilities. This task is collectively known as **research and development (R&D)**. Companies that produce only to clients' designs use R&D primarily to maintain a modern operation. On the other hand, R&D efforts are targeted at both design and production when operating extensively with internal designs. With high-tech or novel products, the design function shares a higher proportion of R&D efforts. R&D activity is relatively more important in companies that produce to in-house designs, since both engineering and production facilities are to be kept in line with the available technology. Figure 2.9 compares the role of R&D with the other three major functions as products progress through various stages of manufacturing.

R&D is an ongoing activity. Manufacturers that underestimate its importance are likely to lag behind in generating new technologies and exploiting existing ones. A lack of investment in R&D becomes evident later rather than immediately. Because of this, management often may reduce the R&D budget during financially difficult times. Today, when manufacturing is both pulled by the market and pushed by technology, R&D is undoubtedly of utmost importance.

Product Life Cycle

Besides the business cycle discussed earlier, another factor stimulates manufacturing: the **product life cycle**. A manufactured product is not in demand forever; it has a life. It is born, it matures, and it dies if the market no longer needs it. In exceptional cases, a product is so successful that it becomes a commodity and lasts for a long time. A product dies for two reasons: First, consumer tastes change over time; and second, a better-value product enters the market as a result of new technology. Sometimes, an altogether new and novel way of fulfilling a consumer need is invented—for example, the microwave oven as an alternative to conventional ovens.

The Ubiquitous Computer

The number of computers used in industry and at home has increased incredibly since the introduction of personal computers (PCs). Today, more than 300 million PCs are in use worldwide. With their continually improving performance-to-price ratio, PCs are becoming ubiquitous. They are changing the way goods are produced and marketed. Since computers are the basic tool for CIM, a full chapter (Chapter 3) has been devoted to them.

Computer-Communications Link

Besides computers, CIM also requires communications technology. If a computer represents the "brain" of CIM, then communications represents its "nerve system." In recent years, these two technologies have been merging into one under marketing pressure. The resulting bond between computer and communications is accelerating the development of CIM. Chapter 4 discusses communications in detail.

TRENDS

❑ Batch production continues to benefit from CIM, especially from flexible automation technologies.

❑ The changes in manufacturing, which began in the 1970s in the name of computer-aided technologies, will continue into the twenty-first century.

❑ The challenges for manufacturers in the 1990s are global competition, consumer demand for better product performance and quality, and judicious application of newer technologies. These forces continue to encourage higher levels of performance and productivity.

❑ Manufacturing is fast evolving toward a development stage similar to present-day farming. Fewer companies employing fewer people are likely to produce all the tangible goods we will need; CIM is a catalyst in this evolution.

❑ The progress toward CIM depends directly on developments in computers and communications.

❑ As CIM advances, other functions normally considered external to manufacturing, such as customer feedback on product quality, will become computerized and fall within the broader meaning of manufacturing.

SUMMARY

Chapter 2 discussed the various issues of manufacturing relating to CIM. Included were the importance of manufacturing to the economy; the usefulness of Standard Industrial Classifications; and manufacturing's role in creating wealth for companies, their stockholders, and the nation.

The complexity of discrete-parts manufacturing, which involves myriad interacting functions, was discussed in relation to the concept of CIM. Manufacturing is so complex that its treatment as a system is warranted. CIM can be thought of as computerized implementation of the system approach to the shop floor and the office.

The term manufacturing is used in two contexts. In the narrow sense, it represents the functions that follow product design but precede actual production, such as selection of cutting parameters and die design. In its broader context, the term includes all the functions and activities of a manufacturing enterprise. These include technologies and their management as well as the management of all other resources, such as finance and personnel. It is this broader context to which CIM pertains. CIM effects incredible changes in manufacturing.

KEY TERMS

Batch production

Business cycle

Capital productivity

Coordinate measuring machine (CMM)

Design for manufacturing (DFM)

Gross domestic product (GDP)

Gross national product (GNP)

Information technology

Integrated Computer-Aided Manufacturing Program (ICAM)

Labor productivity

Machining center

Market research

National income

Payback period

Product life cycle

Productivity

Research and development (R&D)

Return on investment (ROI)

Standard Industrial Classification (SIC)

Statistical process control (SPC)

System

Transfer line

Utility

Vendibility

Wealth creation

EXERCISES

Note: Exercises marked * are projects.

2.1   List 10 discretely manufactured products that affect your life most. Indicate for each the two-digit major group number to which they belong (refer to Figure 2.1 or Table 2.2).

2.2   Which of the three types of facilities—mass, batch, or job shop production—is used to manufacture the products you listed in Exercise 2.1?

2.3   Which major groups in Figure 2.1 fall within continuous-flow manufacturing?

2.4   A manufacturing-based economy is superior to a service-based economy. Do you agree or disagree with this statement? Why? Limit your answer to 300 words.

2.5   Describe the company where you work, as a system. Group it into a few sub-systems, list in each at least five elements, and discuss all the interactions. If you do not work for a company, do this for the college you attend or any organization you are associated with.

2.6   Between productivity and automation, which is more crucial to a manufacturing company—and why?

2.7   As a strong advocate of flexible automation, the production manager of XYZ Company proposes buying a robotized plasma cutting machine to replace the six manual cutting stations, each needing one operator. The robotized machine costing $150,000 will cut six times faster than a manual operator. Each manual operator is paid $11 per hour and works eight hours per day, 250 days a year. The quality of cut is better with the robotized system, but it is difficult to assess the benefit in monetary terms. The robotized system will need spare parts, worth $1,000 annually, and only one of the existing operators, with no change in work hours. If the company approves payback periods of up to three years, will the manager's proposal be funded?

2.8   If the decision in Exercise 2.7 is based on ROI and the robotized machine's life is eight years with no salvage value, should the proposal be approved? The expected ROI must exceed 15%.

2.9   The current resources of a job shop are 6 machine operators, 4 material handlers, and 12 conventional machine tools. As a first step toward CIM, the shop is interested in replacing nine of the conventional machines with three machining centers. Although the modernization will reduce the number of machine operators to two, a NC programmer must be hired. Besides, two new pick-and-place robots will replace two of the existing material handlers. The additional cost on the automated material handling system is estimated to be $40,000 and modernization is likely to increase the annual maintenance cost by $20,000. What is the payback period with the following data?

> Operator cost including overhead = $600/week
>
> NC programmer cost including overhead = $40,000/year
>
> Machining center costs = $60,000, $60,000, and $100,000
>
> Robot cost = $25,000 each
>
> Salvage value of machine tools = $8,000 each

2.10* Visit a library to collect data on your country's GNP for the last 10 available years; use the format below as a worksheet. What contribution (%) has manufacturing been making to the GNP? Draw a bar chart to show the results. What conclusions do you reach from this chart?

Year	Total GNP	Value of Manufactured Goods	%
19_____			
19_____			
19_____			
19_____			
19_____			
19_____			
19_____			
19_____			
19_____			
19_____			

2.11* If your library can help find the data, carry out a project similar to the one in Exercise 2.10 for two other countries of your choice from amongst China, Germany, India, Japan, and the United States. Compare your country's manufacturing performance (Exercise 2.10) with these two.

2.12* Carry out a library search to collect data on manufacturing R&D expenditure in your country for the last available year and compare it with those of the two selected in Exercise 2.11.

2.13* With the help of your instructor, visit a local manufacturing company. Following the visit, write in approximately 400 words a report describing how the company carries out its various functions. Include in the report as many issues covered in Chapter 2 as possible. In another 200 words suggest how the company can exploit CIM.

2.14* With the consent of your instructor, discuss your suggestions of Exercise 2.13 with the appropriate company personnel. (You can learn a lot here!)

2.15* Collect data on the total value of manufactured goods in your state or province for the last five available years. How do these compare as a percentage with those of the best and the worst states? (The local library or the chamber of commerce may be a source of data; ask for a manufacturers' association directory.)

2.16* Use the data collected in Exercise 2.10 to show graphically how the total value of manufactured goods in your country has changed over the last decade. Based on the trend, make a forecast for the next five years.

2.17* Repeat Exercise 2.16 for the number of people employed in manufacturing.

2.18   In the following, circle T for true or F for false or fill in the blanks.

    a.   SIC is an acronym for Standard Industrial Code.   T/F

    b.   Lumber mills are a disjunctive type of manufacturing.   T/F

    c.   Discrete products are manufactured mostly in batch sizes
       of less than 50.   T/F

    d.   In the United States, manufacturing accounts for 50% of GNP.   T/F

    e.   Most companies finance proposals with payback periods of up
       to _____ years.

    f.   Manufactured products fall in the SIC major group num-
       bers _____ – _____ .

    g.   Almost _____ million personal computers are in use worldwide.

    h.   ROI is an acronym for _____.

SUGGESTED READINGS

Books

Chryssolouris, G. (1992). *Manufacturing systems*. New York: Springer Verlag.

Greenwood, N. R. (1988). *Implementing flexible manufacturing systems*. New York: Wiley.

Groover, M. P. (1987). *Automation, production systems, and computer integrated manufacturing*. New York: Prentice-Hall.

Harrington, J. (1984). *Understanding the manufacturing process*. New York: Marcel Dekker.

Hitomi, K. (1979). *Manufacturing systems engineering*. London: Taylor & Francis.

Kalpakjian, S. (1989). *Manufacturing engineering and technology*. New York: Addison-Wesley.

Kruppa, J. R., & Lindbeck, J. R. (1985). *Basic manufacturing*. Peoria, IL: Bennett & McKnight.

Lindbeck, J. R., Williams, M. W., & Wygant, R. M. (1990). *Manufacturing technology*. Englewood Cliffs, NJ: Prentice-Hall.

Maistre, C. L., & Ahmed, E. (1987). *Computer integrated manufacturing—A systems approach*. UNIPUB/Kraus International.

Naisbitt, J. (1984). *Megatrends*. New York: Warner.

Salvendy, G. (Ed.). (1992). *Handbook of industrial engineering*. New York: Wiley.

Smith, A. (1937). *An inquiry into the nature and causes of the wealth of nations*. New York: The Modern Library.

Stonier, T. (1983). *The wealth of information*. London: Thames Methuen.

Monographs and Reports

The Automation News Factory Management Glossary. (1984). New York: Grant.

Ecktein, O., et al. (1984). *The DRI report on U.S. manufacturing industries*. Manufacturing Technology Press.

National Academy of Engineering. (1988). *The technological dimensions of international competitiveness*. Washington, DC.

National Research Council. (1986). *Toward a new era in U.S. manufacturing: the need for a national vision*. Manufacturing Studies Board, Commission on Engineering and Technical Systems. Washington, DC: National Academy Press.

U.S. Office of Management and Budget. (1972). *Standard Industrial Classification Manual*.

Journals and Periodicals

Advanced Manufacturing Engineering. Guildford, Surrey, UK:Butterworth-Heinemann Ltd.
Industrial Engineering. Norcross, GA: Industrial Engineering and Management Press.
Manufacturing Engineering. Dearborn, MI: SME.
Mechanical Engineering. New York: ASME.
Modern Machine Shop. Cincinnati: Gardner.
Production. Cincinnati: Gardner.

Articles

Constance, J. (1992, May). DFMA: learning to design for manufacture and assembly. *Mechanical Engineering, 114*(5), 70–74.

Quinn, J. B. (1983, November). Overview of the current status of U.S. manufacturing: Optimizing U.S. manufacturing. In *U.S. leadership in manufacturing*. Proceedings of the symposium at the 18th annual meeting of the National Academy of Engineering, Washington, DC: National Academy Press. pp. 8–52.

Rowe, B. H. (1990, May). MI 1990 speaker urges firms to cut organizational fat. *ASME News, 10*(1), 1.

PART II

Brain, Nerves, and Heart of CIM

C IM represents the logical evolution of manufacturing through computer and communications technology. Though significant strides have been made in this direction, lack of compatibility still makes electronic integration of the various CIM elements difficult. In addition, CIM will not be feasible on a broad basis until computer, communications, and database technologies are available at a cost the manufacturing industry can afford.

These three technologies, which I call the "brain," "nerves," and "heart" of CIM, are the subject of the three chapters in Part II. Chapter 3 presents principles of computers, Chapter 4 discusses communications, and Chapter 5 concerns database technology.

Chapter 3 begins with an overview of computer fundamentals, followed by discussions of computer hardware and software. Types of computers and common languages are presented. In addition to the basics, this chapter discusses the needs of CIM and recent advances in computer technology. Also included are the results of a recent survey that has identified the manufacturing tasks in which computers are currently used.

Chapter 4 presents the fundamentals of communications to illustrate how the various functions of manufacturing "talk" and "listen" to each other. The chapter includes a discussion of the matrix-type communications needs of CIM. The local area network and MAP (manufacturing automation protocol) are also discussed.

Chapter 5 explains the organization and management of data, a basic requirement of CIM. Various types of data and their sources are discussed. The role of relational database in CIM is compared with those of the other types.

CHAPTER 3
Computer Technology and Manufacturing

"When you can tell a computer, 'Oh, you know what I mean!'—and it does—then that's a computer language!"
—R. Holzman, Jet Propulsion Laboratory, California Institute of Technology

"The computer is fast, accurate and stupid; man is slow, inefficient and brilliant."
—Anonymous

"UNIX is no longer a movement—it's a stampede."
—Paul Cubbage, Dataquest

"Hardware is the skeleton of the computer, software is its life."
—S. K. Gaonkar, State University of New York

"During the last decade, there has been a hundred-fold improvement in raw computing power. Yet, according to Lou Podeska, chief executive officer, Stellar Computer Inc., there has been only a ten-fold improvement in data communications and almost no improvement in software technologies. Most software architectures are based on methods and procedures developed in the early 1970s. Most of the performance gains of the last 10 years have come from the increased performance capabilities of the computer alone."
—C. Hayden Hamilton, PDA Engineering, Costa Mesa, CA

Manufacturing within a CIM environment is based on computers. Why didn't CIM come about earlier since computers have been available since the 1950s? The reason is simple: Manufacturers could not afford them in the past. The situation is different today. Not only are computers affordable, they are more capable, too. It is the enhanced capability for every dollar spent on modern computers that makes CIM feasible.

This chapter discusses the fundamentals of computer technology and presents the recent developments affecting CIM. The advent of microcomputers in the early 1980s generated renewed interest in CIM. The impact of such computers on manufacturing is described. The history of computer use in manufacturing is presented along with current trends. The reasons behind increasing computer use at the shop floor are pointed out. Also discussed are the special computer needs of CIM.

3.1   FUNDAMENTALS OF COMPUTER TECHNOLOGY

Computers represent the most basic element of CIM. These electronic prime movers are driving more and more factories. It is essential, therefore, that the CIM professional understands the principles of computer technology. Such a knowledge promotes effective use of computers in CIM environments.

An Overview

Computer technology has given rise to an array of new terms; articles on CIM and catalogues on computer systems are full of them. A knowledge of this terminology is helpful in deciding which computer system to use for a given task.

A computer system consists of **hardware** and **software**. The hardware includes all the physical components of the computer and devices connected to it. The software, on the other hand, refers to all the **computer programs** that direct the hardware to perform a desired task (a program is a collection of instructions for the computer). Thus, software "commands" and hardware "obeys."

The computer does its tasks using only two digits: 0 and 1. These two represent off and on conditions within the circuits. A device or methodology based on two digits is known as **binary**. Thus, a computer is a binary system. In contrast, people use the decimal system based on the digits 0 to 9. In addition to the numbers, we also use the letters of the alphabet and other symbols, such as $ for dollar, % for percentage, and & for and. The 10 basic digits of the decimal system, the letters, and the other symbols are collectively called **alphanumeric characters**. To communicate with computers, we need to code the characters we use into groups of 0s and 1s. This coding has been standardized; for example, in the **American Standard Code for Information Interchange** (**ASCII**), the capital letter P is represented by the binary group 1010000, whereas the decimal number 8 is 0111000.

The computation or processing inside the computer occurs in the **central processing unit** (**CPU**), which is the "brain" of the computer. When the circuits that do the processing are on one small chip, the CPU is called a **microprocessor**—sometimes denoted by μP. Microcomputers are built around microprocessors. The computation circuitry of large computers is too sophisticated for a single microprocessor; several CPUs are housed within a cabinet, hence the term **mainframe** to describe large computer systems.

The CPU communicates with other hardware units of the computer. Since the CPU is the focal or central element, these units are called **peripheral devices**. Peripherals are either **input devices** for communication to the CPU or **output devices** for communication from the CPU. For example, the keyboard is an input device, whereas the printer is an output device. The peripherals are also called **input/output (I/O) devices**.

Inside the computer, electronic space, called **internal memory**, stores information and data. To keep the cost of computers low, the capacity of internal memory, also called **main memory**, is limited. Additional memory is provided on peripheral devices such as **floppy disks** and **hard disks**. Since such devices provide enormous storage capacity we call them **mass storage** (being external to the computer, this is also called **external memory**).

Long programs and large amounts of data are stored in the external memory as **files**. To distinguish the files from each other, we give them unique names. The individual instructions or data values within the files are called **records**. The files are electronically transferred or **loaded** to the main memory for running a specific program by issuing **system commands**.

The software is either an **operating system** or **application program**. The operating system commands the hardware and is the first to be loaded into the computer or permanently resides there. The application program carries out a specific task in concert with the operating system. Thus, the operating system is an **interface** between the hardware and the application program. The popular **disk operating system (DOS)** for microcomputers is an operating system, whereas AutoCAD is an application program. Software needs a certain minimum of hardware and peripheral devices to operate properly. The required hardware and the associated devices are sometimes called a **platform**, since they "support" the software.

Programs are written in different formats called **computer languages**. The level of a language is either high or low. A high-level language is one that humans can easily comprehend. A **low-level language**, on the other hand, is less comprehensible to humans, but computers understand it better. The lowest level—comprising all 0s and 1s—is called **machine language**. Obviously, a program written in high-level language for our convenience must be translated into what the computer can understand. Such a translation in machine language is done by an **interpreter** or **compiler**. An interpreter examines and implements program instructions, one at a time, every time the program is run. A compiler, on the other hand, examines all the instructions together and then translates them into machine language; the translated version is called an **object program** or code. The object program is saved as a file, and it is this file that is processed when the program is run again. The compilation approach saves time by eliminating the need to convert the program every time it is run. The original program is called a **source program** or code to distinguish it from the object program. The relationship between the two versions of the program and the role of compiler or interpreter is shown in Figure 3.1.

Types of Computers

A large variety of computers are available in today's market. In general, most computers fall into one of three categories: **microcomputers**, **minicomputers**, and large **mainframe computers**. A fourth type—the **supercomputer**—is used in specialized applications such as weather forecasting or large operations that design and manufacture sophisticated products. For example, Nissan uses the supercomputer Cray

Figure 3.1
Translation of the source
code written in a high-level
language into machine code

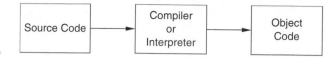

Source: Reprinted with permission from *Microprocessor Architecture, Programming, and Applications* by R. S. Gaonkar, Copyright 1989 Merrill Publishing.

XMP/12 to determine the structural design of the energy-absorbing crumple zone in the Altima model. It conducts more than 15,000 person-years of equivalent crash testing.

The internal working of all computers is based on 0s and 1s. Computers differ from each other in their capabilities such as speed, size, cost, and suitability for applications. Each type is designed with specific characteristics to suit the targeted applications. The three types of computers are compared later in this chapter.

A microcomputer is small and inexpensive. Designed for use by one person, it is also called a **personal computer (PC)**. Its "footprint" is small enough for a desk, hence the term **desktop computer**. Lighter PCs are called **portables** or **laptops** since they can be carried comfortably and placed on the user's lap. They are battery-operated for use in places with no electric power, for example, during a flight or while waiting for one. A typical desktop PC system and a portable model are shown in Figure 3.2. Microcomputers differ from each other significantly in price and capability.

Large computers are called mainframes since their units are transported individually in separate packages and later fixed to a mainframe at the installation site. The units are then tested and commissioned. The mainframe computer is shared by several users who have access locally, or remotely through **terminals**. In either case, the users are physically isolated from the computer. In the remote mode, users may be thousands of miles away with their terminals connected to the computer via satellites or telephone lines, and a **modem** (modulator-demodulator). A terminal with its keyboard and the video display, shown in Figure 3.3, looks very much like a microcomputer, except for the absence of the system unit.

Mainframes are so fast that their instantaneous response may lead a user to believe that the computer is attending only to one person. In fact, the computer takes turns with all the users, but it does it so fast that users do not perceive it. When a mainframe is at its limit in terms of users, its slow response may exhibit time sharing. Such systems are appropriately called **time-shared computers**.

Mainframes are so expensive that only medium or large manufacturing companies can justify their cost. They are purchased primarily for handling the business tasks, rather than for the needs of the plant. But this practice is being questioned as CIM emphasizes the importance of production needs.

In between microcomputers and mainframes lie minicomputers. **Minis'** specifications, capabilities, and prices fall in between those of the **micros** and mainframes. Minis may be the main computer in small or midsize companies. Large companies use minis as plant computers for the needs of production management and operations, while they use mainframes for the business tasks.

Figure 3.2
(a) A desktop personal computer system; (b) A portable or laptop personal computer

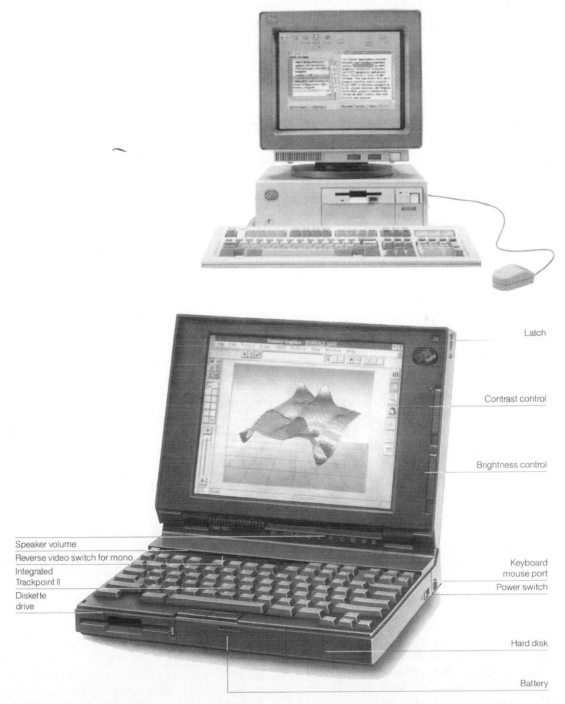

Latch

Contrast control

Brightness control

Speaker volume
Reverse video switch for mono
Integrated
Trackpoint II
Diskette
drive

Keyboard
mouse port
Power switch

Hard disk

Battery

Source: Reprinted with permission from IBM Corp.

Figure 3.3
A typical terminal found in a multiuser computing environment

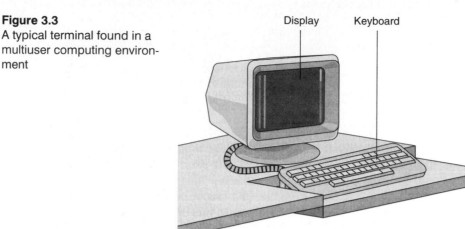

Reprinted with permission from *Using Microcomputers* by D. G. Dologite and R. J. Mockler, Copyright 1988 Prentice-Hall, Inc.

In a typical CIM environment, the mainframe communicates with the minis, which in turn communicate with the micros and other digital devices of the plants and offices. A computer that links, controls, and serves other computers and digital devices is called a **host computer**.

Bit, Byte, and Word

Inside the computer, the two states of off and on are created at an enormously fast rate. These two states represent the binary digits 0 and 1, each called a **bit** (binary digit). A group of eight bits is termed a **byte**. A **word** is a group of bits that are processed together by the computer. Depending on the computer's internal design,

---

☐ **BOX 3.1**   *Computers Make Computers*

At the Fremont (California) Macintosh factory, components are manufactured at rates that produce a new computer every 20 seconds. And this is done using 300–400 Macintoshes helped by a Tandem CLX computer and programmable logic controllers (PLCs) in a CIM facility. The CLX is the factory host that links the shop-floor hierarchy with the site's Tandem VLX-based business system. The Tandem CLX collects data on material movements and assembly status from the shop floor and communicates this information to the

VLX network. The VLX machines handle business tasks such as production control and scheduling, materials procurement and inventory management. According to Mark Oney, the CIM project manager, "This shop-floor-to-business system interface provides improved visibility of the shop-floor activities and improved data accuracy."

Source: From "The Macs That Mac Built" by C. S. Holzberg, March 1990, *Macintosh-aided Design*, pp. 30–35.

a word may be several bits long. Word lengths vary from 4 to 64 bits, sometimes even more. The binary word 1011001000110110 is 16 bits or 2 bytes long. The longer the word, the more powerful—albeit more expensive—the computer. Four-bit words are used in simpler tasks, controlled by dedicated microprocessors, whereas 64-bit or wider words are used in large computers. Modern microcomputers use 32-bit words.

We use so many different characters in our lives that a byte is needed to represent one alphanumeric character and its transmission. Since a byte represents one character, the size of computer memory in bytes equals the number of characters it can hold. For this reason, a byte, rather than a bit or word, is commonly used as a unit. The capacity of computer memory is so large that it is specified in kilobytes or megabytes, and sometimes in gigabytes. The prefix *kilo* (denoted by k) represents a thousand, whereas *mega* (M) represents a million and *giga* (G) a billion. In place of the prefix k, whose value in scientific uses is 1,000, the capital letter K is used in the computer field; it has a value of 1,024. Thus, a capacity of 640 Kbytes means that 655,360 (640 x 1,024) characters can be stored in the memory. To facilitate mental arithmetic, some people round K down to 1,000.

Example 3.1
_____

A manufacturing company uses floppy disks of 1.2-Mbyte capacity to store employee names and addresses. On the average, including the spaces between words, names are 11-letters long; street addresses and city names require 24 and 8 letters each. The state name, a two-letter abbreviation, is followed by the five-digit ZIP code. How many disks are necessary for 552 employees?

Solution

The average number of characters (letters, digits, and spaces) required for each employee's name and complete address is

$$= 11 + 24 + 8 + 2 + 5$$
$$= 50$$

Since each character takes one byte, the required memory space per employee

$$= 50 \text{ byte}$$

The total requirement for 552 employees

$$= 552 \times 50$$
$$= 27,600 \text{ byte}$$
$$= 27.6 \text{ Kbyte}$$

Number of disks required

$$= 27.6/(1.2 \times 1024)$$
$$= 0.022$$

Thus, one floppy disk is more than sufficient. Note that only 2.2% of the disk space is being used.

Example 3.2

A textbook is 400 pages long. Each page contains an average of 30 lines; each line, 10 words. If the words including blank spaces average seven characters, how many 1.2-Mbyte floppies were used to store the manuscript?

Solution

Number of characters in the text

$$= 400 \times 30 \times 10 \times 7$$
$$= 840,000$$

Since each character takes a byte, the required capacity

$$= 840,000/1,024$$
$$= 820.3 \text{ Kbyte}$$

Thus, one (next integer value of 820.3/(1.2 × 1,024)) floppy disk was used to store the entire manuscript. Just one floppy storing the whole book! Also note that computer systems can store large amounts of data in a very small space (a 3.5-inch-diameter floppy disk is extremely small and light compared to the 400-page book).

Example 3.3

The 1987 *Encyclopedia Britannica* is 40,000 pages long. Each page has three columns; each column, 80 lines of text containing on the average (including blank spaces) seven seven-letter words. How many 120-Mbyte hard disks will be required to store the entire encyclopedia?

Solution

Total characters

$$= 40,000 \times 3 \times 80 \times 7 \times 7$$
$$= 470,400,000$$

With each character using one byte, the required memory

$$= 470,400,000/(1,024 \times 1,024) \text{ Mbyte}$$
$$= 448.6 \text{ Mbyte}$$

Number of hard disks

$$= 448.6/120$$
$$= 3.74$$

Thus, four hard disks will be required to store the entire encyclopedia.

Binary System. As letters are grouped to make words in languages, bits are combined to form **computer words**. While our words vary in length, computer words are of fixed length. A computer word can be as short as 4 bits for small microprocessors or as long as 64 bits or more for high-speed computers. Sometimes, the term **nibble** is used for a group of four bits; a byte has two nibbles.

To appreciate how large numbers are formed in the binary system, let us recollect how it is done in the decimal system. Single-digit numbers begin at 0 and end at 9, the next higher number, 10, is formed with a 1 at the left position and a 0 at the right position. The right digit, 0, then progresses through the remaining nine digits to generate the numbers 11 through 19. Next, the left digit, 1, is changed to 2 with a 0 again occupying the right position to form 20. Again the right digit increases, generating 21 through 29. This continues all the way up to 99. The next higher number, 100, is formed by adding a 1 at a new position on the left and changing the two 9s in 99 to zeros. The right two digits are then manipulated the same way to cover the range 100 through 199. Next, the left-position 1 progresses along with the other two digits to reach 999, and a new place is created on the left for numbers with four digits. Other higher numbers are formed the same way.

The binary system follows a similar approach to generate numbers such as 0, 1, 10, 11, 100, 101, etc. Shown below is a comparison between decimal numbers and their binary equivalents.

Decimal numbers	0	1	2	3	4	5	6	7	8	...
Binary numbers	0	1	10	11	100	101	110	111	1000	...

Note in the second row that when all the available positions are filled in by 1s, the next higher binary number is formed by creating a place on the left for an additional 1, then repeating the incrementing process with the digits on the right, exactly as in the decimal system.

The decimal numbers and their largest value that can be represented in the binary system depend on the number of bit positions used in the set. As you can see, with three bit positions, only eight decimal numbers (0 through 7) can be represented. If four bit positions are used, then 16 (0 to 15) decimal numbers are possible. In fact, the total count of decimal numbers (N) and the largest value (M) that can be represented in the binary system is

$$N = 2^n \tag{3.1}$$
$$M = N - 1 = 2^n - 1 \tag{3.2}$$

where

N is the total count of decimal numbers,
n is the number of bit positions, and
M is the largest of the N decimal numbers.

The following illustrates how the values of N and M increase with n:

n	4	5	10	16	20
N	16	32	1024	65536	1048576
M	15	31	1023	65535	1048575

Thus, larger binary numbers are formed in a way similar to that in the decimal system. Binary representation, however, takes more space than the decimal system, since only two bits are used. For example, the number 65535 requires just five places in the decimal system, whereas its binary equivalent (1111111111111111) requires 16 places. In computers, the more the bit positions—the wider the computer word—the more costly the system. However, a technique, called the floating point method, exists to represent large numbers with fewer bits.

Computer words are used to represent both numeric data (consisting of numbers) and character data (consisting of letters and special symbols such as %). Numbers are stored in fixed-point format, floating-point format, or both. The largest positive number that can be represented in the fixed-point format depends on the number of bits used. From Equation 3.2, a 16-bit binary system, for example, can represent in the fixed-point format decimal numbers only up to 65,535 ($= 2^{16} - 1$). This limitation is circumvented by using the floating-point format. Consider, for example, a small job shop's quarterly profit of $72,089.60. This data can be expressed in the popular scientific notation of base 10 as 7.20896×10^4, in which 7.20896 is called the mantissa and 4 the exponent. The floating-point format is similar to the scientific notation except that, instead of base 10, it uses base 2 of the binary system. Noting that $2^{16} = 65,536$, the mantissa for 72,089.60 expressed as a binary number would be $72,089.60/65,536 = 1.1$. Thus 72,089.60 could be expressed as 1.1×2^{16}. Any large decimal number can be expressed this way, as a mantissa part and an exponent part, the exponent being with respect to base 2. A certain number of bits of the computer word are used to specify the exponent part, while the remaining bits represent the mantissa part.

Besides numbers, other characters also need to be represented in the binary system. Seven bit positions are sufficient to represent in the binary system all types of characters we use. The seven bit positions yield a total of 128 (= 2^7) different combinations. These bit groupings have been standardized in ASCII code so that each alphanumeric character is represented by a unique set of bits. ASCII codes for the uppercase letters of the English alphabet and the 10 digits of the decimal system are given in Table 3.1.

When your boss responds to your request for a salary raise by typing N and O on a computer keyboard, the computer reads it as 1001110 for the letter N and 1001111 for O. When these bit patterns reach your monitor, your computer reconverts them into NO. A reply of 10011101001111, rather than NO, might have upset you more!

The computer industry has standardized ASCII for microcomputers. Mainframes use another code, the **Extended Binary Coded Decimal Interchange Code (EBCDIC)**, pronounced eb-see-dic. While ASCII is a seven-bit code, EBCDIC is an eight-bit code permitting 256 (2^8) different combinations.

For ensuring the accuracy of data transfer an additional bit, called a **parity bit**, is added on the left, resulting in an eight-bit transferable ASCII character. Since an eight-bit group is called a byte, we can say that a byte represents one transferable character.

The parity bit could be either 0 or 1. It is placed at the most significant place (on the left of the bit combination). Depending on whether the 1s in the byte add up to an odd or even number, the system is either **odd parity** or **even parity**. Consider, for example, the letter N, represented in ASCII by 1001110. Since it has four 1s, in the odd parity system a 1 will be added to the left to make the total count of 1s five, an odd number. Thus, in this system N would be **1**1001110. The even parity system would add a 0, since N's binary equivalent already has an even number of 1s. Thus, in the even parity system N is **0**1101110.

In any numbering system, the rightmost position has the lowest value and the leftmost, the highest. For example, in decimal number 258, 2 represents two 100s while 8 represents eight ones. This holds true for the binary system, too. In the

Table 3.1

ASCII codes for 26 letters and the 10 decimal digits

A	1000001	H	1001000	O	1001111	V	1010110	2	0110010
B	1000010	I	1001001	P	1010000	W	1010111	3	0110011
C	1000011	J	1001010	Q	1010001	X	1011000	4	0110100
D	1000100	K	1001011	R	1010010	Y	1011001	5	0110101
E	1000101	L	1001100	S	1010011	Z	1011010	6	0110110
F	1000110	M	1001101	T	1010100	0	0110000	7	0110111
G	1000111	N	1001110	U	1010101	1	0110001	8	0111000
								9	0111001

Table 3.2
Numbers represented in the four numbering systems

Decimal	Binary	Octal	Hexadecimal
0	0000	0	0
1	0001	1	1
2	0010	2	2
3	0011	3	3
4	0100	4	4
5	0101	5	5
6	0110	6	6
7	0111	7	7
8	1000	10	8
9	1001	11	9
10	1010	12	A
11	1011	13	B
12	1100	14	C
13	1101	15	D
14	1110	16	E
15	1111	17	F

binary system, the rightmost bit is appropriately called the **least significant bit** (**LSB**) and the leftmost the **most significant bit** (**MSB**). The two four-bit sets in a byte are called the **lower nibble** (the right four bits) and the **higher nibble** (the left four bits). For example, in the binary number 10111110, 1110 is the lower (less significant) nibble, while 1011 is the higher (more significant) nibble.

Octal and Hexadecimal Systems. Since binary representation requires more places than the corresponding decimal number, the binary system is not concise. Octal and hexadecimal systems have been developed to accommodate this limitation. In the **octal system**, eight digits, namely 0 through 7, are used and higher numbers are formed the same way as in the decimal or binary system. The octal system is not as concise as the decimal system. The **hexadecimal system** is based on 16 characters: the first 10 decimal digits and the first six letters of the English alphabet, A through F. Table 3.2 shows the first 16 numbers of the decimal system represented in the three alternative numbering systems.

The hexadecimal system has been developed to facilitate direct interaction between computers and users having specialized knowledge of the hardware. As you can see from Table 3.2, the hexadecimal system requiring the fewest positions is the most concise of all. It is a type of shorthand for the binary system. The octal and hexadecimal systems are used in developing microprocessor-based control devices. The associated digits are provided on special keyboards used in such applications.

A list of graphic and control characters in ASCII hexadecimal is given in Table 3.3. Note that the decimal digits 0 to 9 are represented by ASCII hexadecimal numbers 30 to 39, uppercase English letters are 41 through 5A, and the lowercase letters are 61 through 7A. The other characters have their own **hex** (short form of hexadecimal) equivalents; for example, $ as 24 and * as 2A. Besides the graphic characters, there are control characters such as 0D for a carriage return and 20 for a space.

A character can be converted into its binary equivalent using Tables 3.3 and 3.2. First, convert the character into hex using Table 3.3, then use Table 3.2 to convert the hex equivalent into binary. For example, the letter M is 4D in hexadecimal. From Table 3.2, hex 4 is 0100 in binary and D is 1101. Thus, M is 01001101 in binary.

To avoid confusion between the hex and decimal numbering systems, a common practice is to add the &H prefix to the hex notation. Thus, &H235 is hex 235, which is 565 in the decimal system. Since hex is not a decimal system, it is read differently; for example, &H235 is read as hex-two-three-five and not as hex-two-hundred-thirty-five.

The number of digits available in a system is called its base. Thus, the binary system has a base of two and the octal a base of eight. The generalized format for the decimal value of a number represented in any system is

$$V = A_n B^n + A_{n-1} B^{n-1} + \ldots \ldots + A_1 B^1 + A_0 B^0 \ldots \tag{3.3}$$

V = value of the number in decimal system
A_n = digit at the nth place (counting starts from the right as zero and proceeds leftward)
B = base

Example 3.4

Find the decimal value for the digits 10524 expressed in (a) the decimal system, (b) the octal system.

Solution

(a) For the decimal system, the base B = 10. The MSB 1 in the given set of digits 10524 is at position four (counting zero for the rightmost position occupied by 4), hence $n = 4$. From Equation 3.3, the value (V) of the group of digits 10524 in decimal system is, therefore,

$$
\begin{aligned}
V &= 1 \times 10^4 + 0 \times 10^3 + 5 \times 10^2 + 2 \times 10^1 + 4 \times 10^0 \\
&= 10000 \quad + \quad 0 \quad + 500 \quad + 20 \quad + 4 \\
&= 10524
\end{aligned}
$$

(b) In the octal system, B = 8. Therefore, the decimal value (V) of 10524 expressed in octal system is

$$
\begin{aligned}
V &= 1 \times 8^4 + 0 \times 8^3 + 5 \times 8^2 + 2 \times 8^1 + 4 \times 8^0 \\
&= 4096 \quad + 0 \quad + 320 \quad + 16 \quad + 4 \\
&= 4436
\end{aligned}
$$

Table 3.3
Graphic and control characters in the ASCII hexadecimal system

	Graphic or Control	ASCII (Hexadecimal)	Graphic or Control	ASCII (Hexadecimal)
NUL	Null	00	*	2A
SOH	Start of Heading	01	+	2B
STX	Start of Text	02	,	2C
ETX	End of Text	03	–	2D
EOT	End of Transmission	04	.	2E
ENQ	Enquiry	05	/	2F
ACK	Acknowledge	06	0	30
BEL	Bell	07	1	31
BS	Backspace	08	2	32
HT	Horizontal Tabulation	09	3	33
LF	Line Feed	0A	4	34
VT	Vertical Tabulation	0B	5	35
FF	Form Feed	0C	6	36
CR	Carriage Return	0D	7	37
SO	Shift Out	0E	8	38
SI	Shift In	0F	9	39
DLE	Data Link Escape	10	:	3A
DC1	Device Control 1	11	;	3B
DC2	Device Control 2	12	<	3C
DC3	Device Control 3	13	=	3D
DC4	Device Control 4	14	>	3E
NAK	Negative Acknowledge	15	?	3F
SYN	Synchronous Idle	16	@	40
ETB	End of Transmission Block	17	A	41
CAN	Cancel	18	B	42
EM	End of Medium	19	C	43
SUB	Substitute	1A	D	44
ESC	Escape	1B	E	4
FS	File Separator	1C	F	46
GS	Group Separator	1D	G	47
RS	Record Separator	1E	H	48
US	Unit Separator	1F	I	49
SP	Space	20	J	4A
	!	21	K	4B
	"	22	L	4C
	#	23	M	
	$	24	N	4E
	%	25	O	4F
	&	26	P	50
	'	27	Q	51
	(	28	R	52
	)	29	S	53

Graphic or Control	ASCII (Hexadecimal)	Graphic or Control	ASCII (Hexadecimal)
T	54	j	6A
U	55	k	6B
V	56	l	6C
W	57	m	6D
X	58	n	6E
Y	59	o	6F
Z	5A	p	70
[	5B	q	71
\	5C	r	72
]	5D	s	73
^	5E	t	74
–	5F	u	75
`	60	v	76
a	61	w	77
b	62	x	78
c	63	y	79
d	64	z	7A
e	65	{	7B
f	66	⌐	7C
g	67	}	7D
h	68	~	7E
i	69	DEL Delete	7F

Example 3.5

Find the decimal value of &H8B5.

Solution

In the hexadecimal system, B = 16. The value of n in this example is 2, since the MSB 8 is at the second place from the right (remember that 5 is at position zero and B at one). Thus, the decimal equivalent of &H8B5 is

$$V = 8 \times 16^2 + B \times 16^1 + 5 \times 16^0$$

From Table 3.2, hex B is equivalent to decimal 11. Thus,

$$
\begin{aligned}
V &= 8 \times 16^2 + 11 \times 16^1 + 5 \times 16^0 \\
&= 2048 \quad + 176 \quad + 5 \\
&= 2229
\end{aligned}
$$

Example 3.6

What is the decimal value of the binary number 110111?

Solution

Noting that the base in binary is two and that the value of n corresponding to the MSB 1 in the given number is five, the decimal value V becomes

$$V = 1 \times 2^5 + 1 \times 2^4 + 0 \times 2^3 + 1 \times 2^2 + 1 \times 2^1 + 1 \times 2^0$$
$$= 32 \qquad + 16 \qquad + 0 \qquad + 4 \qquad + 2 \qquad + 1$$
$$= 55$$

Example 3.7

A manufacturing supervisor has typed on a keyboard the message: Part # 619. What binary number group has been generated inside the computer?

Solution

Each character of the given data including the two blank spaces is first converted to hex equivalents using Table 3.3. The hex numbers are next converted to their binary equivalents using Table 3.2. For example, for the uppercase letter P the hex value from Table 3.3 is 50. From Table 3.2, hex 5 in binary is 0101 and hex 0 is 0000. Thus, P is 01010000 in binary. The complete conversion is as follows:

Character	Hex	Binary
P	50	01010000
a	61	01100001
r	72	01110010
t	74	01110100
space	20	00100000
#	23	00100011
space	20	00100000
6	36	00110110
1	31	00110001
9	39	00111001

Thus, the manufacturing data Part # 619, quite meaningful to us in this format, is understood by the computer as the following stream of 0s and 1s:

010100000110000101110010011101000001000000
001000110010000000011011000110001100111001

Computer Systems

A computer is a tool that helps its users carry out tasks faster and more efficiently. Early computers were expensive, bulky, unreliable, and required an air-conditioned room. With the advent of transistors, and later integrated circuits, they have become less expensive and enormously powerful, resulting in their widespread use.

With its peripherals and software a computer represents a complex device, and hence is called a system. A system comprises several elements that operate interactively to carry out a given task. In the case of a computer system, the various elements can be grouped into hardware and software. As explained earlier, hardware is the physical part of a computer and software is the nonphysical (programs) part. The hardware and software are strongly interdependent, with the selection and performance of one affecting those of the other.

Hardware.   The hardware of a computer system includes four main parts: the keyboard, display or monitor, printer, and the system unit containing the disk drive. Figure 3.2 shows the hardware of a typical desktop personal computer and a laptop or portable computer. Inside the hardware are a large number of electronic components that have been assembled on individual interconnected boards. Called **printed circuit boards** (**PCBs**), they are the basic building blocks of computer hardware.

Figure 3.3 shows a terminal that includes a keyboard, display, and modem for connection to remote computers. A terminal links its user to the main or host computer. Note that the terminal has no system unit.

The hardware consists of all the physical components that make up a computer. These components and units can functionally be categorized into four groups: central processing unit (CPU), memory, input, and output, as shown in Figure 3.4. The CPU consists of an **arithmetic logic unit** (**ALU**) and control unit. The categorization of the hardware in these four groups is appropriate to all computers, including minis and mainframes.

Central Processing Unit (CPU)

Also known as a processor, the CPU is the "heart" of the computer and does all the work for the system; it eventually processes all the data. The capability of a CPU is characterized by the number of bits (often expressed as bytes or words) that can be processed simultaneously. The word lengths for PCs vary from 4 to 32 bits in multiples of four. The latest PC processors use 32-bit words. A 32-bit processor allows large memory locations, since up to four billion (2^{32}) combinations are possible with 32 bit positions as compared to only 256 (2^8) with 8-bit processors.

Figure 3.4
Functional categorization of
computer hardware

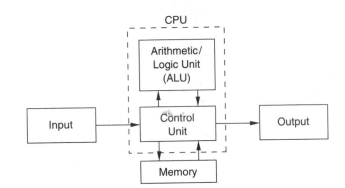

Source: Reprinted with permission from *Microprocessor
Architecture, Programming, and Applications*, by Ramesh S.
Gaonkar, Copyright 1989 Merrill/Macmillan Publishing.

Besides, a larger word size permits more powerful instruction sets and operating modes. The word length is like highway lanes: more lanes (bit positions) allow more traffic.

The words may not be equally wide throughout the computer. Sometimes, information flow between the CPU and memory is 32 bits wide, but on the **bus** it is only 16 bits wide (highways have more lanes near cities!). A bus is the group of wires that connect the devices—say, a printer to the CPU—to transmit the bits. Calling the connecting wire a "bus" is quite appropriate, since the bits can be thought of as "riding" over it.

In a microcomputer, a microprocessor performs the CPU functions. Figure 3.5 shows the four components of a microcomputer: microprocessor, input, output, and memory. These components share a common bus for communication.

The rate at which a processor carries out its tasks is called input rate of data through the computer system. It denotes the number of instructions executed per second and the rate at which the processor can "talk" to the connected devices or interfaces. Even PC systems operate at speeds as high as 66 MHz; high-speed systems are expensive. (Note: Because of rapid advances in technology, this figure may already be low; check current computer magazines or manufacturers for speeds

Figure 3.5
The microprocessor as a
major component of a microcomputer

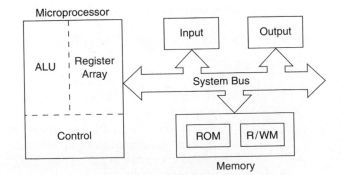

Source: Reprinted with permission from *Microprocessor
Architecture, Programming, and Applications* by R. S. Gaonkar,
Copyright 1989 Merrill/Macmillan Publishing.

available.) The processor can be interrupted from whatever it might be doing, since a computer system needs to keep track of several simultaneous tasks. All processors support interrupts to some extent. The way the processor handles interrupts affects the type of applications it can be used for. Most processors operate at three to five volts dc (direct current).

Memory

In general, computer memory falls into two categories: internal and external. Internal memory is physically very close to the CPU, whereas the external memory is not. Data and programs stored in internal memory can be accessed faster. Due to cost, however, the size of internal memory is limited. In applications involving large amounts of data, as in designs of complex products using finite element techniques, external memory is a necessity.

Internal memory may be either **random access memory (RAM)** or **read-only memory (ROM)**. RAM lets the user both read and write in a random way, whereas ROM can only be read off and not written onto. Though the terminology does not seem to suggest it, ROM is also accessed randomly. RAM is volatile, so everything stored there is lost when the computer is switched off or power fails. In contrast, ROM is nonvolatile (it holds the information permanently). Several types of ROMs are in use, including the standard ROM, PROM (programmable ROM), EPROM (erasable PROM), and EEPROM (electrically erasable PROM) or EAROM (electrically alterable ROM). The main difference among the various types of ROM is in the way data can be altered.

The internal memory may be static or dynamic. Static memory's limitation is that its contents are retained only as long as the power is on. Static memory is designed to store information without needing "refreshing." Refreshing means that the basically intermittent display is repeated at a rate fast enough to make it look continuous, as with ac-powered light bulbs. Slow refreshing gives rise to a flickering display. Dynamic memory requires refreshing and is used for high-density applications, where large memory is needed at a reasonably low price. Static column memory is a compromise between static and dynamic memory. Its main advantage is short access time, but it costs more than the other two types. Bubble memory, a pseudorandom magnetic memory, is the only type of RAM that preserves its contents even after a power loss. Slow access time and relatively high cost are the major disadvantages of bubble memory.

External memory refers to storage media outside the computer, though it may physically be close and under the same cover. It is also known as mass storage, since it can store larger programs and data than the internal memory. The three main types of external storage devices are floppy disks, hard disks, and magnetic tapes.

Interfacing

The CPU connects the three peripheral functional units shown in Figure 3.4—input, output, and external memory. Being physically inside the system unit, these connections are said to interface *within* to distinguish them from interfacing computers with external devices such as machining centers or robots. In nonintegrated environments, such as offices, users are not too concerned with interfacing, since the computers are usually stand-alone. In the CIM environment, however, for data collection or process control, computers are connected to a variety of machines and processes as well as to computers of different makes. Interfacing needs of this type may be termed external interfacing.

Almost all interfaces fall into two broad categories: serial or parallel. **Serial interfaces** allow communication of one bit at a time. An example is the **RS-232C** connection, a standard of the **Electronic Industries Association (EIA)**. The advantage of the serial technique is fewer required connections for interfacing two devices, resulting in easier installation. But, it has the drawback of lower speed of communication, and thus slower data throughput. A **parallel interface** can communicate an entire **packet** (one or several bytes properly packaged) simultaneously, resulting in higher transmission rates. Its main shortcoming is that it requires more data lines. Parallel interfacing is like a multilane highway, while serial interfaces offer a single lane. Most devices, such as modems or terminals, communicate serially while data acquisition and some printers require parallel interfacing.

Software.   As mentioned earlier, software is another name for computer programs that enable the computer to perform its tasks. According to the **American National Standards Institute (ANSI)**, "Software is a set of computer programs, procedures, and sometimes associated documentation concerned with the operation of a data processing system." Software allows the user to interact with the computer hardware, and through it to the outside world if required.

In general, software falls into one of two groups: **system software** or **application software**.

System Software

System software are programs that make the computer operate. The term system (refer to Section 2.6 for its definition) is used to describe the fact that such software may consist of several modules (subsystems or parts) that interact to achieve a well-defined goal. System software is developed to the specifications of a particular computer by its manufacturer or third-party software houses (companies specializing in software development). As users, manufacturing companies are usually not concerned with the intricacies of the operating system. The system software is procured along with the hardware; in fact, the two are purchased together as a computer system.

System software is an "umbrella" term for a collection of programs that may include an operating system, **utility programs**, or programming **language processors**. Of these three, users interface directly only with the utility programs.

The operating system is a special software that manages whatever happens inside the computer. It readies the hardware for the application programs. Like the traffic police, it controls the direction and flow of data over the electronic highways inside the computer. Trained personnel, called systems analysts, administer the operating systems of mainframes. For the microcomputers, DOS is the most popular operating system; UNIX and Windows are other common systems. In the case of microcomputers, the operating system is the major element of system software, and therefore the terms system software and operating system are used synonymously.

Utility programs, sometimes shortened to **utilities**, help to keep track of disk files and perform disk "housekeeping chores." Some useful utility commands are copy, format, directory, and erase. In the case of microcomputers, utility commands are often called DOS commands.

Language processors are programs that perform such functions as translating (source program codes into machine language), interpreting, and other tasks required for processing a specific programming language.

Firmware is a special program that has been stored permanently in the computer system. It is either hard-wired or stored in ROM, and therefore cannot be altered easily.

Application Programs

Application programs apply the computer system to user-specific tasks. It is the application program that produces the result. Application programs are critical in CIM, since they integrate the various functions of the manufacturer. They are so crucial that hardware and associated system software are selected only after the application programs have been chosen.

Application programs are written with a specific function or application in mind. They are usually available commercially. Sometimes users develop their own application programs or get help doing so from companies that specialize in custom software design. CIM operations may require both commercially available and specially developed application programs.

For developing application programs in-house, manufacturing companies employ computer programmers and/or hire outside consultants. Because off-the-shelf application programs are written for a large market, they may not serve the needs of individual users very well. But rather than develop a completely original application program, some companies may find it more cost-effective to procure and adapt a generic package to their needs. When a modification to the purchased program is planned, permission to access and modify the code must be obtained. Alternatively, the software supplier can be approached to do the modifications, usually for a fee, to suit the application.

Should one buy the hardware first or the application programs? Consider an analogy with the farmer's horsecart, where the hardware and system software represent the horse and the application programs represent the cart. Does the farmer buy the horse first or the cart? Normally, the farmer buys the horse first and then makes the cart, right? Not really. Before buying the horse, the farmer has already determined the cart size to suit his needs. Computer systems should be selected the same way. Decide first on the application programs (cart) and then proceed to select the computer platform (horse). Some questions relevant to hardware decision are:

Which tasks are suitable for computer applications?
Are application programs for these tasks available commercially?

If yes, can the cost be justified?

If no, should the application programs be developed in-house or contracted out?

Have the required resources, such as systems analysts, computer programmers, and consultants, been budgeted for?

Some software programs are not copyrighted; they are available in the **public domain** at almost no cost. Companies that specialize in this area distribute public domain software for a nominal fee, say $5. Authors of this type of software choose not to seek formal rights and royalties, rendering their work free to use and alter with few or no restrictions.

At times, software authors and users share their programs; such software is called **shareware**. Shareware distribution is a unique approach to software retailing in which consumers purchase directly from the authors at a fraction of the commercial cost, since marketing, promotion, and packaging efforts have been eliminated. Shareware authors allow their programs to be copied and distributed with few restrictions. Shareware users are encouraged to examine a program, copy it, and pass it on to other users. If a user finds the program useful, he or she is asked to register with the author for a nominal fee, usually under $100. In return, registered users receive the right to continue to use the software and other benefits such as full documentation, technical support, and future updates of the program. Registration generates funds for further development of the program. Thus, shareware is the try-before-you-buy method of software marketing. Not all shareware is in the public domain.

Several useful software packages for both technical and business functions in manufacturing are available in the public domain or as shareware. Examples are PCWrite for word processing, and **Personal APT (Automatically Programmed Tools)** providing two-and-a-half axis NC part programming capability for just $50.

Computer Languages. Software is developed using one of the computer languages. The language in which system software has been written is immaterial since application users normally do not need to make any change in the system software. Language becomes important, however, when a new application program is developed or an existing one adapted to suit a specific task.

Computer languages are either of low level or high level. The main difference between the two is in the extent of programming effort, with high-level languages requiring less effort. In a low-level language, the user programs the computer in 0s and 1s. The low-level language is also called machine language or **assembly code**, depending on how it is entered. In either case, the program "talks" directly with the processor, without any intermediate interpreter or compiler. This has both merits and limitations for application programming. The merits are faster response time, increased program flexibility, and better control. The limitations are excessive programming time, lack of data structures, required detailed knowledge of the hardware system, lack of programming standards, and no upgrade option.

Programming in a high-level language, on the other hand, can use English-like commands. The keyword PRINT used in **BASIC**, a high-level language, is simple to remember. High-level languages allow powerful operations with fewer programming steps, making them ideal for applications where speed is not important but data processing and other considerations are. Applications such as bookkeeping, accounting, database operations, and other complex tasks are well suited to high-level languages. The main differences between a high-level language and machine or assembler language are:

1. High-level languages do not require that users be aware of microprocessor features such as registers, I/O channels, internal representation of data, and so forth.
2. Programs written in a high-level language run on any computer provided appropriate compilers or interpreters are available.
3. High-level language programs are more concise.
4. Programs written in high-level language can be designed to use problem-oriented terms. In other words, they may let users express the problem in familiar words. An example is APT's word PARTNO; as a keyword for part number it is extremely user-friendly with NC programmers.

Computer languages are numerous. Some have been in use for a long time; the popular **FORTRAN** (FORmula TRANslator) developed in 1954 is almost as old as computers. More than 200 programming languages have been developed since the 1940s. Which of these are important for CIM? Will one single language be sufficient for all the tasks of CIM? These are difficult questions; probably no single language will be sufficient. The CIM environment requires both high- and low-level languages—the low-level by technical personnel for individual processes or workstations based on dedicated microprocessor control, and the high-level for other uses by both technical and nontechnical personnel. Table 3.4 describes some languages appropriate for CIM. As yet, no language has been developed specifically for CIM, though several can handle some of its major tasks, such as APT for programming NC machines.

The **C** language is quite popular today. It is a competitor of **LISP**, the original artificial intelligence language. One advantage of C is that it is closer to the machine language, enabling the processor to execute more efficiently. Moreover, programs written in C are easily transportable between computers, because C does not require the special processor board necessary for programs written in LISP.

Comparison.   Now that we have discussed the principles of computer technology, we can compare the three types of computers. Table 3.5 shows the results of this comparison, based on 11 different criteria. A close study of this table reveals that minicomputers overlap low-end mainframes and high-end micros. Modern microcomputers can be upgraded to be multiuser, in which case they may fall within the minicomputer category. With so many factors involved, it is clear that selecting a computer system for a given application is not a straightforward task.

Table 3.4
Computer languages for CIM

Language	Features (year, developer)
APT	Acronym for <u>A</u>utomatically <u>P</u>rogrammed <u>T</u>ools. A pre-processor that translates geometrically described tool movements into instructions understood by NC/CNC machines (1957, MIT)
BASIC	Acronym for <u>B</u>eginners <u>A</u>ll-purpose <u>S</u>ymbolic <u>I</u>nstruction <u>C</u>ode. An easy-to-learn popular language for general use. Originally designed to facilitate the learning of FOR-TRAN (1964, Dartmouth College)
C	So named because the letter C follows B which stands for Bell Laboratories (its developer). A compact, expand-able language. Portable between computers. Structured and popular with micros. Commonly used in developing application programs (1970s, Bell Labs)
COBOL	Acronym for <u>CO</u>mmon <u>B</u>usiness <u>O</u>riented <u>L</u>anguage. Suitable for business and financial tasks, excellent in file-handling, although wordy and at times confusing (1959)
FORTRAN	<u>FOR</u>mula <u>TRAN</u>slator. First high-level language, designed for scientific calculations. Contains an impres-sive library of engineering subprograms, lack of I/O commands (1954, IBM)
LISP	<u>LIS</u>t <u>P</u>rocessing. Symbolically oriented, AI and expert systems, interpreters, text-oriented filters (1958, MIT)
Modula-2	<u>MODULA</u>r language <u>2</u>. Designed as a replacement of Pascal. Subprogram based, easy to build libraries of repetitive tasks. Pascal's strength and C's flexibility (1977, Niklaus Wirth)
Pascal	Named after Blaise Pascal. Subprogram designed for ease of maintenance. Scientific and system use, effec-tive in teaching structured programming (mid-1970s)
PROLOG	<u>PRO</u>gramming <u>LOG</u>ic. French origins. AI, rule-based systems, database, interactive systems, text-oriented fil-ters. Native language for fifth-generation computers (mid-1970s)
VAL	A proprietary robot-oriented language (1970s, Unimation)

External Interfacing

Computer use in controlling processes and in collecting data continues to expand. It is achieved by interfacing computer systems with the various devices of the plant and office. Devices can range from sensors monitoring process parameters to those detecting sight, sound, or touch. Interfacing is also a prerequisite for networking computers and other CIM equipment for data transfer. Since Chapter 4 is fully devoted to the communications needs of CIM, computer interfacing is only briefly

Table 3.5
Computer types: A comparison

Criterion	Mainframe	Microcomputer	Minicomputer
Volume of data	High	Low	Medium
Application	Manufacturing	Manufacturing	Manufacturing
	Governments	Restaurants	City halls
	Banks	Personal finance	Insurance
	Universities	Schools	Colleges
User	Hundreds	Single	Tens
Cost (up to)	$7 million	$5,000	$300,000
Mobility	Low	Very high	Average
Mode	Batch	Real-time	Both
Main memory (up to)	100 Mbytes	10 Mbytes	50 Mbytes
Storage capacity	80 Gbytes	500 Mbytes	8 Gbytes
Speed (MIPS)*	500	20	100
(MFLOPS)**	100	15	75
Interfacing	Complex	Simple	Moderately complex
I/O Ports	Over 200	Under 20	15–200

*   Million instructions per second
**   Million floating point operations per second
Note: Supercomputers are multiuser, can cost up to $20 million, and are used in space
      exploration by NASA (National Aeronautics and Space Administration), in weather
      forecasting, and in scientific research involving large volumes of data at extremely
      fast speeds. Their use in manufacturing is limited to large corporations involved in
      extensive and sophisticated product designs and manufacture.

The data in this table are based on 1993 technology. Since computer hardware is changing rapidly, these data keep improving. For current data, refer to computer trade magazines or consult computer vendors.

described here. Again, communications involves two subsystems: hardware and software.

Communications Hardware.   Hardware designed and developed primarily for communications is called **communications hardware**. Since this type of hardware is in principle similar to computer hardware, the latter can serve the purpose of communications hardware, provided appropriate links to the outside world are made. This requires three basic elements: sensor, **ADC** (**analog-to-digital converter**), and cable.

Sensor

The data collection needs of CIM are fulfilled through sensors that pick up the signals from machines and other devices. The signals relate to the physical activities of interest. For example, a counter may count the parts as they are machined. Sensors are either physically attached (contact type) to the equipment or in close proximity to it (noncontact type). Various types of sensors are available for applications in manufacturing. The general categories are transducers, data-acquisition sensors, and data-conversion sensors. Transducers are used to measure the physical attributes of a process, which can be visual, aural, or based on any other physical phenomenon. Data-acquisition sensors are used to gather data (input) from computers, machines, processes, or other electrical sources (transducers).

Analog-to-Digital Converter (ADC)

Most inputs from transducers and outputs to controllers or actuators are analog signals, which must be converted into digital form for processing by the computer or analyzer. This conversion is ADC's job. Data conversion may be analog-to-digital or digital-to-analog depending on whether the process is monitored or controlled. Normally, manufacturers of data collection or control devices also provide the interface for linking the computer with the device. A recent trend has been to build the sensor and the digital conversion circuitry in a single package, called a chip card, that is inserted in a computer slot. Alternatively, the ADC circuitry is encapsulated within the sensor and the connector is plugged directly into a computer I/O port.

Cabling

Cables connect the various devices to each other—for example, the plant computer to a machining center. Various types of metallic cables are available to suit almost any task. A recent development in cabling and communications technology is the fiberoptic cable, which offers better noise immunity, clearer communication, and long-term affordability. Fiberoptic cables consist of glass fiber strands having protective coverings. Fiber-ready communications products are the latest in this technology. Further discussion on fiberoptic-based communications is presented in Chapter 4.

Communications Software.   Besides the hardware, communications software is essential to enable the various devices to "talk" with each other. The general discussion on software earlier in this section holds true for communications software as well.

3.2   RECENT ADVANCES

Advances in computer technology are taking place at a phenomenal rate. Some of the main developments of the recent past in hardware and software are discussed in this section.

Hardware Developments

Developments in microprocessor technology continue to enhance computers' performance-to-price ratio. On the basis of calculations for every dollar spent on computers, the improvement is phenomenal. As illustrated in Figure 3.6a, every dollar of investment in 1994 on a 486 machine (a PC based on Intel's 80486 microprocessor) translates into 100 trillion calculations; in 1980, this figure was only 10 billion (on Digital Equipment Corporation's VAX 11/780). During the 1980s, the number-crunching power of computers increased 10,000-fold. Furthermore, as shown in Figure 3.6b, the cost per million of PC instructions per second decreased tremendously. A typical microcomputer today is as powerful as the mainframe of 10 to 15 years ago. Their enhanced task-handling capabilities, greater speed, and smaller size make PCs more suitable for an increasing number of CIM applications.

Microprocessors.   At the root of all developments in computer hardware is the rapidly changing microprocessor technology. The trends have been higher speed, more powerful instruction sets, and lower power consumption. Lower power consumption results in decreased heat generation within the circuitry, which increases its life. Other important recent developments are parallel processing and neural learning networks. Such networks are large-scale integrated systems that can learn from examples of what an expert would do in a given situation.

In recent years, the following microprocessors have been introduced:

1. Motorola MC68030 32-bit microprocessor. Its main advantage over previous designs derives from incorporation of new technologies and some external functions into the CPU itself. One such technology is the **high-speed complementary metal oxide semiconductor** (**HCMOS**), which permits faster and more streamlined instruction execution over ordinary processors based on **complementary metal oxide semiconductors** (**CMOS**). The two main elements of a microprocessor—data and instruction cache memory, and inboard memory management—are on the same chip.
2. Motorola 88000 32-bit microprocessor. It is a three-chip set that features **reduced instruction set computing** (**RISC**), enabling more efficient instruction processing (see Box 3.2). One chip contains the processor itself and the other two the data and instruction caches. This arrangement yields a performance of approximately 17 MIPS (million instructions per second).
3. Intel 80386 32-bit microprocessor. Marketed in December 1987 as an extension of Intel's 80286, this processor can run at a clock rate of up to 33 MHz. The clock rate in number of pulses per second represents the speed at which processors access information. The 80386 offers both instruction and data prefetch that enable it to run at near-maximum speed with slower memories. It is effective in multitasking, which permits users to transfer information easily between applications, a clear benefit in a CIM environment. Nearly all new desktop computers incorporate the 80486 chip—the latest version of the 80386 microprocessor; such computers are known as 486 machines.
4. 64-bit microprocessor. The first 64-bit microprocessor built by Intel is also the first chip containing 1 million transistors. Code named N10, it was in 1993 the world's fastest chip, packaging the performance of the CRAY-1 supercomputer. Designed for high-speed multiprocessing tasks for three-dimensional use on workstations and in robotic applications, it integrates on a single chip the capabilities of a supercomputer with three-dimensional graphics.

Figure 3.6

(a) Growth in number of calculations per dollar of investment on computers, 1960 to the present; (b) Decrease in cost per million PC instructions per second

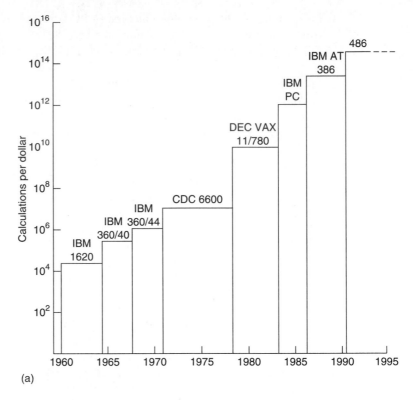

(a)

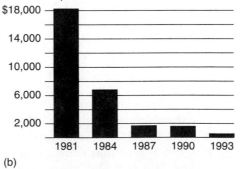

(b)

Source: Reprinted with permission from *U.S. News and World Report*, 26 July 1993, p. 48.

	BOX 3.2	*What's a RISC?*

There are two basic microprocessor architectures, or road maps, leading to the brain—the CPU. One is called CISC (complex instruction set computer), the other is called RISC (reduced instruction set computer). In CISC, lots of instructions are run through at once. In RISC, the instructions that are called upon most often go through faster than the rest. RISC provides more processing power for every dollar.

Source: From "What's a RISC?" September 1989, *Production*, p. 94.

Memory.   Most recent developments in memory technologies and associated products have been aimed at increasing the size and speed of the processor. Some dynamic RAMs available today are in the 4- to 16-Mbits range. This represents a 10-fold increase in size and two to three times the speed of earlier RAMs. Moreover, the newer chips have smaller real estate (physical size) and operate at lower voltage (Wollard and Pricer, 1988). These chips enable data reading in groups of four bits as well as one byte at a time. The 16-Mbit chip can read and simultaneously check the data for parity.

Another new technology for storing large amounts of data is based on optical principles. With more than 600 Mbytes of erasable storage (an improvement in the thousand range) and a transfer rate in the 5- to 10-Mbits per second range (almost unchanged), this technology is pushing existing hard disk capacity further. Optical discs are related to the current crop of **write-once-read-many** (**WORM**) and **CD-ROM** magnetic disks that are based on optical technology. The main attraction is the erasability of new drives, which allows their use in storing and recalling data (Freeze, 1988), similar to a hard disk drive.

A recent development in mass storage is the compact disc (CD) based on optical technology. Originally such discs were not erasable, which limited their use. In 1986, an erasable disc called **CD-I** (**computer disc-interactive**) was introduced. CD-I ROM hardware is now available (Bruno, 1987) and the associated technology is being perfected.

Software Developments

A significant change in software development is also taking place, primarily due to improvements in multiprocessor architectures. A high performance-to-cost ratio and high capacity make parallel computing commercially appealing. Parallel computing allows more higher-bandwidth graphic interaction between humans and machines.

As in the case of hardware, language processors that enable program development have gone through generations. The first-generation software was written in machine language using 0s and 1s, while the second generation used assembly language and mnemonic words. Software called an **assembler** translated the mnemonics into 0s and 1s. The second generation represented a 10-fold improvement in program development time. The third generation saw the introduction of high-level languages such as BASIC, COBOL, and C. Program development in these languages takes one-tenth the time that it takes in assembly language. The fourth-generation language processors, such as dBASE and similar specialized packages, have been used in database management and other new applications of computers. Currently, fourth-generation languages are in wide use. A fifth generation of language processors is cropping up to drive the newer computers based on parallel processing. These offer software development environments that can easily handle the complexities of manufacturing, thus accelerating the progress toward more-automated factories.

Multiuser Multitasking. Small computer systems such as a PC are designed for use by one person at a time. Users can carry out a single task, such as word processing or spreadsheet analysis, at a time. Such systems are called single-user **single-tasking**. Their operating systems are structured and developed with such an environment in mind. The most popular single-tasking operating system for microcomputers is DOS. It is also called MS-DOS, MS being an abbreviation for Microsoft Corp., which developed it around Intel's 8086 and 8088 microprocessors. When IBM put this operating system on its personal computers, it was called PC-DOS; it is now an industry standard due to the large number of IBM PCs and compatibles in use today.

A single-tasking operating system accommodates one user who works on one application at a time. Such a system is effective in carrying out jobs that are isolated from each other, i.e., when the associated computer files are unrelated. While several activities in manufacturing are single-tasking, most are interrelated—especially within a CIM framework. Examples are cost estimating and due date tasks that require information from process planning and **capacity requirements planning (CRP)** modules. Manufacturing professionals typically do several tasks at the same time. Obviously, **multitasking** systems that allow the user to access several files simultaneously are a real need in CIM.

Recently developed operating systems for PCs can handle several tasks at the same time. The impetus for such a development came from advances in microprocessors, such as 32-bit words, which can cope with the sophistication of multitasking. With a multitasking system, users can run several applications (also called programs, tasks, or sessions) concurrently. For example, a multitasking controller can be running production while allowing CNC programmers to develop the next part program, resulting in significant savings from increased machine utilization.

Along with the advent of Intel's 80286 processor, **operating system/2 (OS/2)** was developed to work with this microprocessor in a single-user multitasking mode. The OS/2, created by Microsoft under a joint marketing agreement with IBM, allows 16 Mbytes of memory to run programs. This represents a significant improvement in microcomputer capability over the 640-Kbyte working memory limitation with DOS. Moreover, the extended edition of OS/2 provides users database and communications capabilities, a further attraction from CIM viewpoint.

In spite of their capabilities, single-user multitasking operating systems such as OS/2 serve the needs of CIM in a limited sense only. What CIM needs is a multiuser multitasking operating system such as UNIX. With such a system, several users can share files, application programs, and other computing resources by having everyone's program loaded into the memory. Multiuser multitasking operating systems are already available on mainframes that were designed as multiuser multitasking processors. With multiuser multitasking capabilities on PCs, a new tool is available for advancing CIM.

Development of UNIX

True multiuser operation of computers requires a high-speed CPU, several megabytes of RAM, and an operating system designed to attend to several users running programs simultaneously. Unfortunately, the operating systems available in the market for multiuser operation are massive in size, more than a megabyte compared to 70 Kbytes of the MS/PC-DOS. Since a single CPU must do everything, these operating systems have tended to be rather slow. Operating systems such as UNIX or PICK, especially designed for multiusers, were too large for earlier PCs. This limitation is no longer there with the availability of 80386 and 68030 CPUs. Operating systems designed for standard PCs have also been upgraded to exploit the memory partitioning and multiuser capabilities of these CPUs. An example is the Microsoft-IBM operating system designed for OS/2 machines. Several companies such as Sun Microsystems have capitalized on the existing technology to develop their own networked super-microcomputer workstations operating under versions of UNIX.

In the microcomputer arena, UNIX is the leading software with multiuser multitasking capability (see Box 3.3). Developed by Bell Laboratories and promoted by AT&T (American Telegraph and Telephone company), it has several derivatives such as Xenix. Although fundamentally superior from a CIM viewpoint, UNIX is handicapped by the fact that a large number of application programs written for DOS are already commonly used in manufacturing. This makes the changeover from DOS to UNIX less attractive. Nonetheless, UNIX and other similar multiuser multitasking operating systems are better suited to CIM.

A special feature of UNIX and its derivatives is that it can be used with relatively minor modifications on all types of computers. UNIX is targeted at a level above the PC and below the mainframes. Its growth has been phenomenal. Until recently, the UNIX operating system was a hard-to-use, research-oriented system that had been around for years without attracting much commercial attention. In 1980, there were an estimated 10,000 computer systems running versions of UNIX; by 1990, the number had increased to around 2 million.

UNIX Structure

UNIX operating systems are interactive computer programs designed to support several users in multitasking operations. Essential components of all versions of UNIX system are the following:

Kernel. The **kernel** interacts directly with the hardware and responds to requests from services, such as I/O, and other parts of the operating system. It also

BOX 3.3   *UNIX Is Born*

Computer engineers at AT&T's Bell Laboratories in the 1960s had run aground in the development of a new, time-sharing operating system called MULTICS. Two of the contributors, Ken Thomson and Dennis Ritchie, began to design in 1968 another operating system that would preserve some of the positive features of MULTICS. To distinguish the new system from MULTICS it was called UNICS which was shortened to UNIX later. UNICS incorporated hierarchical filing, special files for devices and directories, a command interpreter called "shell," and the capability to support more than one user. The original designers of the system included a number of subroutines, called "pipes," that take the standard output of a process and make it the standard input of other processes. Another powerful subroutine is called "filter"; it sorts through the output of one process and delivers only selected results as the input of other processes. Using pipes and filters, it is possible to connect small, precise programs/activities into one large, complex program/activity very efficiently. The 1970 version of UNIX, which was granted a patent, had several limitations that were addressed by rewriting the kernel, shell, and utilities in a new high-level programming language called "C," completed in 1973. Some universities, notably Columbia University, obtained the UNIX's source code with the agreement of not disclosing

it. In the meantime, after refinement, Version 6 (the sixth edition) of the UNIX operating system was released in 1975. To capitalize on its popularity, AT&T started to license it. University users were quick to add features and write new applications.

In between 1975 and 1977, input/output subroutines were written by Mike Lesk to expand UNIX's acceptability to computers other than PDP-11 as long as a compiler for C was there. In 1977, a private commercial company, Interactive Systems Corporation, began to resell UNIX to third parties, backing it with installation and service. The same year University of California at Berkeley released their version of UNIX, which made UNIX more popular. However, UNIX did not catch attention outside research, government, and academia until 1980. In 1981, a version of UNIX for PCs was developed by Microsoft Corporation. The development of faster, smaller, more powerful microprocessor chips in the early 1980s fueled the popularity of workstations that exploited the capability of the UNIX operating system. In 1983, AT&T introduced the 'System V' version that contains a million code lines. With the RISC (reduced instruction set computer) architecture available, UNIX is bound to grow.

Source: From "The evolution of AIX" 1988, *AIX manual*, IBM Corp., pp. 35-38.

manages memory, enforces security, and monitors the multiuser environment. The kernels are tailored to the specific computer platforms they are running.

Shell.   The **shell** provides a flexible interface between users and the rest of the operating system. When a command is entered, it is the shell that interprets it and calls upon the right program or utility to perform the desired task.

File system.   Individual files and the directories make the file system. Files are defined simply as strings of characters; the UNIX file system knows nothing about record sizes. Directories consist of a number of files organized hierarchically, like the roots and branches of a tree. Security is achieved by defining files in terms of whether they can be read, written, or executed, and whether the user has permission to do all these functions or only one or two.

Programming tools.   The C and FORTRAN languages are offered with most versions. Also included are highly developed libraries of precompiled codes to aid in program development. Several user tools are also offered with provision for updating the operating system, performance monitoring, and so forth.

Communications.   As a minimum, programs for terminal emulation and file transfer are included. At its best, UNIX provides for distributed processing and other communications functions.

Versions of UNIX are riding two strong trends that are already reshaping the computer industry.

1. First is the networking that has become synonymous with computing. Since CIM may require a variety of hardware from different vendors and run a broad spectrum of application programs, there is a growing need for an operating system that can embrace all elements of the network and leverage their individual values. In manufacturing, computer systems with some version of UNIX are being bought at the department level rather than the corporate level, as was the case a few years ago. The 1980s saw a rapid growth in this trend, which continues in the 1990s.

2. Closely allied with the rise of networking is today's focus on open systems. At their best, open systems offer more options for expanding the networks and more ways to preserve the existing data and applications. UNIX and its versions offer the fastest path to realizing the benefits of software portability, **scalability**, and **interoperability**. Portability refers to users' freedom to run the same application program on computers from different vendors without having to rewrite the program codes. Scalability refers to users' ability to move applications and data among larger and smaller computer systems to meet the changing needs. This is especially crucial to companies that outgrow their startup facilities. Large manufacturing companies face scalability when they replicate application solutions at plants or divisions having different hardware needs. Interoperability refers to the ability to run programs on a network containing various kinds of computers of different makes.

Until an ideal UNIX becomes the common operating system (see Box 3.4) for the CIM environment, the current UNIX and its versions allow continual use of a huge number of programs written for the DOS operating system. They make it possible by providing a DOS server software that converts DOS files to UNIX and vice versa.

Expert System.   As an application of artificial intelligence, expert systems are becoming popular in several industries; manufacturing is no exception. More and more expert systems are being marketed for manufacturing use, which will only accelerate the growth of CIM. Some manufacturing tasks that are appropriate for expert systems are maintenance, process planning, and scheduling.

Interfacing

The upcoming processors and memories based on parallel processing and neural networks require new types of computer architecture. This in turn necessitates new designs for internal interfacing. The latest interfacing techniques are centered around fiberoptic technology. Some of the resulting advantages are noise immunity, low signal attenuation, and long-distance capabilities. To discourage fiberoptic's haphazard use in interfacing, a new protocol—fiberoptic distributed data interface (FDDI)—has been implemented as a backbone network over a fiberoptic link. FDDI works like the token-ring network in which a "disc" is passed around to control the access. Fiberoptic technology and FDDI can currently support transmission rates of up to 100 Mbit per second, with expected future speeds in excess of 1 Gbit per second.

Popularity of PCs

Over the years, microcomputers have become enormously capable. For example, in comparison to the DEC's VAX 11/780, which boasted one **MIPS (million instruction sets per second)** and one **MFLOPS (million floating-point operations per second)**, the PC/AT was in the 0.3- to 0.5-MIPS range. The 1980s will be remembered in the annals of CIM as the period when PCs started challenging their "big brothers," for a simple reason: more power for less money. A close look at Figure 3.6 shows why the present 486 machines are such a good value on the basis of calculations for every dollar spent. Although PCs were not originally designed for the hostile environment of the shop floor, the reduced prices and increased flexibility of micros have opened a floodgate. Today, industrial versions of the PC can operate in temperatures of 0 to 50 degrees Centigrade and 5% to 95% relative humidity. With 32-bit CPUs such as the Intel 80386, PCs are more and more finding a place on the shop floor.

The number of networked PCs continues to increase in the 1990s at an annual predicted rate of 30%. For CIM, more PCs networked together is a logical step forward. But lack of real-time capability of the PC is a serious limitation in the plant. Much of the manufacturing software currently available can operate only under MS-DOS, which inhibits PCs from meeting the real-time control requirements. A promising development is the emergence of operating systems that run concurrently with both MS-DOS and the multitasking UNIX (Boxes 3.3 and 3.4).

Due to their smaller size and weight, portable PCs (or simply portables) are becoming popular, especially with marketing and service personnel who travel frequently. Portables enable such people to communicate with the company mainframe directly from remote places using modem-based communication. This broadens the scope of CIM, since it allows salespersons to instantly link the customer with the manufacturing enterprise.

☐ **BOX 3.4**   *Observations on Operating Systems*

An operating system is the essential software that enables any computer to function. And nowhere is function more important than in manufacturing.

MS-DOS, OS/2, and UNIX are clearly the leaders in the world of PCs. UNIX and OS/2 are competing for the future, while MS-DOS continues evolving to meet present needs. MS-DOS users can find the operating system upgrade choices confusing. The options range from OS/2's "DOS compatibility box," which runs one MS-DOS application, to the 80386 UNIX implementations, which run multiple MS-DOS programs concurrently via the virtual 8086 mode provided by the 80386.

The number of people using the PC and the way they use it are important issues when deciding on an operating system. If only one person will use the system, or if several people will use several PCs in serial fashion, consider a single-user operating system (MS-DOS). If many people will require simultaneous access, pick a multiuser or networked environment. For connection to another computer, the operating system must be able to communicate with that computer's operating system. Investigate the communications protocols supported by the host system and determine what the link requires.

Once you've addressed these issues, it's time to look at the advantages and limitations of each operating system. "UNIX offers the best support for accessing large amount of memory," says Chatha. "It can directly address 32 MB of physical memory along with 4 GB of paged virtual memory, 3 GB of which are accessible by the user. UNIX also allows selective locking in of certain applications in memory. For real-time applications, all time-critical applications must be locked in memory, thus increasing the need for virtual memory for other applications that may also be running.

OS/2 can access 16 MB of directly addressable (segmented) memory. MS-DOS is by far the worst, with only 640 KB of directly addressable memory.

According to Chatha, UNIX is the only multiuser operating system in the group. This is an advantage for real-time data acquisition and control applications that frequently require multiple operator stations, each with its own CRT and keyboard. UNIX is unique in its ability to handle many RS-232 ports simultaneously. This gives an edge for many real-time applications because these ports are the access path for CRT terminals, data entry keypads, bar code wands, and intelligent sensors.

"UNIX systems also have excellent networking capabilities," says Chatha. "Some offer transparent sharing of data files and distributed processing over a network. OS/2 networks are becoming more useful, challenging UNIX with capabilities such as client/server relational database systems. MS-DOS networks are typically difficult to operate because of memory limitations."

Also keep in mind that UNIX supports higher resolution color graphics than OS/2 and MS-DOS. UNIX-based systems can represent 1280×1024 pixels, while OS/2 handles 1024×768 and MS-DOS, 640×480.

Chatha says that even with the limitations of MS-DOS, it will continue to dominate stand-alone systems for a long time for three reasons:

1. MS-DOS has a large installed base of existing applications that users have learned to work with and will continue to use as long as they can. This is especially true in manufacturing, where most users don't like to disrupt something that works.

2. Most programmers find MS-DOS easy to work with, so new applications will continue to flow.
3. MS-DOS systems are much less expensive than OS/2 or UNIX systems.

"UNIX still is the best setting for many PC-based manufacturing applications, however," concludes Chatha. "POSIX compatibility will be extremely important in all future operating systems and applications software. I favor the OSF version of UNIX with POSIX compatibility for manufacturing."

Source: From "Observations on Operating Systems" by J. R. Coleman, March 1990, *Manufacturing Engineering*, pp. 84–85.

Industrial Computers

In the beginning, computers were designed for use in the office or home environment. They could not operate for long under the harsh conditions of the shop floor. This prompted the development of industrial-grade computers (see Box 3.5), which are more rugged. Shock, vibration, airborne particles, chemical fumes, and rough handling can easily damage conventional disk memories. Bubble-memory boards and cartridges may be an answer.

Industrial computers combine the durability of **programmable logic controllers (PLCs)** with the processing power of PCs. The result is a breed of general-purpose computers that bring reliable and affordable computing power to manufacturing plants. Industrial computers are designed to work as manufacturing workstations, cell controllers, and gateways to **MAP** (manufacturing automation protocol) and other in-plant communications networks. They are also used as communications nodes between plant computers, operators, and PLCs. They can operate for hours in plant environments without air conditioning and have battery backup for protection against power vatiations, a common threat in production shops.

Industrial computers are real-time, distributed processing computers specifically designed to fill the gap between management information systems and shop-floor operations. They feature an open, distributed architecture and a real-time operating system that interfaces with shop-floor computers, sensing systems, and data-gathering devices. Thus, industrial computers "bridge" the gap between transaction-processing management information systems (MIS) and the real-time shop-floor operations that "speak" different languages. They are designed with scalable architecture so that the system can be updated later.

Industrial personal computers are increasingly being accepted for shop-floor use. Their lower cost, enhanced power due to 32-bit CPU chips, and improved graphics capabilities are the reasons behind the use in a variety of tasks, such as engineering analysis, cell control, tool management, and inventory control. IBM calls its PC/AT-compatible industrial computer (7552) an "electronic gearbox," probably to emphasize that it is a basic "building block" for plant computerization. The 7552 can operate within temperatures of 0 to 60 degrees Centigrade and relative humidity of 5% to 95%. It looks like a PLC and is housed in a vertically mountable chassis. To minimize the harm from electromagnetic and radio frequency interference, each plug-in board is housed in a metal shroud. One major difference

BOX 3.5 *PCs Tackle Industrial Applications*

Personal computers are playing an expanding role on the factory floor, as a variety of improved microcomputer boards, portable systems, workstation-like platforms, and software products illustrate.

On the factory floor, the personal computer has come a long way in a short time. Equipped with high-performance microprocessors and packaged into rugged enclosures, today's PCs have the power to tackle industrial applications and the toughness to withstand harsh factory environments. At the same time, powerful software specifically designed for industrial applications is appearing.

As a result, the industrial PC is assuming a broadening role. The PC serves as a tool for data acquisition, a controller of other instruments, and an interface to programmable logic controllers. The PC can itself function as an instrument, performing high-resolution readings of flow, temperature, pressure, level, or other phenomena at resolutions of a microvolt or less. It can perform sophisticated on-line data analysis and can port data to other applications programs for spreadsheet analysis, graphing, and mathematical manipulation. It can readily act as an interface between remote factory locations and control rooms, mainframe computers, and other PCs.

In terms of both applications and manufacturers, variety has come to the industrial PC arena. New offerings include microcomputer boards that plug into the PCs and turn them into powerful industrial systems, small portables that can be moved around the factory, and larger units that rival workstations in power.

Microcomputer boards, also known as board-level products, are one manifestation of the trend towards decentralized control in the factory. Some boards are intended to function as stand-alone computers. Others are built to be installed in PCs by plugging them into expansion slots. As PCs are typically equipped with several slots, this allows each PC within a factory to be customized to perform different embedded control, data acquisition, and monitoring functions. Typically, such systems can be put together at a fraction of the cost of equivalent mainframe or minicomputer systems.

Portability can be an asset in operations such as setting up temporary applications around the factory or performing off-site data entry. In this category, the small units available weigh less than 10 pounds.

Many industrial PCs pack so much power that their performance borders on that of larger workstations. In industrial applications, the systems are sometimes used remotely from their keyboard and monitor. For applications such as quality control, the computer, keyboard, and the monitor may be located together. For harsher environments such as process control, the computer is located in a safe place and the hardened keyboard and the monitor are located remotely on the plant floor, connected to the PS/2.

In the industrial environment, PLCs also find themselves distributed about the factory. The PLC, which preceded the PC into the factory, is a hardened industrial computer that is somewhat limited in function. It performs basic relay switching and contact-sensing functions. PCs are now often used to program PLCs.

The most important reason for the PC's invasion of the factory floor continues to be the availability of good software packages. Hundreds of MS-DOS software packages are available. New packages are fault-tolerant.

Though the majority of industrial PCs use MS-DOS as their operating system, according to many experts MS-DOS has a limited lifetime for industrial applications and will go the way of CP/M. For one thing, it has a 640K limit, too little for many industrial tasks. For another, it can only be used in a single task, and many industrial applications are multitasking. Although some multitasking has been tried with MS-DOS, no such standard exists. And MS-DOS does not match the power of newer hardware such as the powerful 80386 microprocessors. Furthermore, available application software is outstripping the capabilities of MS-DOS. These experts foresee MS-DOS being replaced with OS/2, Unix, and the Macintosh O/S operating systems. The first two are multitasking; the last one is expected to become so.

Source: From "PCs Tackle Industrial Applications" by D. Horn, December 1989, *Mechanical Engineering*, pp. 62-65.

between the electronic gearbox and other industrial computers is that its processor is not on a motherboard (another term for the system board, the main circuit board inside the computer), but on a plug-in daughterboard. This architecture has been designed to minimize the downtime that could result from failure of the motherboard circuitry. The current status of industrial computers, vis-à-vis manufacturing, has been presented by Horn (1988).

Workstations

A recent trend in computers are systems designed and built for special applications. Such computer systems are called **workstations**. An example is a CAD/CAM workstation that comprises hardware and software best tailored to CAD/CAM tasks. As an alternative to the workstation, users may adapt a general-purpose PC to suit their needs. Such an adaptation is less expensive than a comparable workstation. Horn (1989) published an article on the pros and cons of these two choices.

Desktop workstations have brought about some change in technology too. They are used not only for design application, but also for software development. Another trend is the development of software as building blocks, which makes specialized system software less expensive and reusable. The aim is to make a smooth and continuous integration of complete systems so that components can be implemented interchangeably either as hardware or software.

Fifth-Generation Computers

Most computers in use today are in their fourth generation. Fifth-generation computers, currently under development, are based on parallel processing, not on Von-Neuman's sequential design. They emphasize new architecture using advanced fabrication technologies and materials other than silicon. They also use languages having enhanced symbol manipulation and logic programming facilities, such as PROLOG and LISP. An increasing number of recent microprocessors implement the principles of fifth-generation computers whose development continues in the United States, Europe, and Japan.

3.3   COMPUTERS IN MANUFACTURING

As mentioned earlier, until recently, computer technology has been slow in addressing manufacturing needs. The earliest application of computers in manufacturing was in financial and administrative areas such as payroll and general ledger, primarily because these are well-delineated, repetitive tasks best suited for computers. In the early days, computers were used in small departments with no thought to integration. Each department was free to select and procure the system that best suited its needs. This practice of computer acquisition gave rise to data processing departments, which reported to the management, especially to the finance function. During the late 1970s—the period when CIM was just a concept—companies began to realize that compartmentalizing computer use in manufacturing was the wrong way to go.

During the punched card era of the 1950s and 1960s, computer use in manufacturing was limited to preparing job or order packets for material requisition, labor reporting, and job tracking. Routing sheets, parts listing, and other functions required data to be punched on cards. Around this time, FORTRAN debuted and IBM 1620 was the landmark computer. Computer use had begun to extend into engineering, but computer architecture was ill-suited for the needs of manufacturing, and computers had fewer I/O capabilities.

Next came manual data collection systems that allowed numeric data entry from different departments, such as production control, to the central computer. The keypunch devices were not computers, but they did demonstrate computer's capability for on-line job tracking and centralized dispatching activities. The next milestone was the use of computers in inventory management. It was spurred in the late 1960s by minicomputers as stand-alone systems for islands of automation. Some companies started developing their own software for tasks such as inventory accounting, materials requirements planning (MRP), operations scheduling, and capacity requirements planning (CRP).

In the 1970s, IBM introduced a **communications oriented production inventory and control system (COPICS)** in eight volumes for overall production and inventory control. COPICS was more than software. It provided a detailed view of the data flow in a manufacturing organization with an integrated system consisting of sales forecasts, engineering design, inventory control, requirements planning, shop-floor control, and more. The development of software began in various functional areas, the main one being the **bills-of-materials processor (BOMP)** task. Thus, COPICS contained the "seeds" of CIM and contributed significantly to development of CIM.

Further impetus to computer applications in manufacturing came from the introduction of COBOL. Before that time, the real impact had mostly been in production and inventory control. Many manufacturers introduced computers in their operations in the late 1970s and early 1980s due to developments in minicomputer hardware and software. During these two decades, computers' performance-to-price ratio improved annually by 25%. Computer vendors seized the opportunity by introducing their products in manufacturing. During this time, terms such as CNC, DNC, and CAD/CAM were coined. Computer use spread to design and shop-floor automation. This growth, resulting in numerous broken promises about computers' usefulness in manufacturing, led to the realization of the need for integration. Soon it was discovered that the once sought-after islands of automation defied integration, primarily due to the lack of attention paid while automating piecemeal. Today, almost everyone agrees that the right way to move forward in computerizing manufacturing is to "keep the forest in mind while planting the trees—one at a time."

Figure 3.7 summarizes the various application areas of computers in manufacturing. Note that the response time needed in manufacturing varies from seconds

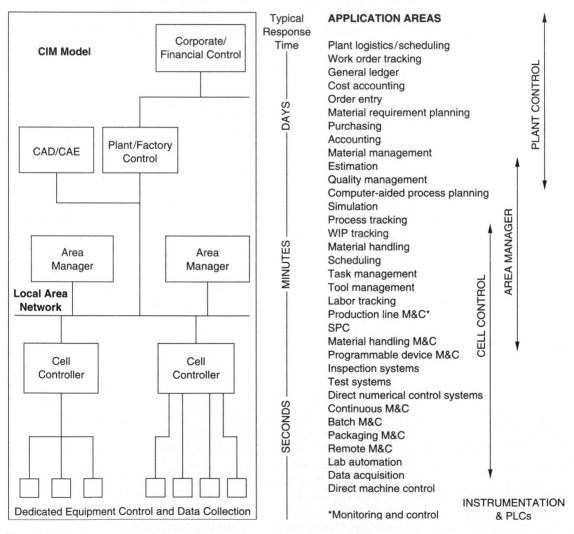

Figure 3.7
Factory tasks for computer integration as perceived by Hewlett-Packard
Source: Reprinted with permission from "Industrial Computers: Tough Enough?" by G. S. Vasilash, July 1988, *Production*, p. 43.

(sometimes fractions of a second when operating in the real-time mode), to days or weeks. Thus, CIM needs a hierarchical arrangement of computers with varying capabilities.

Present Status

The best way to assess the extent of computers' current use in manufacturing is to find out who the users are and what they are using the computers for. *Modern Machine Shop* (Gettelman, Marshall, Nordquist, & Herrin, 1989) surveyed a random sample of firms. The responses were grouped under the categories of NC and non-NC users. Companies were classified as small (less than 100 employees), medium (100-499 employees), and large (more than 500 employees). The survey results showed that U.S. manufacturers use computers in a wide variety of applications, as shown in Table 3.6. The table shows the percentages of manufacturers using computers.

A large variety of software is available for shop-floor use. Packages fall into one of the following five categories:

1. Cell control. Software for cell control is in demand due to the popularity of PCs and specialized cell controllers from control and machine tool manufacturers. Another reason is that the cell controller can also be used as a platform for accessing shop-floor information and other factory management functions.
2. Simulation. Simulation software allows computer analysis of existing facilities to enhance productivity or reduce costs. It is even more useful in the case of "greenfield" (new) facilities as a tool for planning appropriate acquisitions and their layouts.
3. Scheduling. Scheduling can be simplified significantly using well-developed commercial software rather than rudimentary programs or scheduling manually.
3. Process planning. Based on group technology and expert systems, process planning software represents a growth area.
4. Maintenance. Maintenance is crucial to the operation of modern facilities. Current software in this area is aimed at on-line diagnostics and predictive maintenance of equipment. It is available stand-alone or as a module within inventory management software.

An increasing number of computer systems are being introduced in manufacturing. Irrespective of whether the products are hardware or software and who the suppliers are, the questions of compatibility and integration with the current facilities are raised at the time of acquisition. More and more knowledge-based expert systems are aiding manufacturing.

Some current developments are:

Solid-model-based CAD systems with the advantage of unambiguous geometry representation are becoming popular.

Developments in hardware technology are making PCs more attractive to manufacturing.

Table 3.6
Percentage of manufacturers using computers

Functions	NC users			Non-NC users		
	Small	Medium	Large	Small	Medium	Large
Accounting	74	98	98	64	91	100
Bill of Materials	39	74	96	37	76	75
Capacity Requirements Planning	22	62	78	21	62	75
Database Management	37	74	87	28	67	75
Design/Drafting	28	74	91	17	48	50
Digitizing and/or Scanning	11	17	44	5	15	25
Drawing/Engineering Retrieval	19	67	70	17	33	25
Finite Element Analysis	11	24	44	4	19	0
FMS/FMC Control	9	17	26	1	0	25
Geometric Modeling	14	26	30	7	14	0
Group Technology	1	26	26	1	5	25
Inspection/Quality Control	19	57	57	19	33	50
Inventory/Purchasing Control	43	93	87	41	90	100
Job Costing/Estimation	37	67	70	32	67	50
Machine Shop	28	36	43	13	19	50
Machining Database Use	19	29	35	8	14	25
Management Data Reporting	39	74	74	31	76	100
Material Requirements Planning	18	79	87	11	71	75
NC Programming (< three axes)	48	60	87	0	0	0
NC Programming (> three axes)	31	45	61	0	0	0
NC Postprocessor Generation	24	45	60	9	0	0
Sheet/Plate Nesting	10	10	30	4	10	0
Order Entry, Scheduling	35	90	74	40	95	100
Plant Layout/Organization	12	43	44	4	24	50
Process Planning	10	43	30	5	43	50
Production Scheduling/Control	31	83	83	25	80	100
Proposal Generation	19	26	44	16	14	25
Shopfloor Control	14	43	52	6	24	25
Simulation Exercises	8	12	30	0	10	25
Tool Crib/Tooling Control	11	33	39	3	14	50
Work Measurement/Standards	14	52	61	12	38	0
Word Processing	61	71	83	55	95	100

Fourth-generation software is integrating more and more functions.
Fifth-generation hardware and software based on parallel processing have begun to provide CIM the necessary momentum.

3.4   NEEDS OF CIM

Computer-integrated manufacturing represents a complex environment, embodying all activities of designing, making, and marketing a product. The complexity of CIM is obvious from Figure 2.8. The functions represented by individual blocks in this generic model will vary in significance from plant to plant, depending on size and product variety. Nevertheless, most of them will need to be computerized to achieve a true CIM environment. This section discusses the needs of CIM for computer systems.

Hardware

Recent developments in hardware technology are having a profound impact on CIM. Functions such as manufacturing resources planning (MRP II), payroll, and other business data-processing tasks require a mainframe or minicomputer depending on the size of the company. If the company is also involved in product design and the product is complex, then its engineering needs such as finite element capability for stress analysis will be better served by a mainframe. This would also be the right computer for company databases.

With microcomputers becoming less expensive and simultaneously more powerful, PCs and PLCs can computerize most of the functions shown in Figure 2.8. Again, the size and complexity of the business determine what will work best.

More parallel-processing-based hardware and fifth-generation computers are needed for CIM to become widespread. Because of the interaction among numerous computers and other devices, a **local area network (LAN)** is needed for each geographical location. If the manufacturing business is carried out at different locations or in different countries, the LANs must be connected to each other via wide area networks (WANs).

CIM Software

CIM software is an integrated package containing as many individual programs functionally amalgamated into one as possible. The CIM environment requires application programs that can be easily integrated. While a small company may be able to use standard application programs, medium and large companies usually need to develop in-house software, which requires programmers with manufacturing insight. Large companies may have a software department to ensure that both the bought and developed programs are compatible with the hardware as well as with each other.

Typical CIM software should be able to handle all major production tasks in conjunction with other business tasks. It should, for example, be capable of:

1. Managing NC programs and their distribution,
2. Shop-floor data collection,
3. Interfacing to any NC/CNC system,
4. Handling manufacturing management functions such as prescheduling job flow and saving shop-edited programs,
5. Electronic mailing throughout the plant and office,
6. Creating database-accessible files using shop-floor data, and
7. Working in both DOS and UNIX environments.

CIM's progress is being expedited by developments in expert systems as well as in voice recognition and image processing technologies. Whether these will be implemented in the form of hardware or software will of course depend on the volume of the market. Voice recognition, for example, will allow a machinist to operate the machine verbally, thus eliminating the need for a part program. Expert systems are already automating the maintenance diagnostics in manufacturing.

CIM Workstations?

Although the PC is an excellent tool, workstations are likely to be the desktops of the 1990s. Workstations are specialized computer systems that have tailored hardware and software to carry out specific tasks more efficiently; the CAD/CAM workstation is an example. Workstations are fast and have sufficient memory to handle manufacturing tasks more efficiently. Workstation monitors offer higher resolution than typical PC screens, which facilitates display of multiple images and windows.

Once CAD/CAM workstations mature, the next step is the development of first-generation CIM workstations. Initial CIM workstations may be industry, product, or size specific. But by the turn of the twenty-first century, they may become generic to benefit from economies of scale, resulting in affordable CIM workstations (see Box 3.6).

BOX 3.6   *Workstations: The Next Computer for Manufacturing*

Computer workstations are about as common in engineering offices as coffee cups. One of the reasons why there is this proliferation of hot boxes in this office environment, an environment where plenty of personal computers are found behind other doors, is that they offer outstanding price/performance rates (measured in dollars and MIPS). They offer engineers a lot of bang for the buck.

Dr. Martin Piszczalski of the Yanki Group sees three ways to go. One is to link PCs in a network. They are inexpensive and fairly familiar devices for all those involved. But Piszczalski describes making the connections—the linking and coordinating—as "an awful task." Another approach is to base a system on a minicomputer and tie terminals into it. Among considerations: minicomputers are comparatively expensive and their user interfaces, while improving, are still dull stuff compared to those provided by workstations. The

third tack Piszczalski suggests is to use workstations. They are meant to be connected, so the hassle is comparatively minimal.

So let's go through all this.

1. Workstations are powerful computers. This permits computing on more than one front at a single time.
2. Workstations have highly graphical screens. This allows not only the display of wonderfully clear, three-dimensional images, but to put multiple windows on view simultaneously. Minicomputer output is alphanumeric intensive. Workstations can show and tell.
3. Compared to what's available for the PC, there isn't a heck of a lot of software available for the workstation—yet. (The situation will improve only when the independent software developers—the companies that manufacture software—get to work.) According

to George Rybeck, "There's so much talk about UNIX as being ideal that it's very easy to miss the fact that it's an ideal that doesn't have many applications yet." He points out that the independent software developers are waiting until there is a "core standard" version of UNIX before they'll start going gung-ho on writing applications.

4. Networking and multitasking are key to an understanding of the usefulness of workstations. Oak comments, "Manufacturing management are looking for decision support because that's what they do: they make decisions, and it involves knowing far more than what's available through Lotus 1-2-3. Their question is 'How do I get access to data, examine it, and compare it with other data?'" The answer is "Through a workstation."

Pitzczalski points out, "MS-DOS has a huge established base, and a company can do about 80 percent of all the functions on a shop floor with one" —SPC, operator instructions, order routing—things like that. "MS-DOS has well-known weaknesses"—single-tasking, networking snags, comparative low computing power, no graceful recovery after a crash, security problems, etc.— "but it can do many of the things that plants need. And it's dirt cheap—for all practical purposes, free," Piszczalski says.

But what about the future? "For a greenfield plant," Ted Rybeck answers, "DOS is not the operating system of tomorrow." He believes that OS/2, which runs on the IBM PS/2, is a viable candidate for that scenario. As is UNIX, which means workstations. "As applications are offered, UNIX will become a major contender, along with OS/2, for the future of the shop floor."

One of the barriers to workstation implementation within a manufacturing organization may be that there is already a considerable investment in alternative systems, such as networked PCs or minicomputer and terminals. But the primary consideration ought to be whether making the switch would provide a competitive advantage, and what it would ultimately cost the organization if that advantage is not realized.

Workstations hold the promise of truly integrated manufacturing. That is, designers and engineers depend on them right now. And workstation manufacturers are themselves developing manufacturing applications or are supporting this development. Because one of the primary advantages of workstations is networking, going back and forth from CAD to CAE to CAM is facilitated.

Source: From "Workstations: The Next Computer for Manufacturing" by G. S. Vasilash, April 1990, *Production*, pp. 34-38.

Security and Computer Viruses

The term **computer virus** is used to describe the situation in which an unauthorized person makes a computer system replace its memory contents with something of no value. This type of risk can be a serious threat in a CIM environment. PC-based systems and LANs, normally less security-controlled than the mainframes, are more vulnerable. Companies need to ensure that end-user systems that link PCs to the corporate computer cannot enter undesirable information, especially if it can replace existing information. In addition, a crisis management plan should be developed to help the CIM system recover in case of a viral attack. Databases and other erasable components must be backed up.

TRENDS

❑ Manufacturers in the United States and other developed countries continue to expand their use of computers; more than 90% of all plants use computers. The top five areas of application are CAD, production control, NC and CNC programming, inspection and quality control, and MRP/MRP II.

❑ Developments in computer technology are accelerating the growth of CIM. The 32-bit microprocessors are becoming the norm rather than the exception.

❑ PCs and their industrial versions are increasingly used in CIM operations. Fifth-generation computer systems, both hardware and software, aid CIM. Knowledge-based expert systems are entering into the realm of manufacturing.

❑ Workstations and PC-based computing are becoming common platforms for the needs of CIM. Newer microprocessors make this happen.

❑ It is possible to provide personal computers with parallel processing capability by installing accelerator boards containing transputers, which are CPUs executing in parallel. This makes PCs behave as minisupercomputers to handle large tasks such as finite element analyses.

❑ UNIX is not becoming as popular as predicted; this is delaying the progress of CIM.

❑ RISC (reduced instruction set computing) technology is being implemented in newer microprocessors. It allows more MIPS (million instructions per second) to be squeezed from current CPU technology. RISC-based computer systems provide next-generation processing speeds on current machines.

❑ Computers with footprints small enough for the user desk can handle 10 to 15 MIPS; they are as powerful as the mainframes of a few years ago. This trend should continue.

❑ **Optobus** is being developed to increase computing speed even further. Based on fiberoptics, optobus is an optical data bus embedded in printed circuit boards. A single optical fiber carrying serial data is significantly faster than 64 wires carrying parallel data. Its advantages include increased computation speed, immunity to electromagnetic interference, thermal insensitivity, and system reliability.

❑ High-performance programs that once ran on mainframes and dedicated terminals are being implemented on workstations.

❑ Rapid progress is being made in high-performance graphics including high-speed integrated 3-D graphics of surfaces and solids.

❑ The use of touch-screen technology in CIM products such as cell controllers and MCUs enhances human interfacing.

❑ CAD/CAM workstations continue to mature to evolve as CIM workstations.

SUMMARY

The basic principles of computer technology have been discussed in Chapter 3. Common terminologies have been explained so that readers, CIM engineers, technologists, and managers may feel comfortable with computer-related matters. The salient features of computers have been described. A brief history of computer use in manufacturing has been presented along with its current status. Recent advances in hardware and software have been discussed. Finally, the computer needs of CIM have been examined. The popular CAD/CAM workstations are expected to broaden in scope to become CIM workstations. There is no doubt that developments in computers are a boon to CIM.

KEY TERMS

Alphanumeric characters

American National Standards Institute (ANSI)

American Standard Code for Information Interchange (ASCII)

Analog-to-digital converter (ADC)

Application program or software

Arithmetic logic unit (ALU)

Assembler

Assembly code

Automatically programmed tools (APT)

Beginner's all-purpose symbolic instruction code (BASIC)

Bill-of-materials processor (BOMP)

Binary

Bit (Binary digit)

Bus

Byte

C

Capacity requirements planning (CRP)

CD-ROM

Central processing unit (CPU)

Clock rate

Common business oriented language (COBOL)

Communications hardware

Communications oriented production inventory control system (COPICS)

Compiler

Complementary metal oxide semiconductor (CMOS)

Computer disc—interactive (CD-I)

Computer language

Computer program

Computer virus

Computer word

Desktop computer

Disk operating system (DOS)

Electronic Industries Association (EIA)

Even parity

Extended binary coded decimal interchange code (EBCDIC)

External memory

File

Firmware

Floppy disk

Formula Translator (FORTRAN)

Hard disk

Hardware

Hex

Hexadecimal system

High-level language

High-speed complementary metal oxide semiconductor (HCMOS)

Higher nibble

Host computer

I/O device

Input device

Interface

Internal memory

Interoperability

Interpreter

Kernel

Language processor

Laptops

Least significant bit (LSB)

List processing (LISP)

Loaded

Local area network (LAN)

Lower nibble

Low-level language

Machine language

Mainframe

Mainframe computer

Main memory

Manufacturing automation protocol (MAP)

Mass storage

Microcomputers (micros)

Microprocessor

Million floating-point operations per second (MFLOPS)

Million instructions per second (MIPS)

Minicomputers (minis)

Modem

Modular language 2 (Modula-2)

Most significant bit (MSB)

Mutitasking

Nibble

Object program

Octal system

Odd parity

Operating system

Operating system/2 (OS/2)

Optobus

Output device

Packet

Parallel interface

Parity bit

Peripheral device

Personal APT

Personal computer (PC)

Platform

Portables

Printed circuit board (PCB)

Programmable logic controller (PLC)

Public domain

Random access memory (RAM)

Read-only memory (ROM)

Records

Reduced instruction set computing (RISC)

RS-232C

Scalability

Serial interface

Shareware

Shell

Single tasking

Software

Source program

Supercomputer

System commands

System software

Terminal

Time-shared computer

Utility programs or Utilities

Word

Workstation

Write-once-read-many (WORM)

EXERCISES

Note: Exercises marked * are projects.

3.1   A manufacturing company's inventory contains 10,000 different parts. Each part is given a six-digit identification number, a four-digit plant number where it is manufactured, a 10-letter product name in which it is assembled, and a 25-letter additional description. How many 1.2-Mbyte floppy diskettes will be required to store the inventory data?

3.2   Based on the 1990 U.S. census, there are 248 million people in this country. If each person is provided with a nine-digit social security number, how many 120-Mbyte hard disks will be required to store these numbers, along with individual names and addresses. Assume that each name and address requires an average of 70 characters.

3.3   The United Nations wants to develop a directory of every person living on this planet. The required data on each person and the necessary character sizes are:

Name and address	70
Weight	3
Age	3
Sex	1
Education level	3

Race	5
Nationality	10

If the world has 6 billion people, how many 120-Mbyte hard disks would be required to hold this information?

3.4 If 12 bit positions are available in a computer system, what is the largest decimal number that can be represented in fixed-point format? How many different decimal numbers can be represented?

3.5 Explain the difference between odd and even parity systems.

3.6 Write your first name in ASCII binary codes.

3.7 Write in ASCII hex codes your full name, ending with a period. Include the code for blank spaces between words.

3.8 Convert &H2A3E into its decimal equivalent.

3.9 Convert &H6C into binary form.

3.10 Write the binary number 10111101 in hex notation.

3.11 Write a technical specification of your or your friend's PC.

3.12* Visit a nearby mainframe computer installation. Talk with the system manager about its capabilities and prepare a technical specifications summary.

3.13* Collect data on any three mainframe computers and compare them on the basis of cost, memory size, and five other important criteria.

3.14* Repeat the last exercise for PCs.

3.15* Visit a local manufacturing company and write a 500-word report describing its computing facilities, including the applications categorized as in Table 3.6. How much of computer integration exists in the company?

3.16 Circle T for true or F for false or fill in the blanks.
  a. Some recent microcomputers use 32-bit words.   T/F
  b. A transferable byte is 8-bits wide.   T/F
  c. A byte and a computer word are the same thing.   T/F
  d. CIM is impossible without a mainframe computer.   T/F
  e. RISC is British spelling of the word risk.   T/F
  f. A floppy disk is hardware.   T/F
  g. BASIC is a high-level language.   T/F
  h. ASCII is a seven-bit code.   T/F
  i. ASCII is an acronym for _____
  j. RISC is an acronym for _____

SUGGESTED READINGS

Books

Bagadia, K. (1988). *How to select and justify manufacturing software*. Dearborn, MI: SME.

Dologite, D. G., & Mockler, R. J. (1988). *Using microcomputers*. Englewood Cliffs, NJ: Prentice-Hall.

Gaonkar, R. S. (1989). *Microprocessor architecture, programming, and applications*. New York: Merrill/Macmillan.

Halevi, G. (1980). *The role of computers in manufacturing processes*. New York: Wiley.

Heaton, J. E. (Ed.). (1986). *The expanding role of personal computers in manufacturing*. Dearborn, MI: CASA/SME.

Rembold, U., Armbruster, K., & Ulzmann, W. *Interface technology for computer-controlled manufacturing processes*. New York: Marcel Dekker.

Stocker, W. M., Jr. (Ed.). (1983). *Computers in manufacturing*. New York: McGraw-Hill.

Monographs and Reports

AIX. (1988). IBM Corp.

Automation Research Corporation, Metfield, MA. *PCs in manufacturing: Issues, trends, and opportunities*. The handbook of manufacturing software. Madison, GA: SEAI Technical Publications.

Gettelman, K. M., Marshall, H., Nordquist, W., & Herrin, G. (Eds.). (1990). *Modern machine shop 1990 NC/CIM guidebook*. Cincinnati: Gardner.

Journals and Periodicals

Computers and Industrial Engineering. London: Pergamon.

Popular magazines on personal computers, such as *Personal Computing*, *Desktop Computing*, *Computerworld*, and *Byte*.

Modern Machine Shop. May 1990 (emphasis on systems and software).

Articles

Bruno, R. (1987, November). Making compact discs interactive. *IEEE Spectrum*, pp. 40–45.

Diesslin, R., & O'Connor, F. (1990, June). A software selection checklist. *Modern Machine Shop*, pp. 98–103.

Freeze, R. P. (1988, February). Optical disks become erasable. *IEEE Spectrum*, pp. 41–45.

Horn, D. (1988, July). PCs gain a foothold in factories. *Mechanical Engineering*, pp. 44–49.

Horn, D. (1989, December). PCs tackle industrial applications. *Mechanical Engineering*, pp. 62–65.

India abroad. (1989, 16 June). *New York*. p. 10.

Martin, J. M. (1989, June). Personal computers in manufacturing. *Manufacturing Engineering*, pp. 44–46.

Vajpayee, S., & Hajjar, I. (1990). Recent developments in computer hardware and their effect on computer-integrated manufacturing. *Computers and Industrial Engineering*, 18(2), pp. 201–209.

Vasilash, G. S. (1988, July). Industrial computers: Tough enough? *Production*, pp. 36–44.

Vasilash, G. S. (1990, April). Workstations: The next computer for manufacturing. *Production*, pp. 34–39.

Wollard, K., & Pricer, W. D. (1988, January). Solid state expert opinion. *IEEE Spectrum*, pp. 44–46.

Yarnish, Rina. (1989, March). Enhancing the BASIC language: Implications of the new ANSI standard. *Academic Computing*, pp. 28–29, 41–43.

CHAPTER 4
Fundamentals of Communications

"Computer and manufacturing are like the bread of a sandwich. The meat is 'integrated'! Without the integration we have kidded ourselves that we are clever."
　　　　　　　　　—Leo Roth Klein, Manufacturing Control Systems

"Integration means the direct transfer of information between two independent systems. If an intermediate datafile such as the Initial Graphics Exchange Specification (IGES) or the Product Data Exchange Specification (PDES) is needed, there is no true integration. While interfacing is important as the common denominator between two systems, it does not provide the performance and completeness required for tomorrow's environments."
　　　　　　　　　—C. Hayden Hamilton, PDA Engineering, Costa Mesa, CA

"Information processing represents at least 70 percent of the effort expended in the operation of a manufacturing enterprise. Key to full-scale CIM is the ability to harness this information at the right time, in the right format, to make the right decisions."
　　　　　　　　　—Computer-aided Manufacturing International (CAM-I)

"It's less a matter of how fast the computer can compute than it is of how fast it can get the information to where it is wanted."
　　　　　　　　　—Gary S. Vasilash, executive editor, *Production* magazine

"Just as computer technology is moving toward massively parallel architectures, network technology should be moving toward massively parallel networks. Network performance must be kept in balance with computer and software performance. All three areas— computer, network, and software—form the basis for performance-tuning the (CIM) environment."
　　　　　　　　　—C. Hayden Hamilton, PDA Engineering

"Move the expertise without moving the expert. With remote control software, you can access any PC, whether around the block or around the world."
　　　　　　　　　—David Angell and Brent Heslop in *Portable Office*, August 1991

I n success with computer-integrated manufacturing, the name of the game is integration. All elements of the manufacturing enterprise—machines, operators, computers, managers, and consumers—are integrated so that each can communicate with all the others it needs to.

Integration facilitates communication. This chapter covers the basic principles of communications and related developments. It also shows why information is probably the most important resource in modern manufacturing.

4.1   INTRODUCTION

CIM requires that the engineering and manufacturing information be communicated to proper points within the operation, as illustrated in Figure 4.1. As you can see, this task is quite involved, since several "players" participate in producing a tangible product.

Communications comprises three conceptual components: representation of data, the **medium** (through which data flow), and networking (of the devices). This chapter discusses each component in detail.

A lack of standardization among the devices used in manufacturing has been the biggest stumbling block to the proliferation of CIM. This limitation is being addressed under the **Manufacturing Automation Protocol** (**MAP**) and also through developments in local area networks (LANs). Both MAP and LAN, as well as the International Standards Organization's (ISO) **Open Systems Interconnection** (OSI) model, are discussed in this chapter.

All the data and information that flow through the CIM enterprise are generated, acquired, or collected prior to their storage in databases. From there they are retrieved whenever needed. Chapter 5 covers database technology and management.

4.2   INFORMATION: THE NEW "M"

Until recently, manufacturing was thought to require four basic inputs—man, machine, material, and money. These inputs have classically been known as the four "Ms" of production. Operators process materials using machines to produce specific tangible goods, and all these require money (capital). Today, another input—data, or information—has become critical. We can call the new input "message" so that it becomes the fifth "M" of production. Messages are either stationary or mobile. The stationary data are stored in databases, the subject matter of the next chapter. In this chapter, we consider the fundamentals relating to the mobility of data, that is, communications.

Mobile data are either (a) commands that instruct the recipient—machine, operator, or material handling system such as an **automated guided vehicle** (**AGV**); or (b) reports from the sources to the central computer or upper management. Commands flow downstream to lower levels, whereas reports travel upstream in the reverse direction. In a hierarchical system, the communicating pairs are connected bidirectionally as downlinks for downstream messages and as uplinks for upstream messages.

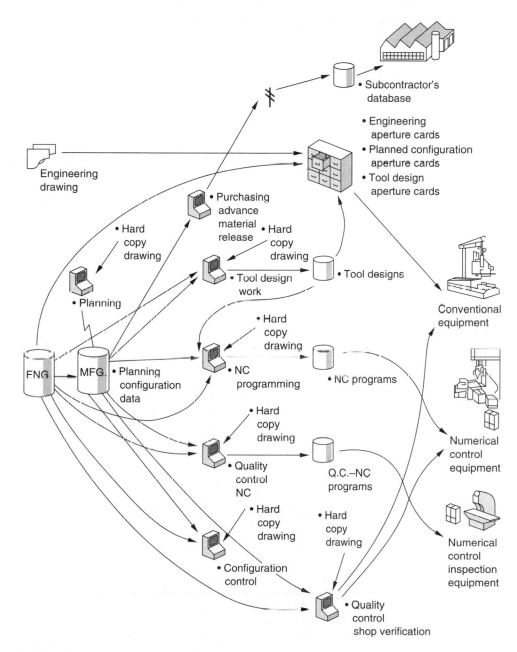

Figure 4.1

Data flow from engineering to shop floor via manufacturing

Source: Reprinted with permission from *Computer-Integrated Manufacturing Handbook* by E.
Teicholz and J. N. Orr (Eds.), 1987, New York: McGraw-Hill.

In real-world manufacturing, information may not be available at the right place in the right format at the right time. CIM attempts to improve this situation by moving the information from the person or device possessing it, called the source, to where it is needed, the sink. This is what communications is all about in the context of computer-integrated manufacturing—for that matter, in any context.

4.3   COMMUNICATIONS MATRIX

From the CIM viewpoint, those that need to communicate can be grouped into three "families": computers, personnel, and processes or machinery such as machine tools. These families need to communicate in an error-free, as-and-when-needed basis.

All the computers and digital devices the company uses are represented in Figure 4.2 by block C. Similarly, all the machines and the personnel are represented by the other two blocks, M and P. From a communications viewpoint, these three blocks or subsystems comprise the manufacturing system. Within each subsystem could be tens or even hundreds of visibly independent units, i.e., sources and sinks.

The communications needs of CIM are basically two-fold. First, any unit within a block or family should be able to communicate with another of its kind. For example, any computer of the enterprise should be able to communicate with any other computer, or a machine with another machine. Second, every unit of a family should be able to communicate directly or indirectly with any other unit of the other two families. Obviously, the first need is easier to fulfill than the second one, since the units of the same family share certain traits.

In light of the preceding discussion, the informational needs of manufacturing can be represented by a three-by-three source–sink matrix, as in Figure 4.3. The three blocks of Figure 4.2 generate this matrix to yield nine possible combinations. From the matrix, the various communication needs of CIM, therefore, are:

1. Person-to-person (P-P)
2. Computer-to-computer (C-C)
3. Machine-to-machine (M-M)
4. Person-to-computer or computer-to-person (P-C and C-P)
5. Person-to-machine or machine-to-person (P-M and M-P)
6. Computer-to-machine or machine-to-computer (C-M and M-C)

Figure 4.2
The three "families" of communicators under CIM

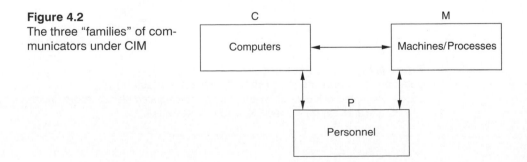

Figure 4.3
Communication needs of CIM as represented by a 3×3 matrix, yielding nine possible combinations

	P	C	M
P	P – P	P – C	P – M
C	C – P	C – C	C – M
M	M – P	M – C	M – M

The communications needs of CIM are thus diverse. It is this diversity that prevents implementation of full-blown CIM to achieve totally automated factories.

Person-to-Person Communications

The need for person-to-person communications has existed since the very beginning of industrialization, when one person could not carry out all the tasks of manufacturing. Person-to-person communication has its merits and limitations; for example, the use of gestures to make a point (or a threat!), ambiguity, the mood element in communicating with the boss, and so forth. Even true CIM plants may not be able to operate without this mode of communication. However, rather than communicate directly or through hard copies (information on paper), employees communicate more and more via keyboards and "soft" copies (video display units).

Computer-to-Computer Communications

The earliest attempt to promote computer-to-computer communication was made in 1969 by the U.S. Department of Defense's **Advanced Research Projects Agency** (**ARPA**). This agency initiated the development of a nationwide experimental computer network, called **ARPANET**, to promote computer resource sharing and to pioneer packet switching technology. Although developmental, the network proved so successful that it was soon adopted for use by various defense research establishments as well as researchers at major universities. It encouraged communication of three types: person-to-person, person-to-computational resources, and computer-to-computer. Three basic end-user functions, namely electronic mail, remote log on, and file transfers, became possible. These functions are based on Transmission Control Protocol (TCP) and Internet Protocol (IP). The TCP/IP had a profound effect on ISO's architecture for **Open Systems Interconnection** (**OSI**), on which MAP is based. Over the years, the network has benefitted from improvement in the capabilities of telecommunication lines. From 56 kilobits per second (Kbps) in the beginning, the line speed has increased to provide a data flow rate of 1.544 megabits per second (Mbps); this decade, that speed is likely to get as high as one gigabit per second (Gbps).

Computer-to-computer communication has grown tremendously over the last two decades. Most developments took place in the 1970s in response to the need to access large mainframe computers for jobs that minicomputers could not handle. In addition, newer microcomputers were designed with communications in mind. Point-to-point communication between computers using micros is quite common today. In the name of CIM, this technology is being extended to include machine tools and other equipment used in design, inspection, and other tasks. Newer machine tools have built-in capabilities that permit direct communication with computers. Most discussions in this chapter are targeted at computer-to-computer communications, since this forms the basis of all communications in CIM.

Machine-to-Machine Communications

"Machine" here refers to any device or equipment other than computers. Production machines of various types may communicate with each other through automation. In an automated machine, the operations are capable of communicating with each other without any help from the operator. Machine-to-machine communication takes place using mechanisms—mechanical devices such as gears or cams that transfer motions. In several situations, a conversion of mechanical input to electrical signals, and back again to mechanical output via actuators, using electromechanical devices is the basis of communication. Mechanism-based automation is "hard" communication. Today, the trend is for "soft" or programmable communication as embodied, for example, in a robot or **automatic tool changer** (**ATC**).

Person-to-Computer or Computer-to-Person Communications

Person-to-computer communications is the most obvious modern type in a typical plant. The increasing need for this type of communication is behind the development of high-level computer languages. So is the trend for user-friendliness of equipment and software. Ergonomic considerations in the design of keyboards and other hardware units are meant to render person-to-computer communication less tiring.

Person-to-Machine or Machine-to-Person Communications

The amount of direct communication between machines and their operators is significantly low under CIM. For example, the operator does not need to remember the appropriate values of cutting speeds and feeds for a given machining job. This information is retrieved from the database and appended to the part program that directs the machine operations. CIM, on the other hand, demands better communication between machines and the maintenance crew to maximize the machine utilization.

Computer-to-Machine or Machine-to-Computer Communications

Machines suitable for a CIM environment have computers or built-in microprocessors so that computer-to-machine communication is direct, involving minimal interfacing. Lack of computer-to-machine communication represents a bottleneck in the development of CIM.

4.4   FUNDAMENTALS OF COMPUTER COMMUNICATIONS

Some common terminology relating to communications are:

Data—entities that convey meaning
Information—the content or interpretation of data
Signals—electric or electromagnetic **encoding** of data
Signalling—the act of propagating the signal along a medium
Transmission—propagation of data by processing of signals

Note that data and information are not the same thing. Data is raw, whereas information is processed data. Information displays some useful message, whereas data merely contains it. Nonetheless, as is commonly practiced, this text uses the terms data and information interchangeably. In manufacturing, as elsewhere, data and information are generated, collected, manipulated, transmitted, stored, retrieved, plotted, and shared.

Data communications using computers began with minicomputers. Later, when microcomputers became inexpensive and powerful, data communications applications proliferated. In fact, microcomputers were developed with communications in mind; that is obvious from their serial asynchronous communication capability through RS-232 as well as the parallel ports. Despite this standard, however, there are incompatibility problems. RS-232 ports limit the distance between the communicating devices. Paper and magnetic tapes were developed as alternatives to direct on-line communication. While they are used to move large amounts of data between devices separated by any distance, their drawback is the requirement for physically moving the data.

Regardless of the types of source and destination, the computer or terminal usually serves as a communications link in CIM environments. Consider, for example, the case of a machine tool operator who requires input, say a part program, from the plant host computer. The operator sends a request through the machine control unit (MCU) to the host, which downloads the program. In effect, then, communication between the operator and machine occurs via a computer (the MCU).

As with human conversation, computer communication involves three levels: cognitive, language, and transmission. The cognitive level requires that the devices have enough intelligence to take part in communication. The language level demands that both understand a common language. The transmission level requires a physical mechanism for information transfer. Each of the three levels of the sender must be compatible with that of the receiver. Human communication follows some rules (etiquette or **protocol**); even when they are broken, they allow communication. For example, a speaker who does not possess high proficiency in English or has a difficult accent can still be understood. The protocols of computer communications do not allow such variations, however.

Most computer communications take place in a way very similar to communication between two individuals. Communications technology principles are easier to comprehend if we keep in mind the situation of two persons communicating with each other. Consider, for example, what is happening at this moment between you as a reader and me as an author. Whatever I have to say about communications relating to CIM is here in the pages you are reading. We are communicating with each other, albeit in a one-way mode (only from author to reader). In a two-way mode, called bidirectional, the communicating partners or devices change their roles of receiver and transmitter as and when required.

Irrespective of the mode, whether unidirectional or bidirectional, communication between two participants (persons, computers, machines, or any combination) involves these three components:

Transmitter or sender—the source of information
Receiver—the person or device needing the information
Medium—the path through which information flows

In addition, a language—the protocol—must exist for communications to take place. In our example of communications between you and me, I represent the transmitter, you the receiver, the book the medium, and English the protocol.

The process of transferring information within CIM involves three basic issues:

1. Representation and signalling of data
2. Medium
3. Networking the devices

Representation of Data

Before data can be transferred, they must be represented in some suitable format. Data representation is analogous to having the thought process mentally work out what to say (or write) and how to say (or write) it. When two persons talk with each other, they describe data or messages (what is said) using the conventional rules of grammar.

Representation of data is a function of the mode of its transfer. For example, in preparing a resume you would use formal English, not informal conversational expressions you use while talking to a friend. There are four different ways to transfer manufacturing data (for that matter, any data): in spoken, written, analog, or digital form. Since the Industrial Revolution, most manufacturing data have been transferred either orally or written on paper as memos, letters, reports, drawings, and the like. These methods of information transfer are unsuitable for CIM. Data can be lost on the way, they could be misunderstood by the recipient, or their storage may be cumbersome as with designers' drawings. CIM can't afford these deficiencies; it requires more reliable and faster information transfer.

The other two formats, analog and digital, are based on whether digits are used to represent data. Historically, we have used analog data in manufacturing. An example of analog representation is the physical model of a component. In such a case, the data is embodied in the model rather than described verbally or written (expressed through drawings). Other examples of analog data representation are forming tools, jigs, fixtures, and prototypes. For CIM purposes, digital representation proves superior. Paper tapes, floppy diskettes, and direct communication between machines and computers are all based on digital data representation.

Data represented in digital form are most convenient for the needs of CIM. Since integration is achieved through computers, digital representation is, of necessity, binary rather than in the decimal system of the human world. Thus data conversion back and forth between the computer's binary and our decimal system is essential. CIM should convert nonbinary data to the binary format at the very first opportunity—at the source of entry or generation—and keep it binary throughout. Its conversion back to nonbinary format is done only to facilitate human interaction. Such conversions cannot be avoided, since computer processing is all binary, whereas human processing is nonbinary. We are just not capable of comprehending and processing binary data.

Coding

Whenever the prevalent practice of data representation is unsuitable, inefficient, or both, some sort of coding is done. For example, the Greek letter ϕ is used in drawings as a code for the diameter of a shaft or hole. This symbol saves the time and effort of expressing the term diameter. Since we use this term quite often in design and technical communications, coding could be justified. An acronym used to shorten a term, such as CIM for computer-integrated manufacturing, is another example of coding.

Communication requires two types of coding: data and control. While data coding translates the data to be transmitted, control coding is essential for controlling the communicating devices.

Characters. Manufacturing data or information is expressed using decimal numbers and the letters of the English alphabet (or of other languages). In the case of English, both uppercase and lowercase letters are used. The letters are combined to form words which, following the rules of grammar, are used to form sentences, paragraphs, and complete documents. Similarly, the 10 digits of the decimal system (0, 1, . . ., 9) are used individually or as groups to express larger numeric data. An example of manufacturing data is the expression PART # 619, which contains both letters and numbers. Besides the letters and digits, we also use other symbols such as #, %, and math notations such as + and –.

The various letters, symbols, and the decimal digits used to express manufacturing data are called characters. When both upper- and lowercase English letters are used, the total number of characters required to express manufacturing data is in excess of 62 (52 letters plus 10 decimal digits), since symbols are also needed.

Binary System

Of the three groups of participants in manufacturing communications, as illustrated in Figure 4.2, the computer system is the most efficient in handling data. It remembers better than we do, it stores much more data than we can, and it transfers data faster than we do. For these reasons, CIM requires expressing manufacturing data in a format suitable for computers. Since there are only two characters, bits 0 and 1, in the vocabulary of computers, all CIM data must be coded into groups of 0 and 1—that is, in the binary system.

The maximum number of different combinations that are possible using the two bits depends on word lengths (the number of 0s and 1s used). For example, since six-bit words yield a maximum of 64 (= 2^6) different combinations we can code only 64 characters using six-bit words. The two popular coding systems for binary representation of data are the ASCII and EBCDIC systems.

ASCII and EBCDIC Codes

CIM needs to handle data that are expressed in English (or any other language) and decimal numbers. This requires that enough combinations of bits 0 and 1 be available to code all possible characters used in our day-to-day life. The question is: How many bits should be grouped to represent one basic character? The answer depends on how many alphanumeric characters we need to code. As mentioned earlier, we have the 52 letters of the English alphabet (26 uppercase and 26 lowercase) and 10 decimal digits (0, 1, 2, . . . ,9). Besides these, we would need the mathematical operators such as +, -, and symbols such as ( ), „ , $, and so on. Let's consider using six bits to represent one character. With six-bit words, we can have only 64 (= 2^6) different combinations—not sufficient, since the letters of the alphabet and the decimal digits themselves add up to 62 (52 + 10) characters. What about seven-bit words? This results in 128 (= 2^7) different combinations, which are sufficient to represent all possible characters we are likely to use.

The preceding is the basis of the ASCII code developed by the American National Standards Institute (ANSI). The ASCII coding system is widely used in communications with microcomputers, minicomputers, and other digital devices. It has standardized every character of our world into a unique seven-bit binary word, as shown in Tables 3.2 and 3.3. Referring to the ASCII codes in Table 3.3, the first 32 codes from hex (short for hexadecimal) 00 to 1F, termed low codes, are used as control codes. The commonly used characters of our lives begin at 33 (hex 20) with the blank space. The decimal digits begin at hex 30, uppercase letters at 41, and lowercase letters at 61. ASCII contains a code for every communication need; there is even a buzzer, represented by hex 07 (binary 00000111), to ring a bell at the receiver end.

With the addition of a parity bit, ASCII becomes an eight-bit code. The parity bit is appended to the binary code at the most significant place (leftmost position) and checks against transmission errors.

Let us consider an illustration. How will the manufacturing data JOB # 2468 look in the binary system in ASCII? Using Table 3.3, we code each data character in hex, then use Table 3.2 to translate it into binary. This gives 01001010 for J, since J from Table 3.3 is 4A where hex 4 (from Table 3.2) is 0100 and hex A is 1010. Similarly, we get 01001111 for O, 01000010 for B, 00100000 for the blank space between B and #, 00100011 for #, 00100000 for the other space, 00110010 for 2, 00110100 for 4, 00110110 for 6, and 00111000 for 8. Thus, the data JOB # 2468 will be represented in ASCII as

01001010010011110100001000100000000100011
00100000000110010001101000011011000111000

While the seven-bit ASCII codes are sufficient for most purposes, for communications involving mainframes, an eight-bit code giving 256 ($= 2^8$) different combinations is used. This code is called EBCDIC for Extended Binary Coded Decimal Interchange Code.

Octal and Hexadecimal Systems

Besides the binary, there are two other systems: octal and hexadecimal. These have been developed for use primarily in computer hardware and its operation and programming. Manufacturing data may sometimes be expressed in octal or hexadecimal systems. CIM communications, especially those involving dedicated microprocessor-based control, may entail these two systems.

The octal system is based on the eight digits 0 through 7. To express any decimal number greater than 7, these digits are combined in a way similar to that in the decimal system. Thus, the other higher numbers in the octal system are 10, 11, 12, 13, 14, 15, 16, 17, 20, 21, . . . , 76, 77, 100, . . . and so on. These numbers are respectively equivalent to the decimal numbers 8, 9, 10, 11, 12, 13, 14, 15, 16, 17, . . . , 63, 64, and 65. Chapter 3 provides further details on the octal system.

In the hexadecimal system, the letters A, B, C, D, E, and F are used along with the 10 decimal numbers (last column of Table 3.2, Chapter 3). Thus, the hexadecimal system is based on 16 digits. For decimal numbers greater than 15, the rules of multiple place number generation are used. For example, the decimal number 20 is expressed by 14 in hexadecimal, while 27 is 1B. For details on the hexadecimal system, see Chapter 3.

Note that the binary system has a base of two, decimal of ten, octal of eight, and hexadecimal of 16. To avoid ambiguity in expressing numbers, it is a good practice to express the base, too. For example, 15_{10} means number 15 of the decimal system, whereas 15_8 means number 15 of the octal system.

Irrespective of the coding system used, all data are eventually converted into binary codes prior to transmission. Since more than one code may be in use, code conversion is occasionally a necessity. Code conversion is embedded somewhere in the data communication system and is generally based on translation tables. The conversion of ASCII codes to EBCDIC is straightforward, since ASCII has fewer codes (128, exactly half that in EBCDIC). The reverse is difficult, however, since 128 codes of the EBCDIC will not have an equivalent code in ASCII. For these codes, ASCII uses two characters.

Baudot Code

Baudot code is based on five bits and is used commonly in telex. It was used extensively in the 1950s on paper tapes and punched cards in which a hole represents 1 and a blank (no hole) represents 0. The early teletype terminals also used five-bit codes. Baudot code served its purpose in the early days, but is limited for CIM communications; ASCII and EBCDIC codes developed in the 1960s have proven to be more useful. Baudot code is insufficient for CIM because: (a) the 32 combinations possible with five-bit words are not enough, since manufacturing data involve significantly more characters; and (b) it does not provide for error checking, so essential in the noisy plant environment.

BCD Code

A code common with machinists is the **binary coded decimal (BCD)** which represents information by four bits. With four-bit words, 16 (= 2^4) different combinations are possible. Of these, the first 10 combinations are used to represent the 10 decimal digits as shown below:

Decimal Digit	BCD Code
0	0000
1	0001
2	0010
3	0011
4	0100
5	0101
6	0110
7	0111
8	1000
9	1001

An examination of BCD coding shows that it is basically a four-bit binary representation in which the last six combinations—namely 1010, 1011, 1100, 1101, 1110, and 1111—have simply been discarded. Note that this code can handle only numeric data; alphabets and other characters are not coded. Hence BCD code also has only limited use in CIM.

BCD expresses larger numbers by coding each individual decimal digit. For example, coding 147 involves coding 1, 4, and 7 individually, and then grouping the BCD codes together as follows:

Since 1 is 0001, 4 is 0100, and 7 is 0111; therefore, decimal 147 = 000101000111 in BCD code

BCD codes are converted to their decimal equivalents by decoding each four-bit group. Consider the BCD code 10000110, which comprises the four-bit groups 1000 and 0110. Since 1000 is 8 and 0110 is 6, this BCD code is equivalent to 86 in decimal.

BCD code can also be used to handle a decimalized fraction of a number. For example, the decimal number 35.84 is represented by coding each digit on either side of the decimal point. This results in

35.84 = 00110101.10000100

BCD is also known as **8421 code**, since the decimal value of the bit in the word depends on its position. It is equal to decimal 1 for the extreme right position, 2 at the second position from the right, 4 at the third position, and finally 8 at the fourth position (extreme left). Consider, for example, the BCD code 0111.

Position	Bit	Value	Decimal Equivalent
First (rightmost)	1	1	$1 \times 1 = 1$
Second	1	2	$1 \times 2 = 2$
Third	1	4	$1 \times 4 = 4$
Fourth (leftmost)	0	8	$0 \times 8 = 0$
			Total = 7

Thus, by adding the equivalent values of the bits, the decimal value of the BCD code 0111 is found to be 7.

On eight-channel paper tapes, the first four channels from the right are given place values of 8, 4, 2, and 1 so that any numerical value from 1 to 9 can be assigned on a horizontal tape level. For example, decimal 7 (0111 in BCD) is represented by punching holes in the first, second, and third channel (a hole signifies bit 1) so that the positional values 4, 2, and 1 add to yield number 7. Other channels are used to code the 26 letters and some miscellaneous and control characters. Figure 4.4 shows two paper tape standards currently in use. The ASCII subset used in numerical control is EIA 358-B standard, which is even parity, while the original standard EIA 244-B was odd parity.

Gray Code

The **Gray code** is a variation of the binary code. The variation ensures only one bit change for effecting adjacent change in decimal numbers. The advantage of this code can be noted through the following list containing both binary and Gray codes.

Decimal Number	Binary Code	Gray Code
0	0000	0000
1	0001	0001
2	0010	0011
3	0011	0010
4	0100	0110
5	0101	0111
6	0110	0101
7	0111	0100
8	1000	1100
9	1001	1101

Figure 4.4
The original 244-B and the new 358-B EIA punch tape codes

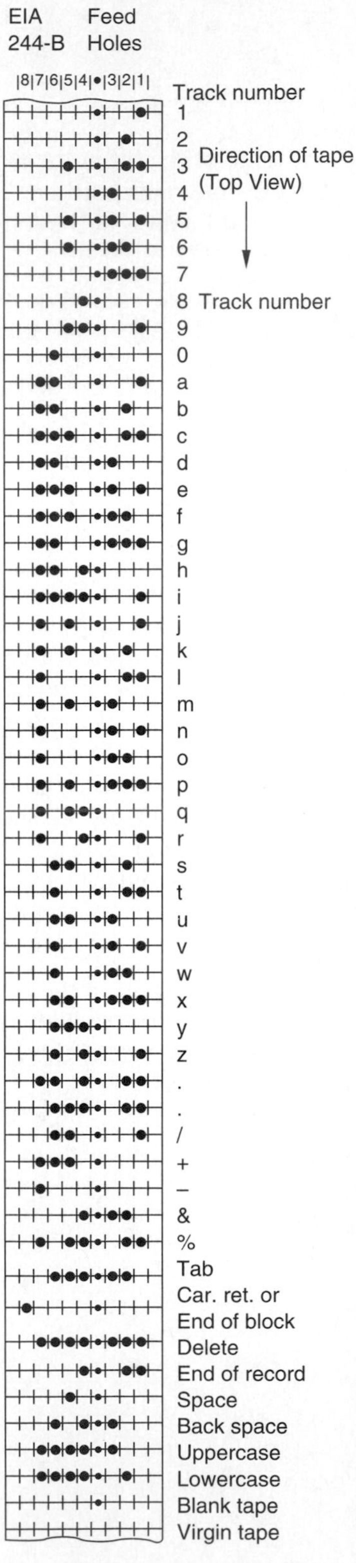

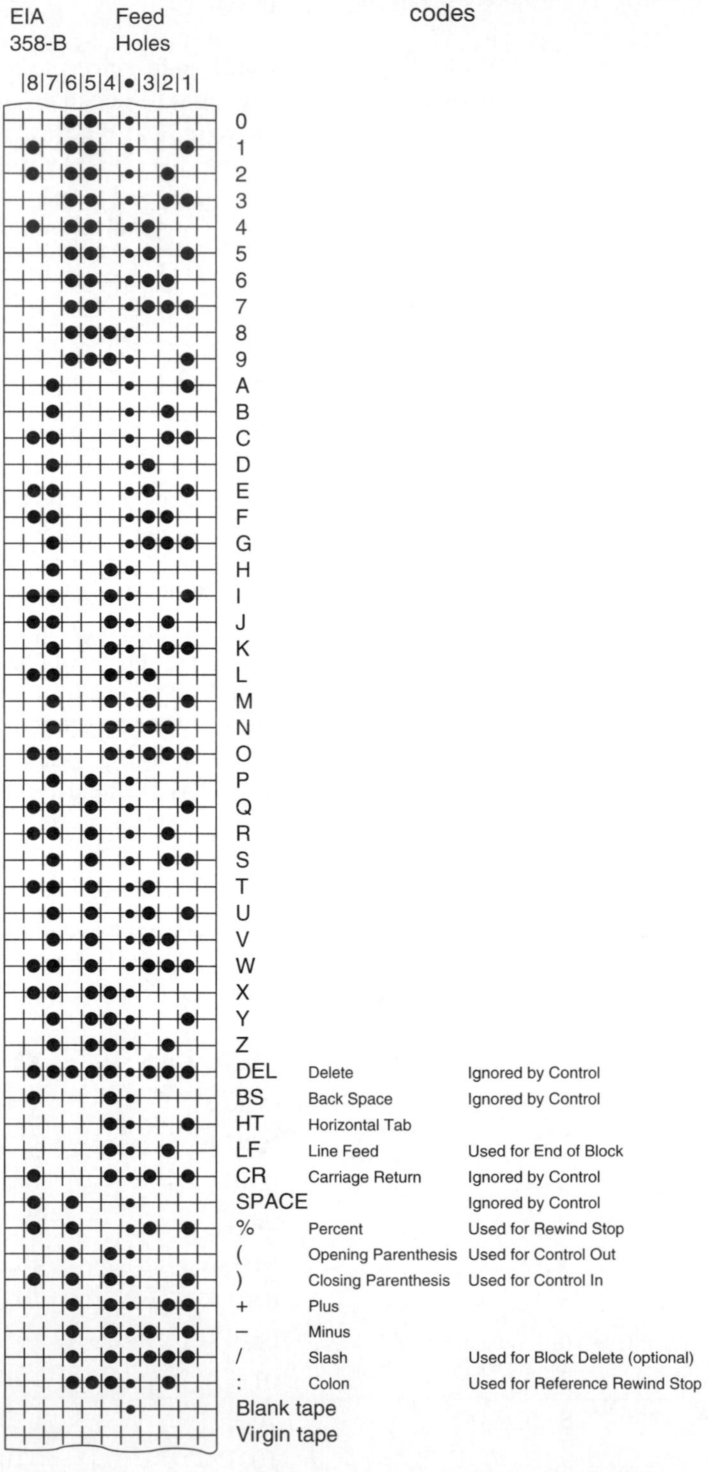

Source: Reprinted with permission from *Modern Machine Shop 1989 NC/CIM Guidebook*, Copyright 1989 Gardner Publications, Inc.

Consider that the decimal number 7 is to be changed to its adjacent value, 8. In the binary system, this requires changing 0111 to 1000, which involves replacing all the four bits—the three 1s to 0s and the leftmost 0 to 1. For the same change, the Gray code 0100 requires only one change—the leftmost 0 replaced by 1—to yield 1100. Such a benefit is derived for all adjacent changes, forward or backward, as seen in the table.

This property of the Gray code helps in many applications, such as in minimizing error susceptibility in shaft position encoders. Error is more likely in non-Gray coding, since more than one bit change may be required for adjacent positioning.

The Gray code is not an arithmetic code. It is merely an unweighted code, so there are no specific weights assigned to bit positions. The Gray code can have any number of bits to make a word.

Transmission

Once manufacturing data have been coded, the next step in effecting communication is to transmit the bits. The bits are transmitted as signals. The two bits 0 and 1 are represented by two different signals, which flow through the transmitting medium. Each signal is at a certain level of voltage. In practice, bit 1 is represented by a negative voltage, usually between -5 and -25 volts, and bit 0 by a positive voltage between 5 and 25 volts.

Signals transmitted through a cable may become distorted. Distortion can take two forms:

Attenuation. The signal gets weaker due to power loss to the medium. Amplifiers are used along the transmission path to correct for this. Attenuation is not the same for each frequency; higher and lower frequencies lose more than midfrequencies.
Delay. The delay arises due to the fact that signals have different speeds of propagation at various frequencies. This effect, which is hardly noticeable in voice transmission except over long distances such as overseas calls, can give rise to errors in data transmission.

Equalizers are used to compensate for distortion; for example, frequencies losing power are boosted, those traveling faster are delayed, and so on. Communication lines supported by equalizers are said to be "conditioned." The lines leased by telephone companies are usually conditioned.

Serial Versus Parallel Transmission.   With data represented as a group of 0s and 1s, the next questions are: How are the bits transmitted from source to destination, and how can one ensure that what has arrived is error-free?

Consider transmission of the letter P, as represented by 1010000 in ASCII. For an even-parity system, bit 0 is appended to the left of this code. Thus, the eight-bit combination 01010000 is transmitted for transferring letter P from one device to another. There are two ways to do this. In one, these bits are sent simultaneously through eight wires running in parallel—one bit through each wire. The bits arrive at their destination together and appear as the letter P. However, the cost of eight wires running in parallel may be high. An alternative is to use one wire and send one bit at a time. Obviously, transmission will take eight times longer. The first technique of transmission is called **parallel transmission**, whereas the second one, using single wire, is called **serial transmission**. Parallel transmission is expensive, although fast, and hence serial transmission is most often used.

In serial transmission, the binary numbers are transmitted from right to left. Thus, the code 01010000 representing the letter P is sent as four zero-level signals, followed by a one-level, another zero-level, another one-level, and finally the last zero-level signal, as shown in Figure 4.5. At the receiving end, these bits are arranged in the order received, yielding the letter P. The parity bit arriving last is ignored after having been used to confirm an error-free transmission.

How does the receiving end know that the code contains four zeros, not one or two? This is achieved through a clock. Depending on the transmission rate, a bit takes a finite time that marks the end of that bit and the start of the next. When the consecutive bits are the same, the clock time is proportionately longer. When they are different, the signal voltage changes accordingly. For example, if the transmission rate is 4800 bps, then each bit takes 1/4800 second, or 208.33 microseconds. Thus, in the above example, the four consecutive zeros would keep the voltage steady at the level corresponding to zero bit for 833 ($= 4 \times 208.33$) microseconds. The time duration of a bit depends on the transmission rate. Transmission rates in common use are 2400, 4800, 9600, and 19200 bps, giving approximate bit times of 417, 208, 104, and 52 microseconds, respectively.

Serial transmission can take place in one of two ways: asynchronous or synchronous. In **asynchronous transmission**, one character is transmitted at a time and the time gap between individual characters is immaterial. The asynchronous method is commonly used in data transmission; an example is the personal computer's asynchronous interface with printers, terminals, and so on. In **synchronous transmission**, a group of characters are sent as a packet in such a way that the time gaps between individual characters are synchronized between the sender and receiver.

Figure 4.5

Transmission of ASCII-coded bits 01010000, representing the letter P

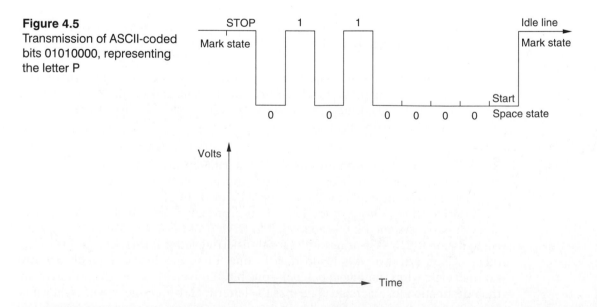

Asynchronous Transmission.   In asynchronous transmission, data is transmitted and received with no control on the time interval between characters. Individual bits travel at a fixed rate, say 4800 bps (see Box 4.1, which illustrates how higher speeds save time). Since characters can arrive at an arbitrary rate, a method of recognizing the beginning of the character is devised. This is accomplished by "framing" the beginning with a START condition and ending with a STOP condition, as shown in Figure 4.5. That is why asynchronous transmission is also called START/STOP transmission. The data travels serially; that is, one bit after another. When no transmission takes place, the line is kept at a voltage corresponding to bit 1, the prevailing condition called MARK state (Figure 4.5), and the line described as "idle." When data is to be transmitted, the line is first switched to the start condition by transmitting a START bit. This is done by changing the line voltage to the value representing bit 0; the resulting line is said to be in SPACE state. The duration of the START condition equals the bit time (reciprocal of the transmission speed); for example, 208 microseconds at 4800 bps. Each character is terminated with a STOP (MARK state) condition that lasts for one or two bit durations, depending on the hardware design.

A clock samples the idle line continuously. The moment it detects a START condition, bit clocking begins and lasts for a duration corresponding to eight data bits. After that, the line is returned to the MARK state. After collecting the bits, the receiver stores them temporarily in its hardware buffer and fetches them later via the communications software.

Synchronous Transmission.   In asynchronous transmission, time is lost between the end of one character and the start of the next. If this time can be saved, it will speed up transmission. This is what synchronous transmission achieves. Instead of sending one character at a time, a block of characters are sent together as one packet. A special bit pattern, called a synchronizing bit pattern (or **header**), is used to synchronize the transmitting and receiving devices. It acts for the whole data block as

---

 BOX 4.1   *Modern Speed Economics*

In remote-control computing, as with PC telecommunications in general, time is money. The biggest cost to stem from remote-control computing is telephone charges. The low purchase prices of older generation 2400 bits per second (bps) modems mask the true costs of using them. The new generation of 9600-bps modems that comply with the V.32/V.42/V.42bis protocols yield speeds up to 38,400 bps over standard telephone lines. These speed demons, while more expensive than 2400-bps modems, can quickly pay for themselves by reducing your telephone bill. For example, transferring a 1 MB file using V.32/V.42/V.42bis modems at each end of the connection takes around eight minutes; using 2400-bps modems to transfer the same file takes over one hour. It doesn't take a financial genius to figure out that's nearly a 90 percent savings in connect time.

Source: From "Computing by Remote Control" by D. Angell and B. Heslop, August 1991, *Portable Office*, p. 46.

shown in Figure 4.6. Following the data, another bit pattern, called a **trailer**, marks the end of transmission. Note that the synchronizing pattern works like the START of asynchronous transmission, while the trailer works like the STOP.

In the synchronous system, too, the receiver continuously searches for the header, which is normally two occurrences of an eight-bit synchronizing character. In EBCDIC, the header is 00110010 (decimal 50). Once found, the receiver and transmitter get synchronized and transmission ensues. The first eight bits received form the first character, the next eight the second, and so on until the end of the data block. The exact format of the data block transmitted depends on the protocol **syntax**.

As long as time can be saved, synchronous transmission is preferred. This means that for any given data, the total duration of the header and the trailer should be less than the overhead in the corresponding asynchronous transmission. The overhead is the sum of all the start and stop bits as well as the time lost in between the characters.

Assuming that

n = length of header and trailer in characters
m = number of characters to be transmitted, and
S = transmission speed in bps

the duration (t) of data transmission in the synchronous case can be expressed as

$$t = (m + n)/ (S/8) \tag{4.1}$$

In this equation, the numerator is the total number of characters actually transmitted and the denominator is the transmission rate in characters per second (8 accounts for the character's eight-bit width).

Consider a case in which $n = 3$, $m = 400$, and $S = 2400$ bps. Synchronous transmission of these 400 characters of data will take 1.34 seconds [(400 + 3)/(2400/8)]. Let us now compare this with asynchronous transmission, where there are two additional bits—one for start and another for stop—associated with each character. Thus, every transmittable character is 10 bits long. Assuming no waiting between the characters, asynchronous transmission will take 1.67 seconds [(400 × 10)/2400]. Thus, with a duration of 1.34 seconds, synchronous transmission is faster in this case.

The header and trailer are together called overhead bits. A high overhead can be justified if the data block is long. In fact, there is a cut-off value at which both asyn-

Figure 4.6
Formatted synchronous transmission with data as a packet

Trlr.	DATA	Syn. Pattern or Header	→

chronous and synchronous transmissions take the same time. Only when data blocks are longer than this value is synchronous transmission faster.

Error Detection.   Electrical noise can create errors in data communications. In a telephone conversation, the receiver can judge whether the message was error-free. In case of an error, the receiver responds: Could you repeat that please? In computer communications with no human reasoning capability, the software is designed to detect errors.

There are two techniques for error detection: parity check and **cyclic redundancy check (CRC)**. In the parity technique, a bit—called a parity bit—is appended to the character bits at the left, in the most significant position. The parity bit is either 0 (zero) or 1. Parity check may be odd or even. In the odd method, the parity bit renders the number of 1s in the character to be odd. In the even parity method, the number of 1s in the eight-bit group is even. Error detection based on parity is effective when one bit has suffered transmission error. If two bits are in error, the transmission error will go undetected. In fact, the parity scheme is able to detect only when an odd number of errors have occurred. In general, this method works well, since the chance of two or more bits out of eight being in error simultaneously is extremely low.

The CRC technique detects errors by performing calculations on the bits. The sender appends the results of the calculation to the message. The receiver carries out the same calculation and compares it with the appended results to determine whether the transmission was error-free.

Occasionally, noise bursts appear in the transmission line. Depending on its duration, a burst may smash a large number of bits. Consider, for example, a burst lasting only for one-hundredth of a second. If the transmission takes place at 9600 bps, then 96 (9600 bps × 0.01 s) bits would be lost.

Modulation and Demodulation.   Prior to transmission over telephone lines, digital signals are "imprinted" on an analog signal—appropriately called a **carrier signal**. The process of imprinting is termed **modulation**. At the receiver end, a reverse process called **demodulation** takes place to recover the digital information. The device that implements these processes is a **modem**, an acronym for modulation and demodulation. Modulation may be of three types: **amplitude modulation (AM)**, **frequency modulation (FM)**, or **phase modulation (PM)**.

The binary bits ready for transmission enter a modulator that generates sine waves. The sine wave is modified depending on the bit pattern. In the AM method, the modulated signal looks like the one in Figure 4.7a, where bit 1 is represented by the high-amplitude and bit 0 by the low-amplitude portion of the signal. A frequency-modulated signal looks like that in Figure 4.7b, where the high-frequency portion represents bit 1 and the low-frequency portion bit 0. In the PM technique, the phase of the sine wave is shifted by the bits so that peaks do not occur where they should. This technique can result in dibits—two bits per signal state, using phase shifts of 90, 180, 270, and 360 degrees.

Figure 4.7
Modulated signals: (a) amplitude modulated, (b) frequency modulated

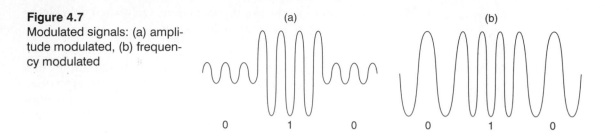

Simplex and Duplex.   When communications is one way, as with a doorbell, it is called **simplex**. The two-way mode, called **duplex**, may be half-duplex or full-duplex. In half-duplex communications, only one device or person can use the line at a time, whereas in full-duplex mode both can do it simultaneously. The CB (citizen band) radio is an example of half-duplex, with the word "over" used to indicate the end of transmission so that the other party may use the line. The common telephone is an example of full-duplex communications. Even when the medium is capable of full-duplexing, the protocol may limit the communication to half-duplex.

Line Speed.   Communications speed is usually expressed in **baud rate**. The baud rate indicates the number of times the line condition changes state every second. When the state can be either 1 or 0, as in AM or FM modulation, the baud rate equals line speed in bps. If there are four states, as in phase modulation, then the baud rate is twice the bps value. To avoid this confusion about line speed, it is better to use bps than baud rate.

Depending on the ratio between signal strength and background noise in the medium, there is an upper limit to the transmission rate. The theoretically possible **maximum transmission rate** (**MTR**) through a medium in bps is expressed by Shannon's Law:

$$MTR = W \log_2 (1 + S/N) \tag{4.2}$$

where

W is the bandwidth, and
S/N is the signal-to-noise ratio

The S/N ratio is sometimes expressed in dB (deciBell), which is given by $10 \times \log_{10} (S/N)$. Thus, a S/N ratio of 1000 is equal to 30 dB.

For voice-grade telephone lines, W is 3000 Hz. Therefore, for a signal-to-noise ratio of 1000,

$$MTR = 3000 \log_2 (1 + 1000) = 29{,}901 \text{ bps}$$

In practice, voice-grade lines can transmit only up to 9600 bps, significantly below the theoretical maximum of 29,901 bps.

Medium

A medium is that through which data is transmitted; the medium links the sender with the receiver. It provides for both off-line and on-line communications. Paper tapes and magnetic disks or tapes are examples of off-line media (these need to be transported physically from sender to receiver). Alternatively, a cable connecting the sender and receiver can be used for on-line transmission.

Communication media include

wire communication path
microwave
fiber-optic cable
satellite

Wire communication paths may be open wire, twisted cable, or **coaxial cable**. Open wires have no covering and are usually attached to ceramic insulators fixed on poles. They are suitable for low traffic (data rates) only. In twisted cables, copper wires insulated from each other are twisted in pairs. There is a risk of **crosstalk** (undesirable signals from one wire to another) when multiple twisted pairs have been warped in the cable. Coaxial cables avoid this risk, since they have a grounded shield around the conductor. Coaxial cables are used in LANs, since such cables can transmit at higher frequencies than twisted-pair cables.

Microwave transmission uses very high-frequency radio waves in the range of 4.6-12.0 GHz. This **broad-band** facility provides line-of-site transmission capability. A **repeater** is used every 30 miles—provided the path is not obstructed—to transmit to the next receiving station.

In **fiber-optic cable**, a strand of glass with a diameter as thin as that of human hair is used to transmit data. A fiber-optic cable can carry high **bandwidth** signals, since it is immune to noise and distortion. A light source—laser diode or **LED** (light emitting diode)—allows modulation of data at high transmission rates. Fiber-optic cables are impossible to **tap**, which makes them a secure medium. However, this makes it difficult to use such cables in bus topology LANs.

Satellites work as relay stations. A station receives signals at one frequency and transmits them at another frequency to avoid interference with the incoming signals. A device called a **transponder** accomplishes this. Propagation delays are longer (about one-fourth of a second) with this method as compared to data transmission through ground links (approximately six microseconds). Manufacturing companies engaged in worldwide operations use satellite communications to implement CIM.

The choice of transmission medium depends on such factors as cost, error rate, security, and speed. For CIM, coaxial cables provide a balance between cost and benefits, though fiber-optic cables are the medium of the future.

Fiber-optic Medium. Fiber-optic-based communications is a relatively new technology. It involves transmission of light and images through hair-thin, flexible, optical, glass fibers and provides capabilities superior to those of conventional metal conductors. Applications of fiber-optic cabling in telecommunication to achieve

Figure 4.8
Cross-sectional view of a typical fiber-optic cable

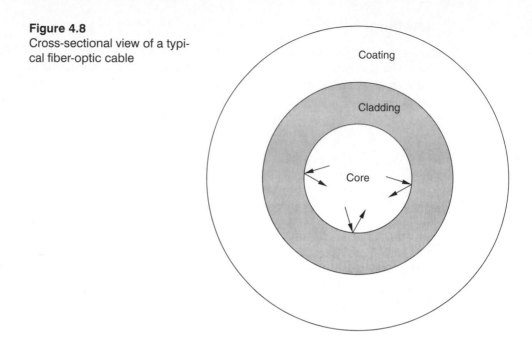

transmission at speeds of 500 Mbps, over an unrepeated distance of 30 miles, are routine. Fiber-optic cables were first installed in 1977. Today, service industries use thousands of miles of such cables. In manufacturing plants, coaxial cables are preferred due to their lower cost. This is changing rapidly, however, as fiber-optic cabling costs continue to drop.

A fiber-optic cable contains three concentric layers, as shown in Figure 4.8. The central core, which actually transmits the light, is usually silica, glass, or plastic. The cladding is made of materials with a refractive index lower than that of the core material. The interface between the core and the cladding acts as a reflective surface, which abates the leakage of light from the core. The external multilayer coating provides strength and flexibility to the fragile fiber. The cable is so designed that photons reflect (Figure 4.8) at the internal cable wall to maintain their forward travel along the cable axis. The fiber guides the light down its length with minimal signal attenuation and pulse dispersion or distortion. The function and construction of fiber cables differ radically from those of metal conductors.

Fiber-optic cable's appeal lies in the fact that it involves photons rather than electrons. Photons, which are light waves of the same frequency, travel 50% faster through glass fiber than electrons through a metal conductor such as copper wire. Moreover, because photons do not carry any electric charge, the signals have no risk of electromagnetic interference. Photons can travel simultaneously in more than one direction. These basic characteristics of photons give fiber-optic cables several advantages, including the following:

Immunity to Electrical Noise.   Optical fibers are dielectrically immune to electromagnetic interference (EMI). They neither emit nor are susceptible to EMI. They are also immune to transients generated during the switching of heavy loads. An optical fiber can be run next to high-voltage power lines without detrimental inductive or capacitive coupling. It does not radiate energy and thus eliminates any crosstalk. It requires no cabling shield or filtering at I/O points.

Manufacturing plants with large electric motor systems and similar equipment generate electromagnetic fields (noise) that interfere with the electrons flowing through copper-cable networks. Fiber-optic cables are free of such problems as crosstalk, ringing, and echoing. Immunity to in-plant electrical noises due to EMI and RFI (radio-frequency interference) from welding equipment, machine tools, and so forth, is very desirable in CIM plants.

Speed.   Transmission speeds through fiber-optic cables are higher, since optical energy travels 50% faster than electrical energy. A single cable can simultaneously carry up to 10,000 digital television channels. Moreover, higher bandwidths are possible.

Efficiency.   Minimal power losses with fiber-optic cabling result in more efficient signal transmission. Signal attenuation through fiber cable is low, in the range of 6 to 7 dB/km. With little degradation in signal quality and strength, cable runs can be long without needing repeaters. For example, nonsilica-based fibers allow signals to travel up to 7000 miles without requiring amplification. The number of repeaters in fiber-optic systems can be as low as one-tenth of that in a copper-based medium. This benefit is very attractive in telecommunications, where source–receiver distance is usually long.

Safety.   Since fiber cables do not carry electrical current, there is no risk of spark hazard. Manufacturing plants involving chemicals find this attractive. Unlike electrons, photons do not initiate explosion in hazardous environments. Optical fiber is inherently safe for use in hazardous situations, since there is no risk of spark, fire, or explosion at contact or break points. In addition, the risk of lightning to fiber-optic-based communication systems is minimal. The risk of nuclear-attack-generated electromagnetic pulses damaging such systems is absent as well.

Absence of Ground Loops.   Since the optical transmitter and receiver are electrically isolated by the very nature of optical fiber, ground loops and associated problems are altogether absent.

Distance.   Optical fibers extend the distance of standard communications channels such as RS-232C, RS-449, and IEEE-488. Unrepeated distances of up to a mile are commonplace; the distance can be extended even up to eight miles.

Security.   Since fiber-optic lines do not radiate electromagnetic energy, they cannot be tapped magnetically and hence are more secure against wiretapping and electronic snooping. However, this becomes a limitation of fiber-optic cabling in multidrop networks.

Easy Handling and Light Weight.   Fiber-optic cables, being relatively light and thin, are easy to handle while laying and transporting. This also makes such cables attractive in avionic applications. Fiber cables are as light as twisted-pair cables. With a weight of 25–150 grams/meter, fiber cables are lighter than the coaxial cables (70–700 g/m for the **baseband** type and 140–1500 g/m for the broadband type). Fiber-optic cables can use existing conduits for copper wires without any additional cost.

Reliability. Unaffected by moisture, heat, and most acids, glass-based fiber-optic cables are extremely reliable. They keep sending data even at 2,000°F.

Installation. The installation of fiber-optic cables is simple and inexpensive. Their light weight eliminates the need for special fixtures for hanging them from walls or ceiling. No special conduits are necessary under the floor, and space is saved. But, installation of fiber-optic cables demands higher skills. Manufacturers are addressing this by developing special tools that make installation easier. For example, preparation of fiber ends using a special tool leaves the surface flat and smooth so it will not require polishing.

The only drawback of fiber-optic cabling is its high cost. At 30 to 80 cents per foot, its 1995 cost is twice that of twisted-pair cables and 50% more than coaxial cables. The additional expenses of special connectors and other accessories make the relative cost of fiber-optic cabling high. This cost difference is expected to narrow as fiber-optic cabling benefits from economies of scale. Use of plastic (rather than glass) fibers further reduces the cost. However, plastic fibers attenuate the signals more than glass fibers.

Fiber-optic cables are used in LANs where connected devices can be as far apart as hundreds of feet. With copper wires, the distance is limited. The fiber-optic cable has another use in CIM plants: as a photoelectric transducer it can measure pressure, vibration, noise, temperature, and other variables.

Since most industrial equipment has standard interfaces (RS-232C ports), the simplest approach to using fiber optics is to opt for add-on devices at these ports. A module with the standard 25-pin RS-232C connection for the equipment and optical I/O ports for fiber-optic transmission is an example. Such a module plugs into a standard serial port and automatically handles all electro-optic signal conversions. It can extend the 15-meter transmission limitation of RS-232C to more than 2 kilometers, while operating at 20 Kbps. A computer terminal or peripheral may have built-in optical interfaces in place of or along with the electronic ones. Alternatively, equipment and devices can be specifically designed for an optical fiber medium.

A fiber-optic system (Figure 4.9) comprises three major elements: modulator or transmitter, fiber cable, and demodulator or receiver. The modulator converts electric signals into light pulses, which are channeled into the fiber cable through an optical coupling for eventual transmission to the receiver. At the receiving end, the pulses strike an optical detector that, along with other components of the demodulator, converts the pulses back into electrical signals.

The modulator carries out two main functions: (a) converting the electrical signals into light pulses (the primary working element is a light source—usually laser or

Figure 4.9
Basic principle of a fiber-optic system

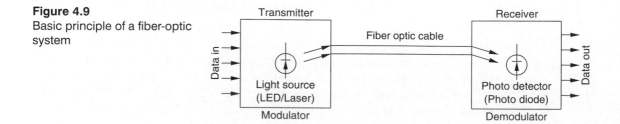

LED) and (b) combining or multiplexing multiple signals (as many as millions), and sending them on a single cable.

The demodulator also serves two functions, which are actually the "undoing" of what the modulator does: converting light pulses back into electrical signals and decoding the multiplexed signal into separate channels. This is achieved with a photodetector (optical detector) having a high-frequency response and wide bandwidth and operating temperature range. The standard photodetector, also called photodiode, is a semiconductor; incoming light strikes the semiconductor and excites the electrons to generate current.

The major problems and limitations of fiber-optic systems are:

1. Connectivity sophistication. A laser or LED, two couplers (one at each end) and a detector are needed (Figure 4.9).
2. Higher risk of damage to fiber from cable bending.
3. Analog signals cannot be transmitted.
4. Miscellaneous: need for special test equipment, additional training of workers, lack of standards for the technology, and a slight risk to the eye.

The manufacture of fiber-optic cable is in itself challenging. The aim is to make the fibers clearer (free of impurities) and stronger. Clearer fibers can send signals over longer distances without needing amplification. For example, silica-based fibers can cover a distance of 400 miles without needing a repeater.

Figure 4.10 and Table 4.1 compare the three commonly used transmission media.

Figure 4.10
Comparison of speeds for
three transmission media

Twisted pair

0.1 Mbits/sec

Coaxial

Broadband
400 Mbits/sec
Baseband
10 Mbits/sec

Fiber optic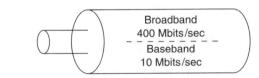

50 Mbits/sec and above

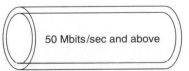

Source: Reprinted with permission from *Computer-Integrated Manufacturing Handbook* by E. Teicholz and J. N. Orr (Eds.), 1987, New York: McGraw-Hill.

Table 4.1
A comparison of the three commonly used transmission media

| | Twisted Pair | Coaxial | | Fiber Optic |
		Baseband	Broadband	
Bit rate	100 kbits/sec	10 Mbits/sec	400 Mbits/sec	50+ Mbits/sec [up to 5 Gbits/ sec in tests]
Type of data	Voice/data	Voice/data	Voice/data/video	Voice/data/video
Usual topologies	Point to point Star	Star Bus	Ring Bus	Point to point Ring
Advantages	Lowest cost Good EMI/RFI immunity	Low cost Easy to install	Relative noise immunity Uses CATV components	Immune to EM interference Very secure Low weight and size
Disadvantages	Poor security High loss	Low noise immunity Usually requires conduit	Higher cost Skill required to install RF modems required	Very high cost Mostly limited to point to point High skill required for installation and maintenance

Source: Reprinted with permission from *Computer-Integrated Manufacturing Handbook* by E. Teicholz and J. N. Orr (Eds.), 1987, New York: McGraw-Hill.

Types of Communication Lines

In the simplest communication line, two devices are directly linked through a point-to-point (PTP) connection, as shown in Figure 4.11. Each device has a direct dedicated link with the others.

The PTP line represents the most basic arrangement of a network. Early networks used PTP lines in star configuration (Figure 4.12), in which the host is the mas-

Figure 4.11
Point-to-point (PTP) connection between two devices

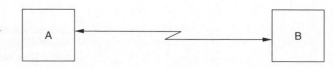

Figure 4.12
Star configuration

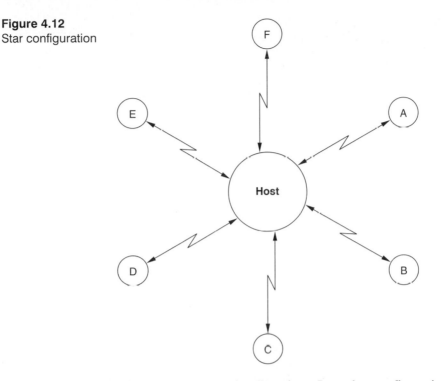

ter **station** and A, B, C, . . . are remote "slave" stations. In such a configuration, each slave needs a line of its own that remains underutilized if the traffic volume is low.

The main drawback of a PTP connection is that the number of lines can become large even in a moderate-size facility. If N devices are connected in PTP fashion, then $N(N-1)$ lines are needed, and each device requires $(N-1)$ input-output ports. Just five devices ($N = 5$) require 20 (5×4) lines, as shown in Figure 4.13. This makes cabling cost prohibitive, especially in a CIM environment that may need to link tens or even hundreds of devices. That is why other networks have been developed.

We can compare a PTP arrangement to constructing a direct road to connect all the houses in a residential community. A better alternative, commonly practiced, is to build houses on both sides of a common road. In communications, such a layout is termed a multidrop line, as shown in Figure 4.14. A multidrop connection enables devices to share the same line and is suitable in cases where they do not need to transmit data constantly. However, in a shared line when two devices try to transmit at the same time, a **collision** may occur that can garble the data. Protocols avoid such accidents by continuously **polling** the devices to find which one is ready to use the line.

Multiplexing.   Sending one stream of information at a time over a line may not be cost-effective. It is possible to support multiple streams of data on a single line through **multiplexing**. The line capacity is divided by **time division multiplexing (TDM)** or **frequency division multiplexing (FDM)**. Referring to Figure 4.15, information from various stations or devices is transmitted over several lines to multi-

Figure 4.13
Just five devices require 20
lines!

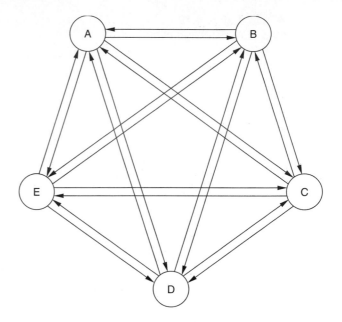

plexer 1, which may be a central computer or host system. Multiplexer 1 sends all the information to its counterpart at the receiving end (multiplexer 2), which directs it to the proper recipients, as shown in the figure. Multiplexer-to-multiplexer transmission takes place at a high speed, say 9600 bps, whereas station-to-multiplexer and multiplexer-to-station transmissions are slower, say 1200 bps. The concept is analogous to vehicular traffic originating on approach roads at slower speeds, say 30 mph, entering an expressway through a junction, covering the expressway distance at higher speeds, say 65 mph, and finally exiting on secondary roads at lower speeds, say 30 mph.

Multiplexing reduces the cost of communications lines, since different data streams can share the leased lines. Multiplexers allow transmission between two clusters of widely separated equipment. A typical multiplexer offers 8 to 32 ports. Multiplexers, adapters, and extenders are available for RS-232C, RS-422, IEEE-488, and other interfaces as well.

Communications Hardware

CIM requires a variety of hardware especially designed for communications. Some of the important ones are described here.

Figure 4.14
Schematic diagram of a multidrop line

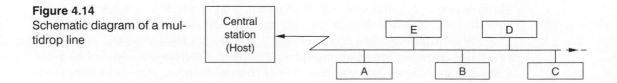

Figure 4.15
The principle of multiplexing

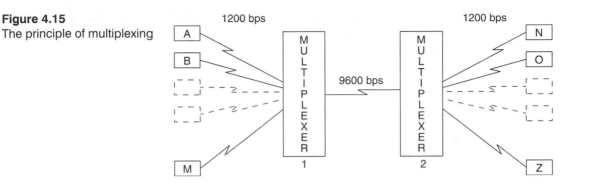

Interactive Terminals. Interactive terminals are basically a keyboard and a display unit, such as a video display terminal (VDT) or printer. The earliest terminals were teletypes (TTY), which were cumbersome because their keys required pressure that tired the fingers. Moreover, their operating characteristics could be changed only through switches (wire jumpers). Modern terminals based on internal microprocessors are sophisticated; their operating characteristics such as speed, number of bits per character, parity, and mode are user-programmable. A setup mode is provided for the user to effect changes as desired. A terminal can even emulate other terminals and allow changes in the protocol. These flexibilities save users the effort of making changes in application programs to suit a given terminal.

Remote Batch Terminals. Remote batch terminals were developed in the 1960s for accessing mainframes. A remote job entry (RJE) terminal can read local data and transmit them to the host for processing; the results of processing can be made available back at the terminal. Manufacturing plants use such terminals for uplink and downlink information transfer. RJE terminal capabilities were added to minicomputers in the 1970s and to PCs in the 1980s.

Microcomputers. Today, microcomputers carry out RJE tasks. The micro option is attractive since, besides serving as terminal, a PC can also run local programs. Micros' low cost has made older RJE terminals virtually obsolete.

Front-End Processors. In the late 1960s, interactive and RJE terminals interfaced with mainframes using hardwired communication controllers. This did not provide much processing capability, since the mainframe had to process most of the data transmitted or received. Eventually, the hardwired controllers were replaced by programmable front-end processors (FEP). FEPs are especially designed to process a high volume of data from several communications lines and to transfer information to and from mainframes efficiently.

UARTs/USARTs. Data received or transmitted are read or written by some interface in the hardware. In the case of micros, communication interfaces use a **universal asynchronous receiver/transmitter (UART)** or **universal synchronous/asynchronous receiver/transmitter (USART)** for this purpose. UARTs process data asynchronously (time intervals between characters vary), whereas USARTs provide for both asynchronous and synchronous (time intervals between blocks of data vary, but between characters there is no time gap) communications. Compared to customized interfaces, UARTs and USARTs are relatively inexpensive.

Modems.   To communicate, discrete digital signals are modified to an analog format that can be transmitted through a communication medium. This is accomplished by a modem.

Acoustic Couplers.   **Acoustic couplers** connect modems to telephone lines for communication. The modems either originate or answer at set frequencies. In Bell-103F compatible modems, originate corresponds to 1070 Hz for bit 0 (space state) and 1270 Hz for bit 1 (mark state). These are 2025 Hz and 2225 Hz, respectively, for the answer mode. **Full duplex** can be accommodated on two-wire facilities, one channel using 300–1700 Hz and the other 1700–3000 Hz. Electronic filtering at each end allows correct frequency detection.

Multiplexers.   Multiplexers manipulate data for line sharing (Figure 4.15) to reduce the cost of communications. There are two approaches: time division multiplexing (TDM) or frequency division multiplexing (FDM). TDM allocates a certain time period to a specific terminal or line. In FDM, all the data coexist on the line, with different frequency bandwidths representing information flow for different terminals. Signals are divided into narrower bandwidths within a given bandwidth. Figure 4.16 illustrates the TDM and FDM techniques.

Multistream Modems.   Multistream modems provide for more than one stream of data on a single line. Also called split stream modems, they work on the TDM principle, where, for example, data at 9600 bps is split into four 2400 bps streams.

Concentrators.   These process the data received from terminals by placing them in a buffer and identifying the source. An identification for the terminal is appended to the beginning of the data. The common buffer is transmitted at higher speed.

RS-232C Connector.   This is a 25-pin connector (Figure 4.17) based on a standard developed by EIA. Which pin carries a certain signal is theoretically standardized, but in practice this is not followed strictly. The pins are neither used nor interpreted the same way. Normally, the pins are designated as:

1—for protective ground
2—for receive (IN)
3—for transmit (OUT)

Other pins are used for auxiliary functions such as carrier detect, request to send, data set ready, and ring. In the simplest cable connection—called null modem wiring—asynchronous communication can take place for short distances of up to 2000 feet. In this wiring, pins 2, 3, and 7 of one device are connected to pins 3, 2, and 7, respectively, of the other. The hardware implementing RS-232C is either data ter-

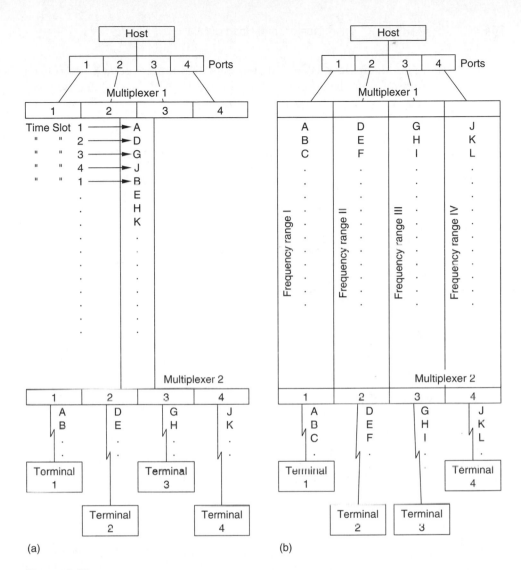

Figure 4.16
Multiplexing types: (a) Time division multiplexing, and (b) frequency division multiplexing

Figure 4.17
A 25-pin RS-232C connector

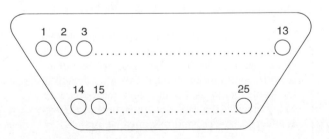

minal equipment (DTE) or data communication equipment (DCE). For parallel transmission, the appropriate standard is RS 422. Box 4.2 describes some basic terms relating to RS-232C communication.

BOX 4.2 *Understanding RS-232-C Communications, Part I*

All current CNC controls allow programs and other data (such as offsets and parameters) to be sent to and received from an outside device such as a computer, tape reader/punch, or floppy diskette unit. The method by which data is transferred is called the RS-232-C port (sometimes called the serial port).

Terminology. Each term will describe one of the "variables" that must be set properly (exactly the same) on the sending device and receiving device in order for the transmission to take place. Some of these variables are "fixed" on some devices. This means that you may be limited with regard to the variables of RS-232-C that you can modify. Before you purchase any equipment/software for communications, always check to be sure that what you buy is "compatible" with your CNC controls.

Baud Rate. The baud rate is the speed at which the transmission takes place. Baud rate can be specified in several "pre-selected" numbers. You cannot simply choose any number you like. The most commonly used baud rates for CNC applications are 110, 300, 600, 1200, 2400, 4800, 9600, and 19200 bps.

Generally speaking, the larger the number you choose, the FASTER the transmission will take place. However, the true speed of transmission is also related to the type of "handshaking" being used.

Even at a very fast baud rate, it is possible that the transmission will take just as long as at a slower baud rate if the receiving device cannot accept the data quickly enough. Also, the length of the cable being used can affect the baud rate. Many times, if the cable length is quite long (over 100 feet), a slower baud rate must be used.

Parity. This "variable" allows both communicating devices to do some error checking during transmission. Your choices for parity will usually be even (ISO or ASCII), odd (EIA), or none (NULL). Of the bits that are being sent, one is reserved for parity. In the case of an even parity system, for example, if either device finds a total number of bits that adds up to an odd number, a parity alarm is generated. Many RS-232-C devices can only use EVEN parity, so usually this is your best choice.

Stop Bits. This controls the number of bits used (per character) to terminate each character. Most CNC devices require two stop bits.

Word Length. This is the number of bits that make up each character. Again, the coding format (EIA, ISO, ASCII, etc.) determines the number of bits to a word. Most CNC devices use a seven-bit word length, and an eighth bit as the parity bit.

Control Codes. Some devices require that a special character be included in the transmission at specified times. For example, some CNC controls require that a "control Z" be used to end a transmission. In this case, the receiving device would be "hung up" until this character is received. Other uses for control codes might be to begin the transmission, pause for a specified length of time, and to delimit multiple programs during transmission.

Some older controls may require non-standard techniques for communication to take place.

Source: From "Understanding RS-232-C Communications, Part I," July 1990, *Modern Machine Shop*, pp. 134–135. (The other parts appeared in subsequent issues.)

4.5   NETWORK ARCHITECTURES

The purpose of a network is to allow error-free data transfer from one end-user to another. Network architecture describes the details of how devices are connected to the network and through it to each other.

The Seven Layers—OSI Model

Computer network architectures are based on the layering principle following a standard, namely the Reference Model of OSI defined by ISO. The model is intended to facilitate writing software that can be transported easily from one network to another. Since each layer performs a specific communications task, an application program written to interact with an upper layer function does so even when changes are made to the lower layers. Thus, layering makes application programs adaptable to various computers with minimum changes.

The OSI model has seven layers, as shown in Figure 4.18a. The end-user or the application "resides" in the highest layer (7).

How is data transferred in this model? Data originates (Figure 4.18b) at the **application layer** of the sender, passes downwards through each layer to the **physical layer**, and then "out" onto the medium. On arrival at the receiver's physical layer, the data moves up through the layers to reach the receiver at the application layer. Each layer at the sender end usually adds a header, and sometimes a trailer, to **frame** the data passed to it by the higher layer. At the receiving end, these headers and trailers are removed by the appropriate layers so that the end user receives only the original data.

Figure 4.18b illustrates the protocol and information flow. Layers 4 through 7 have peer-to-peer or end-to-end protocol (conversation), with a dialogue between similar programs on the source and destination devices. In layers 1 through 3, protocols allow conversation with the immediate neighbor, which may be the source or ultimate destination device. Only layer 1 is involved in the physical transfer of data; no data is directly transferred from layers 2 through 7. Layers 3 through 7 involve information transfer among layers within the same computer system. Layer 2 controls the transfer of information on layer 1. A layer can be implemented either as hardware or software. Moreover, two layers may be "collapsed" into a single layer that carries out the functions of both.

The application layer is user-dependent and serves as a window between the corresponding end users. It provides several services, including the following: (a) identifies the communication partner (name, address, etc.), (b) establishes authority to communicate, (c) allocates costs for communications, (d) determines adequacy of resources, and (e) identifies constraints on data systems, and more.

The presentation layer represents information that applications communicate with or refer to in their dialogue. This layer copes with the variety of syntaxes that the applications use. The presentation layer deals with representation (syntax of the sender, receiver, and that used by the transfer) of data. It provides for "transparent" transfer—a transfer that frees users from concern about its details. The presentation layer provides code conversion and data compression and decompression in which duplicate characters or common words are represented by algorithms. Data com-

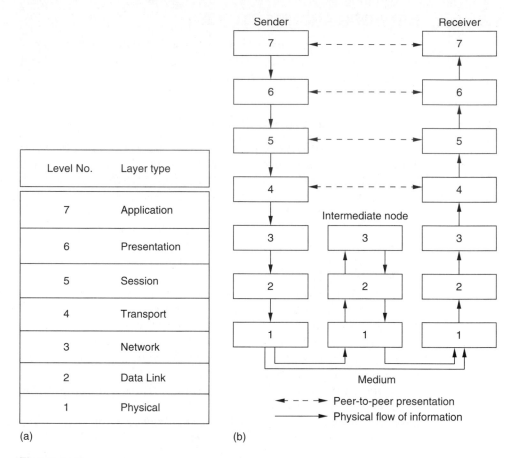

Level No.	Layer type
7	Application
6	Presentation
5	Session
4	Transport
3	Network
2	Data Link
1	Physical

(a)

(b)

Figure 4.18
(a) OSI model's seven layers; (b) information flow through the layers

pression and decompression eliminate some bits from the message without changing the control of information being transmitted. The most common character compressed in ASCII data streams is the blank space, since it occurs so often. The presentation layer also responds to an application's request for network services to transfer information.

The session layer is the actual interface to the network. At this layer connections are negotiated. Session means connecting the two users following the rules of data transfer established prior to connection—for example, whether one-way interaction, or two-way alternate or simultaneous. Services offered by the session layer include logging into a timesharing system, transferring files between two computer systems, accessing records in a file, preventing aborts in the middle of a database update, and so on.

The transport layer creates a transport pipeline, even though the path is unknown. It takes into account any limitation on the message length; long messages are divided into smaller units called packets, which the receiving transport layer reassembles. Its main service is **flow control** of the streams of data units. For higher through-put, several transport connections may be multiplexed.

The network layer provides the routing necessary to move information through the pipeline defined by the transport layer. It decides the best route to avoid local congestion by considering the normal (fixed) route, an alternate route if the normal route fails, or an adaptive route that considers traffic congestion on a real-time basis.

The datalink layer controls the message traffic on the physical medium and provides the rules for moving data between **nodes**. It also frames the data to be transmitted and controls the stations connected in a multidrop environment. The layer also decides when to transmit and/or receive, acknowledges correct receipt of a frame, re-transmits erroneous frames, and more.

It is the physical layer that actually transmits "raw" bits over a communications channel. It converts the bits appropriately so the receiving physical layer recognizes them correctly—0 as 0 and 1 as 1. It is thus concerned with electrical and procedural interfacing.

Each OSI layer is defined by three characteristics:

1. Protocol. The protocol comprises the set of rules for operating a data communications system.
2. Syntax. The syntax of information in a protocol is analogous to the language used in person-to-person communications.
3. Semantics. The **semantics** of a protocol is like the rules that determine when and how people say certain things. In telephone conversation, for example, the answering party's initial response is Hello or Hi—not Bye, which is reserved to signal the end of conversation.

Each layer carries out several functions, which are quite detailed. To appreciate the sophistication involved, we can consider the procedural working of one of the layers—say the datalink layer—and compare it to human conversation. The datalink layer is involved in:

Startup control. This function is similar to "Hello" in a telephone conversation. In computer systems, the computer is always in the listening mode; that is, ready to receive data.
Framing. Framing defines how a message is packaged: where it begins and where it ends. In a phone conversation, the other side can sense when the talker has finished; in computer communications, it must be stated clearly. Framing uses headers and trailers, which are analogous to the pauses in person-to-person communications.
. This function is achieved through rules of protocol that specify when the computer system should be sending or receiving data. The two parties must be in compatible modes, such as send/receive, rather than send/send.
Timeout control. Timeout control is achieved through a set of rules that determine whether the connection has been lost. In telephone conversations, one would say, "Hello! Hello! Are you still there?" In computer communications, a timer

within the computer accomplishes this. It times out the sender if no message is received within a certain interval. After a given number of consecutive timeouts, other actions such as an operator message or system reinitialization are taken.

Error control. If the phone line is noisy and the other party can't be heard, one would say, "What? Can you repeat that?" or something similar. The receiver carries out two tasks—telling the sender that the line is bad and asking him or her to repeat. Likewise, error control performs these tasks in computer communications. It does it in one of two ways: (a) request and get new data or (b) correct the data already received. The former technique is more common.

Sequence control. This function detects whether the message has been lost halfway through the communication.

Why Layered Structure?   A layered structure is desirable for several reasons. Such a structure facilitates modification in one layer without affecting the others. In a modular structure, each layer performs distinct, isolated, but well-defined functions. Layer boundaries are so chosen that information flow across the interface is minimal. There are enough layers so that distinct functions need not be combined in the same layer, which may otherwise have resulted in unwieldy architecture.

The only drawback of layered architecture is the high data overhead compared to a single-level structure. The advantages outweigh this limitation, however.

Local Area Network (LAN)

A local area network (LAN) is an arrangement of hardware and software that permits logically related devices to communicate with each other over distances up to 20 miles. In the CIM environment, these devices may be machining centers, CAD/CAM workstations, AGVs, NC equipment, robots, PLCs, data acquisition systems, bar code readers, and so on. LAN is usually proprietary for a single organization.

Developed by Datapoint Corp., the first LAN, implemented in 1977, was called an Attached Resource Computer (ARC) rather than a local area network. Early promotions for this product prophesied that it would "dramatically alter the way the business world thinks and uses computers." By 1983, 5000 units were in place in the U.S. alone. Early computer networks used star configuration with "dumb" (nonprogrammable) terminals accessing a host's computing power. This centralized approach was cost-effective during the 1970s. As minis and micros evolved and became inexpensive, they replaced dumb terminals. Minis and micros encouraged distributed processing and are now hosts or nodes in a network.

A LAN deploys switching technology to transfer data. Its special features are:

1. Shared transmission medium
2. Peer-to-peer communication (i.e., a device can communicate directly with another one at the same level).

3. High-speed communication, up to 10 Mbps, as compared to a wide area network (WAN), which uses standard communication facilities such as telephone and leased lines.

Thus, LANs permit a large number and variety of computer systems and other digital devices, including machining centers, to share peripherals and exchange information at high speed over limited distances. This fulfills the communications needs of most CIM plants.

LANs have become popular in recent years. The main reason for this is the microcomputer's low cost and enhanced capabilities. As a network node (see next section), a micro can share expensive computer resources such as databases and large printers. LANs are also expandable as the need arises, so investment is not wasted.

Several factors should be considered in designing and implementing a LAN. For example, the speed of the LAN should be approximately equal to the fastest computer or device on the input/output bus. The goals are reliability, maintainability, cost, flexibility, compatibility, and extendibility. LANs should allow networking of a variety of devices, even of different makes. For security, the files and records should be lockable.

A wide variety of LANs are available. Prospective users should discuss their particular needs with application engineers of companies engaged in the networking business. Such a company is sometimes called a **systems integrator**.

Elements.   LANs comprise the following elements:

Nodes. In a CIM environment, the usual nodes are PCs, PLCs, CNC machines, and similar equipment and other digital devices.

Network communication cards (adaptor cards). The network cards provide communications paths between a node and the network.

Network software. The network software is usually a package from the vendor that provides all the necessary protocol to communicate from the node environment, say DOS, to the network environment.

Transmission media (cabling). Depending on the type of network, the transmission media may be a telephone line twisted-pair cable, coaxial cable, or fiber-optic cable.

Server (PC or dedicated). The server is a special node with a hard disk where the network software "resides." It is the junction for file transfers and mass storage. A dedicated server offers several storage and performance advantages over a PC used part time as a server.

Peripherals. Printers, plotters, modems, and so on.

Box 4.3 describes the important tasks carried out in a multivendor network environment.

BOX 4.3   *The Layered Look of Multivendor Networks*

Engineering (manufacturing) networks may be based on local area networks (LAN) or wide area network (WAN) technology. Because WANs were introduced first, many of the older networks may be WAN. Along with the older networks that are planned for conversion or already have been converted to LANs, most new networks are either pure LANs or some combination of LAN and WAN communication.

Compared to WANs, LANs usually offer much higher maximum data rates, called bandwidth, depending on the transmission media. However, they are limited in the distance range to one or more buildings at a single site. Improvement in this limitation has extended this distance to 25 miles, which is sufficient for most manufacturing enterprises operating at one site. Companies with international operations use WAN to integrate the individual sites on LAN.

The principal levels of connectivity are physical connection, terminal emulation, file transfer, and program-to-program links. A physical connection allows the network to transmit data streams. But that is not enough for the computers in a working network. First, for any two computer nodes to communicate, networking software at each node must be able to determine which node is the destination for transmitted message packets. Second, networking software must be available on possible destination nodes to prepare received data for processing by application programs. In many cases, each networking program is unique to a specific type of application software.

The seven-layer OSI reference model represents the evolving standards of the ISO, designed to allow computers made by different companies to communicate by complying with internationally accepted protocols. Each layer of the reference models builds on the functions of the layers beneath it.

Physical Connection. Comprising the two lowest layers in the ISO/OSI model, physical connection provides the actual interface to the transmission medium (or carrier) and establishes an error-free communications path between adjacent nodes in a network. However, there can be no useful communication between nodes without message routing and communication control (layers 3 and 4) and the application-specific functions (5, 6, and 7).

Three aspects of physical connection must be considered when setting up a network: transmission media, medium access control, and physical transmission protocol. The most widely used transmission media today are baseband cable, broadband cable, fiber optics and microwave signals. Baseband cable may be either standard coaxial, standard ThinWire coaxial, or unshielded twisted-pair cable. Most local data networks use standard coaxial cable as the communication backbone but may include other media types as well. Broadband cable is used for simultaneous communication of data, video, and voice over the same medium. Optical fibers and microwave signals are used mainly to connect geographically separated LANs in a metropolitan region. However, use of optical fiber for LANs is expected to grow dramatically with the introduction of ANSI's fiber-distributed data interface (FDDI) standard.

Medium access control has to do with how the network nodes' access to the transmission medium is controlled. The two most common techniques are token ring (IEEE 802.5) and CSMA/CD (IEEE 802.3). In token ring networks, a token—a special bit pattern—passes from node to node around a ring-shaped LAN. Possession of the token gives a node exclusive access to the ring for data transfer; when a node has completed its current transfer, it places the token back into circulation for use by another node.

In CSMA/CD (carrier sense multiple-access with collision detection) networks, all nodes can detect the presence of data traffic on the medium. Those having message packets awaiting transmission send them out on the carrier as soon as network traffic clears. Since more than one node may start to transmit data at the same time, all nodes are equipped to detect any resulting collisions of data. When a collision is observed, each node stops and begins retransmitting after a brief interval. Because the interval differs from node to node, one node soon gets exclusive access to the medium.

The physical connection protocols define the characteristics of the networking hardware and software that implement the two lowest layers of the model, namely physical and datalink. The most widely followed physical connection protocol today is the Ethernet specification. It is identical to the IEEE 802.3 standard. In addition to defining CSMA/CD medium access control, the Ethernet specification covers such network characteristics as topography, allowable number and separation of nodes, message packet format, and data rate, which is a maximum of 10 megabits per second.

Terminal Emulation. The lowest level of useful network connectivity is terminal emulation, which allows one vendor's computer or video display terminal to interactively access applications executing on another vendor's computer. This may be done over a direct connection or through a network. The emulation software is layered on the operating system running on the first vendor's computer. If the two computers are in different networks, additional layered software is needed to furnish network-to-network communication.

Terminal emulation is useful in an engineering (manufacturing) LAN, for instance, when the engineering manager uses a PC mainly to execute spreadsheet and project management programs. He/she may want to be able to log into individual workstations in the department to check design changes or project status.

Terminal emulators allow mechanical/manufacturing engineers to access such information as on-line parts lists, catalogs, and specifications. They also support the use of computer bulletin boards, note conferences, and videotext databases, all of which can contribute to improved productivity by allowing engineers to address many questions and issues as soon as they are raised.

Electronic Mail. E-mail provides for informal one-to-one or one-to-many communication between network users through their video displays. It is offered as an integral function by most high-performance operating systems.

Electronic mail basically allows users to read from or write to a text file under the control of an application program executing on another computer in the network. And, with additional capabilities, electronic mail can provide for transfer of mechanical drawing files between workstations or other types of computer nodes.

E-mail promotes the communication of ideas and information among engineers (employees) in a number of ways. Engineers may make suggestions to project managers concerning the design and manufacture of a product, keep up with design activities and technical issues associated with their specific assignments, and follow related business developments and concerns. Since E-mail is usually time-stamped, their design thinking can be documented as a project progresses.

There is a natural tendency for engineers deeply involved in a single aspect of development to neglect communication with colleagues. With E-mail, they can communicate effectively in very little time, from their own desks and, with a node, from their home at their leisure.

File Transfer. This is probably the most commonly implemented level of the connectivity. The long-term trend in engineering departments is for files to become larger and larger, as the generation of more complex and 3-D drawings becomes standard practice. Design files that are typically 1 to 2 MB today will increase to 10 to 20 MB and more. To avoid long transfer times, network data rates will have to increase correspondingly.

There are three file transfer methods commonly used today: ASCII transfers, direct translation, and neutral file translation. In ASCII transfers, the file consists of a stream of eight-bit binary numbers representing the alphabet letters, punctuation, numbers, and special symbols. ASCII transfers can be accomplished without communication software above the transport layer on the sending and receiving nodes.

In direct translation, software on each pair of incompatible nodes converts bidirectionally between their dissimilar data formats. In neutral file transfers, software on the sending node converts files in its own format into a neutral outgoing format, and software on the receiving node converts the neutral incoming file format into its own. The neutral file method can substantially reduce the amount of translation software that might be needed in a multivendor network and so is popular with both application program vendors and customers. In both direct and neutral file translation, format conversion takes place in layer 6 of the ISO/OSI model but requires the communication functions in layers 5 and 7 as well in order to be comprehensible to application programs.

Typical applications of file transfer include moving structural analysis solution files from a supercomputer to workstations or mechanical drawing images between central engineering databases and workstations. Mechanical design databases are transferred to manufacturing for tool design, NC programming, and development of process sheets.

Program-to-Program Connectivity. Sometimes called task-to-task connectivity, this highest form of computer communication allows an application program on one computer to communicate directly with a program on a second computer. This network capability may be thought of as command transfer, in contrast with file transfer.

Assume, for instance, that a user at node A in the network needs to verify the version number of a drawing file at node B. Application program A sends the drawing number to program B, which searches the appropriate drawing file at node B, retrieves the version number, and sends it back to program A, where it is displayed. Usually, the user would be unaware of the cooperation between the application programs and might not even know where the drawing file is actually located.

In another example, program-to-program connectivity enables computer conferencing. When an engineer strikes a key on his/her keyboard that indicates "get the next reply," the workstation sends a command to the computer where the conference database is located. The computer accesses the next reply, reads it, and transmits it back to the workstation, where it is displayed.

Program-to-program connectivity is also useful for requesting quick calculations by other computers in a network. For instance, an application program at a workstation might ask an analysis program at a remote supercomputer in the network to compute the maximum torsion stress under load in a newly designed component. The maximum stress figure might then be used locally by the application program to calculate an assembly loading parameter. This type of computer cooperation saves a good deal of time by allowing the supercomputer to compute the stress and return the result far faster than could be done by the workstation itself.

Interconnects. The first engineering LANs were homogeneous, based on one computer vendor's proprietary networking architecture, hardware, and software and often including some compatible products by third-party vendors.

Much of the current development work done on networking products is directed at achieving similar capabilities in heterogeneous networks, so that computing resources from different hardware, software, and system vendors can communicate just as freely. But it will be some years before such multivendor networks can be designed and operated as straight-forwardly as homogeneous networks are today. In the meantime, multivendor networks are being created with the help of various types of hardware and software interconnects that allow otherwise incompatible systems to communicate.

There are three types of interconnects between computers in a multivendor environment: proprietary-to-proprietary, partially standards-compliant, and fully ISO/OSI-compliant. The first two have been broadly implemented in engineering networks today. The third is the ultimate goal of many industry standards groups, vendor consortiums, and individual vendors.

Source: From "The Layered Look of Multivendor Networks" by D. A. Joy, September 1989, *Mechanical Engineering*, pp. 60–62.

A comparison between LANs and the OSI model reveals that:

1. LANs correspond to layers 1 and 2 of the OSI model.
2. LANs are connected to WANs at the network layer.
3. Two LANs are connected by **gateways**, which basically packet-convert the source protocol into target protocol. The speeds of the two LANs must be compatible when implementing the gateways.
4. Standards exist for LANs at the physical layer. The CSMA technique is equivalent to IEEE 802.3 or ISO 8802/3. These standards are similar to the **Ethernet** standard developed by Xerox Corp.

The enhanced capabilities of computer networks and distributed processing have increased the challenge of security. Computer "hackers" access information they are not entitled to. In manufacturing, new product designs and financial and personnel data are usually the prime targets.

Network Topologies.   A topology is made up of nodes and links (communication paths) between nodes. LANs can be implemented as one of the three topologies shown in Figure 4.19, where the blocks represent devices called nodes. LAN implementation provides for logical data exchanges, regardless of whether a direct physical connection exists, between the source and destination nodes. It allows switching and routing of data to the receiver.

The following are major characteristics of the three topologies; some of these may be obvious from Figure 4.19.

1. Star:

    The central node (host) routes the message to the outer nodes.

    If traffic is from or to the central node, it is efficient.

    If it is from one outer node to another, it gets heavy, since data must go through the central node, where switching can degrade the network performance.

    The control could be either centralized or distributed.

2. Ring:

    The nodes act as active repeaters.

    If a node fails, the entire network fails.

    The node decides how to route (switch); in star topology, the host does it.

    A **token** is used to give a particular node permission to use the channel.

3. Bus:

    All devices share the same physical channel.

    Similar to a multidrop environment, messages are broadcast to all the nodes. A node must be able to recognize its **address** in a message.

    Bus topology does not require repeating of the message as in ring topology.

    Failure of one node does not shut down the entire LAN.

Figure 4.19
Star, ring, and bus topologies

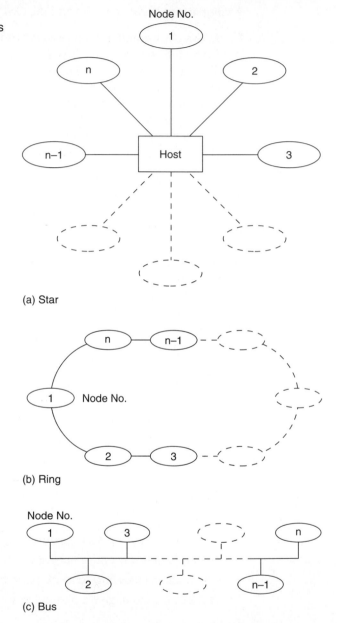

(a) Star

(b) Ring

(c) Bus

Protocols.   A well-defined protocol is essential for data exchanges. The two common types of protocol are: (a) the polling technique, such as token passing; and (b) the collision detect technique, such as **carrier sense multiple access collision detect (CSMA/CD)**. Polling techniques can be centralized, as in a multidrop environment, or distributed, as in the token passing or slotted ring methods.

Polling Technique

The polling technique is based on the token passing or slotted ring method. In token passing only one device can transmit on the LAN network channel at a time. The token is simply a special bit pattern that continuously moves around the network from node to node. Each node or device receives the token in sequence. If it has data to transfer, then it grabs the token to use the channel and on finishing passes the token to the next node. If it has nothing to transmit, then the token is allowed to move forward. Token passing is used mostly in ring topology, but the bus can also use it.

The node holding the token has the right to use the LAN. It continues to hold the token until it has finished. The message to be transmitted includes source and destination addresses. On recognizing its name, the addressee collects the message. It then informs the sender about the receipt of the message so that the sender can pass the token to the next node. To give all nodes equal opportunity, the protocol can impose constraints; for example, no node can use the LAN twice in succession.

The slotted ring is another method for distributed polling used in ring topology LANs. In this case, rather than one token passing around the ring, several slots or frames have space for data and addresses. The node ready to communicate grabs an empty frame and loads the data along with the address. The addressee recognizes its name and unloads the data, thus emptying the frame. Each node constantly checks whether the frame carries any mail and whether the mail is addressed to it. When a node needs to send, it looks for an empty slot. The prevailing situation is like a multicar freight train running around a specific track.

Collision Detect Technique

The popular Ethernet is based on the collision detect technique of CSMA/CD protocol. This protocol handles three tasks: carrier sense (CS), multiple access (MA), and collision detection (CD). Carrier sense means that the nodes are always listening. Once a node finds the channel free and has something to send, it does so immediately. Multiple access means that all the nodes that have something to send are free to try to access the channel. Any of these can send a message immediately on detecting the channel to be free. Thus, a node need not wait as in the polling technique.

What if two nodes try to send at the same time? In such situations, a collision can occur; it is detected by the collision detect function. A collision changes the energy levels the hardware can detect. The hardware and protocol recover from the collision. Following the collision, another attempt is made, again based on listen-before-send. The situation is similar to when two persons begin to speak at the same time and on realizing the oral collision, one says, "I am sorry, you go ahead."

Signalling Techniques.   Besides the transmission media, signalling technique affects the speed of data transfer within a LAN. Signalling technique may be baseband or broadband. The baseband is more prevalent, since it is less expensive. In this, either coaxial or twisted-pair cable is used and the signalling bandwidth is less than 50 MHz. The only signal on the transmission media at any time is that of the LAN itself.

The four popular baseband LANs and their characteristics follow.

Manufacturer	LAN	Media	Access Method
AT&T	Starlan	twisted-pair	CSMA/CD
IBM	Token ring	twisted-pair/coaxial/fiber	Token-ring
Novell	S/net Netware	twisted-pair	Polled
3Com	3+/Ethernet	coaxial	CSMA/CD

The broadband signalling takes place at frequencies higher than 300 MHz. In this case, multiple signals, (e.g., both audio and video) can coexist on the transmission media. Since the data rate is proportional to the bandwidth, the broadband technique provides higher transmission rates. Because of all these features, however, broadband costs more.

While some of the networks originally specified the use of coaxial or twisted-pair cable, all have been adapted for fiber optics.

Networks and Standards.   The following are three of the several networks in common use:

1. ARPANET—Created by the Advisory Research Projects Agency (ARPA) of the U.S. Department of Defense, ARPANET was first used in the late 1960s as an experimental four-node network test bed for basic research
2. SNA—System Network Architecture, a proprietary network of IBM
3. DECNET—A proprietary network of Digital Equipment Corp., DECNET provides most services described in OSI reference models for different layers

The four network standards associated with LANs are:

1. IEEE 802.3—A bus network based on CSMA/CD technique (Ethernet is an example of this network)
2. IEEE 802.4—A token bus; MAP employs this bus
3. IEEE 802.5—A token ring; IBM's ring network is an example
4. ANSI X3T9.5—A token ring used for computer-to-computer or computer-to-peripheral communications; also called a **fiber distributed data interface** (**FDDI**)

Table 4.2 summarizes various IEEE 802.x standards. Box 4.4 describes Ethernet-based communication in manufacturing. Ethernet's importance is obvious from a 1989 survey, which found that 75% of all the installed LANs used Ethernet.

Manufacturing Automation Protocol (MAP)

In the course of computerization and implementation of flexible automation toward CIM, manufacturing companies procure modern equipment, such as machining centers, robots, workstations, computer hardware and software, and other devices. Most user companies face incompatibility of equipment due to lack of appropriate com-

Table 4.2
An overview of IEEE 802.x standards

IEEE 802

Specifies the physical and link layers. Baseband uses the bus scheme with a 50-Ω cable. Devices use RS-232C to interface with network controller. Also runs 1, 5, and 10 Mbytes over 2.5 km. Broadband uses 75-Ω cable with rates of 1, 5, 10, and 20 Mbytes. Broadband uses a single-cable CATV standard for normal use. Operates in half- or full-duplex mode.

IEEE 802.2

Specifies medium access and logical link. Protocol specified is a multipoint peer link. Any compatible physical layer can be used. Supports data rates to 20 Mbytes. Used in all three access methods (802.3, 802.4, and 802.5). Approved in 1984.

IEEE 802.3

Ethenet (CSMA/CD). Uses a carrier sense multiple access with collision detection. The requesting sender is obligated to "listen before using the network." Specified lengths of up to 2500 m can be used. The system has low efficiency for high loads. The specification outlines method, protocol, physical medium (low-noise coax), connections, and cable to I/O device. Works on baseband and broadband. Approved in 1983.

IEEE 802.4

Token bus. Token passing over a bus-type network on broadband cable or on a baseband cable with two allowable types of communications. The specification allows equal access, but devices can transmit only when the token is acquired. The token is not physical, but logical. The system is stable and predictable. The documentation details the token passing, protocols, physical connections, and media (base- or broadband cable). Typical vendors complying are Concord Data, Gould, Allen-Bradley, Western Digital, and 3M. IEEE approved in 1983, and the ISO is reviewing the specification. Fast becoming the de facto standard for the factory floor.

IEEE 802.5

Token ring. Mostly an IBM thrust. Similar to 802.4 but with a physical ring instead of a logical bus. Data can pass in only one direction. One station failure can disable the entire network. Media can be twisted pair or coax. The documentation defines the token management, protocols, physical media, and connections. Approval is expected about one year later than the others.

Source : Reprinted with permission from *Computer-Integrated Manufacturing Handbook* by E. Teicholz and J. N. Orr (Eds.), 1987, New York: McGraw-Hill.

munications protocols. Users have one of two choices: (a) continue to buy from the same vendor, which may not allow competitive bidding, or (b) buy from different vendors to save on equipment cost, but incur additional cost on developing communication interfaces.

◻ **BOX 4.4**   *Ethernet Communication in Manufacturing*

While new installations of MAP systems continue to lag behind the early optimistic forecast, Ethernet installations continue at a heavy pace. A 1989 survey conducted by Computer Intelligence Inc. revealed that 75 percent of all LANs installed used Ethernet.

Ethernet is a baseband LAN which supports one transmission at a time on the line as compared to broadband; which supports multiple transmissions. The advantages of baseband are its favorable cost, speed, and maintainability. Broadband, however, is a more complex technology permitting simultaneous transmission of voice, video, and data over the same cable with multiple transmissions under each mode. But, for this expanded capability there is a higher cost. Each of these network technologies is covered by a set of OSI sanctioned standards: baseband under IEEE 802.3 (CSMA/CD), which Ethernet closely aligns to and broadband under IEEE 802.4 (token bus on broadband). The TOP standard recognizes IEEE 802.3 whereas the MAP standard recognizes 802.4.

Ethernet was developed in 1976 by Xerox Corporation as a prototype network for internal use in their Palo Alto Research Laboratories. Even though the majority of Ethernet installations are in office environments there is no element of its design that prohibits use at the shop-floor level.

Ethernet uses "Carrier Sense Multiple Access with Collision Detection." With this arrangement, all devices share a common data line. Any device wishing to transmit listens to the line first to see if another device is transmitting. If there is existing transmission, the device waits until the line is clear. There is a chance that two devices can start sending information at the same time and thus create a collision. When collisions are detected, the data sent are ignored and each device stops transmitting for a random interval of time. The random interval assures that the same two devices do not create another collision on retry.

Ethernet provides data at a transfer rate of 10 Mbps and considerable transmission flexibility in that it supports four transmission mediums: thick wire coax, thin wire coax, twisted-pair and fiber optics. The distance without repeaters ranges from 230 feet on twisted-pair to 5,000 feet on fiber optics.

To illustrate Ethernet's speed, two tests were made by Cincinnati Milacron's Research and Development Center to compare the speed of an Ethernet link with a serial data link. The first test consisted of transferring a very lengthy part program from a Silicon Graphics Workstation to a VAX host computer. Using an RS-232 serial data line with Kermit protocol running at 9,600 baud, the transfer time was 6 hours. The same part program transferred between an IBM PC and the Workstation over an Ethernet link running at 10 Mbps took 3.5 minutes.

Ethernet interfaces are not available today on CNCs except through front-end computers serving as Network Interface Units or Protocol Converters. There is usually a loss of network capability when this method is employed, especially where a slow speed serial communications port is the only way to communicate to the CNC. Backplane Ethernet interfaces to CNCs do not exist today primarily because CNC builders have been giving development priority to MAP interfaces. Any major shift in communications philosophy at this point will have an impact on MAP and merits close monitoring by both CNC builders and users.

Source: From "Ethernet Communication in Manufacturing," June 1990, *Modern Machine Shop*, pp. 118–20.

General Motors experienced such a problem in the mid-1980s. The company found that only 15% of the 40,000 or so programmable devices then in their plants could communicate with each other. With a projection of such devices increasing to 200,000 units by 1990, GM could see the problem getting out of hand. Until the 1980s, GM was opting for choice (b) above. When it found this option to be both impractical and costly, it decided to develop a standard that has matured into what is called MAP (manufacturing automation protocol). In 1982, GM adopted a standard form of exchange (MAP) recommended by a select task force set up earlier. The MAP was meant to minimize communication difficulties arising from a lack of standards in machines and equipment from different computer and control vendors.

MAP's feasibility was first demonstrated at a trade show, AUTOFACT '85. It was shown that computers from different manufacturers can be linked to exchange data. PLCs from Allen Bradley, Siemens, Gould, and Honeywell could talk to each other, and with ASEA robots, IBM computers, and Honeywell and Hewlett Packard inspection systems. Other participants in this demonstration were Motorola, DEC, and Control Data Systems. The AUTOFACT exhibition attendees were invited to enter on a terminal the order for a wooden assembly in any color combination of a product called "tower of Hanoi." From there on, all functions of manufacturing including final inspection were computer-controlled.

MAP is a specialized LAN designed for a factory environment. It is a hardware-cum-software implementable set of rules that facilitate information transfer among networked computers and computer-based equipment. It has been so designed that computers and associated devices from different vendors can communicate with each other. The alternatives are vendors' proprietary networks whose rules are understood and followed only by their own products. MAP is being developed in conjunction with TOP (technical office protocol) so that plants can communicate with the administrative functions as well. MAP/TOP is a means of consolidating the fragmented information systems prevalent in manufacturing, and thus it is extremely CIM-oriented.

Box 4.5 describes some of the growing pains MAP is currently undergoing.

▣ BOX 4.5   *Application Layer of MAP*

If you have stayed abreast of the development of MAP you are aware that its progress has been slower than predicted. The MAP 3.0 specification was released in June, 1988 with the promise of specification stability for six years. So why hasn't the application of MAP flourished as predicted? One limiting factor is the application layer.

MAP is built on the International Standards Organization (ISO) Open Systems Interconnect (OSI) seven-layer reference model where each layer is assigned a protocol specification. Layer 1, for example, is the physical layer which defines the connection to communications cable. Layer 7 (application layer) specifies the interface for the computer device's software. The computer device may be a CNC, programmable controller, cell controller, or other computer device. Layer 7 is of interest to suppliers and users because it is at this level that the issues are addressed as to what type of data may be transferred in or out of the devices such as part programs, machine/control status, tool data, diagnostic information, probe data and so on.

Suppliers of CNCs, programmable controllers, and cell controllers have started committing resources to integrating MAP 3.0 interfaces into their products. But, they are finding that the specifications for the application layer are still in the development process.

MAP 3.0 defines two specifications for the application layer, File Transfer Access Management (FTAM) and Manufacturing Message Specification (MMS). MMS is the specification expected to be utilized by CNC manufacturers. This standard was jointly developed by Electronic Industries Association (EIA) and International Standards Organization (ISO).

The services provided by MMS RS-511 within the application layer of MAP 3.0 are generic in that they specify how messages are assembled and sent, but do not provide application-specific information. That information is intended to be provided by companion standards, envisioned for each specific class of application. For example, there are currently projects in process to create companion standards for robots, process controls, programmable controllers, and NCs. The NC companion standard identifies the data content of specific NC domain types like part definition, machine part program, tool data table information, tool offset table data, fixture data information, pallet data table, and cutter compensation table which is not covered in MMS.

Writing and debugging the software within the application layer is a sizeable task. One way being considered by some CNC suppliers is to provide a fully compliant MAP 3.0 interface and in addition offer a collapsed MAP version.

Cincinnati Milacron has selected a different approach to maximizing their efforts on MAP. On their Acramatic 950 CNC they are offering a full compliant MAP 3.0 interface and in addition an RS-511 async interface which complies to MAP only at layer 7. This permits the MAP and the non-MAP interfaces to both share the same RS-511 and companion standard implementation. The RS-511 async interface substitutes an RS-423 serial connection in place of carrierband or broadband specified by MAP and runs on an RS-484 line protocol. It is slower than MAP and requires a separate data line for each connection but it is desirable because of its lower cost per connection.

Still another way to maximize the design efforts is to wait until MMS RS-511 and its NC companion standards are published, thus avoiding rework caused by working with premature specifications. This appears to be the case with not only some CNC vendors but also other computer device vendors as well. This may help explain why MAP interfaces are still not readily available on a broad base of computer devices.

Source: From "CIM Perspectives" by G. E. Herrin, May 1989, *Modern Machine Shop*, pp. 134–135.

As in other LANs, MAP allows devices to be interconnected in one of the three ways: star, ring, or bus (see Figure 4.19). In the star network, each device communicates with the other through the host, similar to the way two persons talk with each other over the telephone via the switchboard. In the ring network, the devices are connected through a continuous ring. A message travels from station to station until it reaches the recipient who, on finding its address on the packet, unloads the message from the ring. In the bus network, devices are located on either side of the cable (bus) to which they are connected. The arrangement is similar to the layout of a residential community in which houses are built on either side of the street to which they are connected through driveways.

MAP is designed to allow broadband transmission. Broadband transmission differs from baseband transmission. In the baseband method, one device occupies the entire cable until the transmission is completed. In the broadband method, each device has its own frequency, thus several data channels coexist on the line. The devices can send their signals simultaneously, as in cable TV, which may be audio or video.

Of the two approaches—deterministic and contentional—MAP is based on the former. Access is available only to the device holding the token being passed around the network continuously. The device needing to send information must wait for the arrival of the token and claim it. To ensure that every device gets a chance to use the network, the messages in a token-based system are strictly controlled. This feature is what makes MAP so useful for CIM.

An important consideration in any network is the protocol of access, since several devices may be competing for the line at the same time. As an example, Box 4.6 describes "Kermit" protocol in manufacturing.

 BOX 4.6   *Kermit Protocol in Manufacturing*

As computer-integrated manufacturing expands, users and systems integrator alike continually search for lower-cost, reliable methods of getting computer equipment to talk to each other. Kermit File Transfer Protocol is starting to appear on the manufacturing floor, filling the gap between low-cost serial connections without error checking capability and the high-cost, full-capability local area network (LAN).

There are many interface solutions available including Manufacturing Automation Protocol. The cost of utilizing these various interface methods is generally proportional to their capability and speed with LANs at the top end of the price range. The lowest-cost solution to connecting computer equipment is still a serial connection provided by RS-232, RS-422, or RS-423 running very basic protocols without error checking. Most CNCs and programmable controllers today include one or more of these interfaces as standard equipment. Kermit File Transfer Protocol implemented on these controls upgrades the existing interfaces and data lines to a packet transmission-type system, with check sum and retransmission capability to enhance data integrity.

The name Kermit is derived from "Kermit the Frog." The association of Kermit to the protocol was prompted by a picture of Kermit the Frog on a wall calendar located in the room where the development team was working at Columbia University Center for Computing Activities (CUCCA) back in 1981–82. Initial objective had been the archiving of files on floppy disks from mainframes. Soon it became apparent that Kermit could be useful for more than just archiving. It is estimated that it has reached more than a thousand sites through various user groups. The software is free and available to all provided the user pays a distribution fee of $100 to defray the cost of media, printing, postage, etc.

There may be another cost payable to the CNC or PLC vendor for an interface for Kermit, but this may be a fraction of the cost of MAP interface.

The proliferation of Kermit into the shop is primarily due to the wide variety of computers on which the protocol has been implemented. These computers, when used as DNC (direct numerical control) hosts, cell controllers, front-end processors, programming systems, and status gathering systems, have made available a low-cost interface protocol for connecting into shop-floor equipment. Kermit is desirable since it runs on the existing serial data lines. Another plus is emerging out of the Binary Cutter Location (BCL) environment. Because of its packet transmission capability, Kermit has become a very desirable protocol for transmitting BCL data to BCL input controls.

Even though Kermit is implemented on a wide range of computers, it is still not available on all shop-floor control devices.

The role of Kermit in manufacturing is not clear yet. It could be that Kermit will serve as a short-term solution until MAP products become available, or it could become a long-term low-cost solution coexisting with MAP. Much depends on where MAP interface costs settle out.

Source: From "Kermit Protocol in Manufacturing" by G. E. Herrin, November 1989, *Modern Machine Shop*, pp. 116–118.

MAP has been promoted since 1984, when a MAP Users' Group formed (in 1985, it became a technical group within SME). Currently MAP has strong global support. Major U.S. manufacturers such as General Motors and Boeing, SME, the British Department of Trade and Industry, the Japanese Manufacturing Information Technology Institute, and European ESPRIT/CNMA (a European computer communication effort managed by British Aerospace) are a few of the organizations behind MAP. With the scale economies arising from international commitment, MAP is becoming affordable in comparison with the proprietary network options. It is projected that MAP's cost is likely to drop 25%–30% annually over the next couple of years, with broadband at $750 per node and carrier band under $500.

What is holding MAP back? It is not the technology as much as the lack of human resources and the will of the manufacturing companies. More companies are beginning to choose MAP-compatible equipment over the non-MAP types. In 1988, MAP version 3.0 became a de facto standard for the next six years (see Box 4.7).

BOX 4.7   *The Current Status of MAP/TOP Development*

In late 1988, MAP/TOP version 3.0 of the specifications had been published. It is to remain unchanged until 1994, so that companies get time to implement it in their products.

The transition from Ethernet-based network to MAP/TOP-based network is not taking place fast enough since many plants already have Ethernet-based network; they do not feel a need to change. Computer companies such as DEC and IBM are pushing through their own networks, DECnet and PCnet respectively. The former is offering its users a direct migration to OSI, bypassing MAP/TOP.

During the confusion prevailing at this time, the best strategy for manufacturing companies is to ensure that the network, be it MAP/TOP or a proprietary or a combination of both, they have or intend to have in their plants follows the OSI rules.

Source: From "To MAP/TOP or Not?" by P. M. Noaker, February 1989, *Production*, pp. 60–62.

4.6 TOOLS AND TECHNIQUES

Information can be transferred from one point to another either on-line or off-line. Manufacturing demands both these modes of transfer. For example, since information on tool wear needed by an adaptive control (AC) system cannot be delayed, an on-line fast system is desirable for its transfer. Off-line transfer is acceptable when data can wait without affecting the receiver's performance. An example is the count of parts processed at a workstation, say during the Tuesday morning shift, for review at the next week's production meeting.

Whether on-line or off-line, CIM communication is aided by a variety of tools. The major ones include

1. telephones, including cellular systems
2. facsimile terminals or fax machines
3. satellite dish and videoconferencing
4. personal computers (PCs)

Telephones

For person-to-person communication, the telephone has been in use for a long time. What is new, however, is the use of telephone lines for data communication between devices—computers and machines alike. Plants located hundreds of miles apart use telephone lines for this purpose. Cellular phones enable company personnel, especially those in marketing, to keep in touch with the plant and with each other, if within the cell vicinity. The cell is a geographical area within which transponders connect the caller with the receiver. Cellular phone transmission is through radio waves.

Fax

Using the fax machine, a sender can transmit hard copy information to the receiver over telephone lines. The sender's tele-facsimile (fax in short form) machine transmits photocopies of documents to the receiver's machine. Fax machines allow manufacturing companies to send information to, or receive it from, their subsidiaries, distributors, and customers—thus broadening the scope of CIM.

A fax machine's operation is simple. It is connected to a telephone. The sender puts the original on the machine, which is like a copier, then dials the receiver's number and presses the send key. The copy is scanned electronically to generate specific signals and codes, which are sent across the telephone line to the other fax machine miles away, even continents away. Technical data, such as graphs, charts, and drawings as well as handwritten material—in fact anything on paper—can be sent. The size of the original document does not matter; it is even possible to reduce or enlarge the original.

With current prices starting at under $500, even small manufacturers can afford a fax machine. Besides the cost of the machine, the user must pay telephone charges. To fax a document internationally costs less than $5. In the U.S., more than 10 million fax machines are in use.

Satellites and Videoconferencing

Satellite data distribution and videoconferencing are new technologies for fast business communications. They assist all functions, especially marketing, in both national and international communications. The time saved in business dealings, reduced expenses, and less travel for staff are the factors behind their increasing use. Texts and facsimiles can be transmitted at rates between 50 and 19200 bauds. As an example, GE Information Services has developed a software (Box 4.8) that connects 750 cities worldwide to exchange engineering and manufacturing data.

The videoconference is a live, interactive television program delivered via satellite for a special audience. Videoconferencing can encompass several countries. In it, even a third party can participate. For example, in a videoconference manufacturing engineers may discuss "live" a product defect with the designers who may be located at company headquarters hundreds of miles away. Occasionally, customers or distributors may be called in "live" to clarify a point relating to the defect.

Videoconference technology is similar to that of television programs received in homes on cable networks. It has two common uses in industry. In one, important messages can be delivered simultaneously to all the employees who may be at different locations. In the other, two groups of people, say of marketing and production, can talk face-to-face with each other even when thousands of miles away.

Satellite and videoconference technology are currently expensive for manufacturing. The cost consists of three elements: (a) setting up the network, (b) putting up a studio, and (c) cost of time logging on the network. The first two are one-time costs, which can be high. The logging time represents running cost, which may be $500–$1,000 per hour. However, these costs are decreasing and CIM operations have already begun to benefit from this technology.

□ BOX 4.8   *Manufacturing Network*

GE Information Services has developed a software called "Design Express" that allows companies to exchange engineering and manufacturing data with suppliers and trading partners worldwide using their global teleprocessing network. Available year-round on a 24-hour per day basis, the network connects 750 cities around the world. The Design Express consists of three modules: Design PC, Design PC Encryption/Decryption, and Design Display.

The Design PC accepts and transmits design and manufacturing data in a variety of formats, such as ASCII, binary, IGES, GKS, and protocols (1200, 2400, and 9600 bps asynchronous or 2400, 4800, or 9600 synchronous) for several microcomputers, workstations, and mainframes. The Design Encryption/Decryption provides data security by letting only authorized persons gain access to the information. Design Display module is for creating parts programs and also for displaying CAD/CAM, APT source files and other files.

Source: From "Manufacturing Network" by D. Deitz, March 1989, *Mechanical Engineering*, p. 8.

PCs

Personal computers are used as terminals on a network for accessing central facility's databases that may be miles away. The benefits from networking the micros are:

1. accessing of central data files
2. sharing of expensive peripherals such as graphic plotters, laser printers, and so on
3. electronic mail
4. access to mainframes for "number crunching" tasks
5. sharing of expensive software

Major difficulties in connecting micros to a network are the following:

1. The PCs are designed inherently for stand-alone operation.
2. The basic CPU (Intel 8088) of an IBM-PC or equivalent is too slow to work on a network. However, newer CPUs in 386 and 486 machines have eliminated this limitation.
3. The network software takes much of the memory.
4. I/O ports are insufficient.
5. The cost of a plug-in networking circuit board per PC is high (20%–30% of the PC cost).
6. For adequate performance, a high-speed central network-serving computer system costing $4,000–$10,000 may be needed. A 25-MHz PC based on a 80286 CPU is the minimum requirement.

TRENDS

❑ The merging of computer and communications into one technology—sometimes called information technology—is accelerating the growth of CIM. Today, several companies specialize in what they call computers and communications (C&C) technology. Although developed primarily for service industries, C&C technology has begun to cater to the needs of manufacturing as well.

❑ Fiber-optic cabling is developing rapidly as a communications medium. It suits the factory floor where data transmission is subject to a hostile environment. The lower cost of fibers made of plastics rather than of glass enables fiber-optic cabling to compete with twisted-pair and coaxial cablings.

❑ New techniques to transfer data at faster rates continue to develop. For example, it is possible (Bergman and Tell, 1989) to transfer data synchronously through fiber-optic cable at rates in excess of 100 Mbps.

❑ The MAP 3.0 specification released in 1988 did not adopt fiber-optic cabling. But in the appendix of the document, two fiber-optic standards—802.4H and 802.5—were cited. The ANSI Committee X.3T9.5 is proposing another standard—called FDDI (Fiber Distributed Data Interface)—for high-performance general purpose LANs with 100 Mbps transmission rates.

❑   Personal computers and workstations are being integrated using networks, scanners, CD-ROM devices, and more. A recent trend is to link these with video cameras and recorders. It is achieved through a printed circuit board called live color video digitizer (LVD). Two possible manufacturing applications are (a) taking a live image of the assembly and overlaying it on the computer model to see how the parts fit, and (b) pointing the camera at a moving part to check the clearances and tolerances.

SUMMARY

Communications is fundamental to the integration of personnel and CIM devices. In a CIM operation, almost all communications occur via computers. This chapter covered the basic principles of communications technology. Representation of data, coding and their systems, transmission, and fiber-optic-based communications have been presented in detail. Networking of CIM devices, local area networks, OSI protocol, and MAP have also been described.

CIM's progress is being hampered by a lack of standards for manufacturing communications. One reason for this is that the development of standards is time-consuming due to the complexity of discrete manufacturing. The other reason is the less-than-enthusiastic support of computer vendors and system integrators who fear the demise of their proprietary systems with a universal standard in effect. Further development of MAP/TOP and its wider acceptance by the industry would help the growth of CIM.

KEY TERMS

8421 code

Acoustic coupler

Address

Advanced Research Project Agency (ARPA)

Amplitude modulated or modulation (AM)

Application layer

ARPANET

Asynchronous transmission

Automated guided vehicle (AGV)

Automatic tool changer (ATC)

Bandwidth

Baseband

Baud rate

Baudot code

Binary coded decimal (BCD)

Broadband

Carrier sense multiple access collision detect (CSMA/CD)

Carrier signal

Coaxial cable

Collision

Crosstalk

Cyclic redundancy check (CRC)

Demodulation

Duplex

Encoding

Ethernet

Fiber distributed data interface (FDDI)

Fiber-optic cable

Flow control

Frame

Frequency division multiplexing (FDM)

Frequency modulated or modulation (FM)

Full duplex

Gateway

Gray code

Header

Light emitting diode (LED)

Manufacturing Automation Protocol (MAP)

Maximum transmission rate (MTR)

Medium

Modem

Modulation

Multiplexing

Nodes

Open Systems Interconnection (OSI)

Parallel transmission

Phase modulated or modulation (PM)

Physical layer

Polling

Protocol

Repeater

Semantics

Serial transmission

Server

Simplex

Station

Synchronous transmission

Syntax

System integrator

Tap

Time division multiplexing (TDM)

Token

Trailer

Transponder

Universal asynchronous receiver/transmitter (UART)

Universal synchronous/asynchronous receiver/transmitter (USART)

EXERCISES

Note: Exercises marked * are projects.

4.1 Explain the role of a communications matrix (Figure 4.1) in highlighting the communication needs of CIM.

4.2 What is the limitation of the BCD code?

4.3 What is the attraction of the Gray code?

4.4 Represent the following manufacturing data in hexadecimal and binary codes as per ASCII:

JOB # 1 IS ON MACHINE # 7

4.5 Discuss the limitation of parity check in detecting transmission errors.

4.6 List five major benefits of fiber-optic cabling. Which of these is most beneficial and why?

4.7 It has been said, "What's good for GM is good for America." Comment on this statement in light of GM's MAP efforts.

4.8 A communications line operates at a signal-to-noise ratio of 25 dB and a bandwidth of 4,000 Hz. What is the maximum speed at which data transfer is possible on this line?

4.9*   Conduct a literature search and prepare a 300-word report on the current status of MAP.

4.10*   Invite the person who had been in charge of LAN implementation and development at your college/university or company to give a talk on LAN to the class. If a LAN does not exist at your place of study or work, then invite someone from a nearby company where it has been implemented recently.

4.11   Circle T for true or F for false or fill in the blanks.

a.   In a full-duplex multidrop environment, two slave stations can transmit data at the same time.   T/F

b.   Open wires are still used in communications.   T/F

c.   Multiplexing techniques enable transmission of data from several sources onto one line.   T/F

d.   If you are willing to pay for better quality modems, the speed at which data can be transmitted over voice-grade communication lines is unlimited.   T/F

e.   Remote batch terminals are often found in timesharing systems.   T/F

f.   In asynchronous transmission, data is transferred in blocks.   T/F

g.   Whether it is between persons or computers, the three basic components of communication are _____, _____, and _____.

h.   The rules that govern communications between computers are called _____.

i.   The basic problem in transmitting digital signals over long distances is _____.

j.   The two types of multiplexing are _____ and _____.

k.   A _____ is used by satellites to retransmit the signal received at another frequency. This is done to avoid _____.

l.   _____ Law determines the theoretical limit on the speed of data transmission over noisy communications lines.

m.   Match the term on the left with that on the right by connecting them with a line:

simplex   simultaneous bidirectional

half-duplex   unidirectional

full-duplex   alternating bidirectional

n.   Broadband uses a bandwidth of _____ , while baseband uses a bandwith of _____.

SUGGESTED READINGS
Books

Baker, D. G. (1986). *Local area networks with fiber optic applications*. New York: Prentice-Hall.

Durr, M. (1987). *Networking IBM PCs*. Que Corp.

Fritz, J. S., et al. (1985). *Local area networks: Selection guidelines*. New York: Prentice-Hall.

Halsall, F. (1988). *Data communications, computer networks, and OSI*. Reading, MA: Addison Welsley.

Ranky, P. G. (1990). *Computer networks for world class CIM systems*. Guildford, Surrey (England): CIMware Limited.

Stallings, W. (1987). *Local networks*. New York: Macmillan.

Journals and Periodicals

NASA Tech Briefs
IEEE Proceedings
Manufacturing Engineering. Dearborn, MI: SME.
Mechanical Engineering. New York: ASME.
Modern Machine Shop. Cincinnati: Gardner.
Production. Cincinnati: Gardner.

Articles

Bergman, L. A., & Tell, R. G. (1989, April). *NASA Tech Briefs*. *13*(4), 38.

Horn, D. (1989, July). Fiber optics extend into factories. *Mechanical Engineering*, pp. 94–96.

Jackson, G. (1988, September). Dialing the software. *Mechanical Engineering*, p. 11.

Joy, D. A. (1989, September). The layered look of multivendor networks. *Mechanical Engineering*, pp. 60–62.

Lynch, M. (1990). Understanding RS-232-C communications. *Modern Machine Shop*. In four parts; first part in July issue, pp. 134–135 (other parts in subsequent issues).

Mathews, J. I. (1989). Networking computers: Designs and problems. *Journal of Industrial Technology*, *5*(2), 9–10, 28–29.

Miu, N. T. (1989). Communications with light-laser/fiber optics technology. *Journal of Industrial Technology*, *5*(2), 1–2, 27.

Schenck, J. P. (1989). Manufacturing automation protocol. *Journal of Industrial Technology*. *5*(2), 6–8.

Southard, R. K. (1989, February). Fiber-optics for industrial and control applications. *Manufacturing Engineering*, *102*(2), 85–87.

CHAPTER 5
Database

This chapter discusses the principles of database technology and management. Database technology and management play a critical role in computer-integrated manufacturing. Regardless of the company size and product, an effective system for developing and controlling various databases is essential. In recent years, several computer-based tools for automating database management functions have been developed.

Databases involve two groups of people—their "creators," called system designers or analysts, and users such as manufacturing engineers, customer service representatives, and so forth. The discussions in this chapter concern the user's needs rather than design issues, which are normally the concerns of system designers and analysts.

5.1   INTRODUCTION

Database technology is important for electronically integrating the various manufacturing functions of design and production. Such integration is vital in CIM environments, since the original design data provides input to several downstream functions, such as finite element analysis, NC machining, bill of materials (BOM) processing, or CMM. As an example, two designers may be working on the same file and save their work at the same time. If the files cannot be "locked" properly, saving one may erase the other.

Electronic or optical databases provide a method of filing and retrieving volumes of data. Employees skilled in the use of databases can quickly gather information through a keyword search. The database of a manufacturing company is its storehouse of all the relevant data. The terms data and information are often treated synonymously, as in this book. In reality, **data** refers to the values stored in the database and **information** to the meaning as understood by the user.

Although a database can be generated and maintained manually, in a CIM environment it is computerized. A computerized database is created and maintained by a group of application programs or a **database management system (DBMS)**. DBMS is software, such as the database manager Oracle R, that uses simple forms with rows and columns.

The term *database system*, rather than *database*, is a more appropriate description of what storing and retrieving data and information entails. A system involves operation of interacting elements with a specific purpose. In the case of database systems, there are four interacting elements: data, hardware, software, and users. Between the database and its users is a layer of software, usually called DBMS. This software handles user requests for accessing the database. Thus, DBMS is like the operating system of a computer, similar to DOS. The three user groups of database systems are database administrators, application programmers, and end-users such as design or manufacturing engineers.

A database is a collection of related files consisting of records that contain data. Data are known facts that have implicit meaning and are worthy of storing. A database is designed, built, and "populated" with relevant data. It is a logically coherent collection of data; merely a random assortment of data does not constitute a database. Since manufacturing data vary in type, databases can become complex and large. CIM databases contain design data, employee data, customer data, NC programs, BOMs, and more.

According to Martin (1975), "A database may be defined as a collection of interrelated data stored together without harmful or unnecessary **redundancy** to serve one or more applications in an optimal fashion; the data are stored so that they are independent of programs which use the data; a common and controlled approach is used in adding new data and in modifying and retrieving existing data within the database."

Date (1982) offers a layperson's definition of a database: "Basically, it is nothing more than a computer-based record-keeping system: that is, a system whose overall purpose is to record and maintain information." He adds, "The information concerned can be anything that is deemed to be of significance to the organization the system is serving—anything, in other words, that may be necessary to the decision-making process involved in the management of that organization."

Date (1982) has identified the following seven reasons why companies need databases:

1. Redundancy can be reduced.
2. Inconsistency can be avoided (to some extent).
3. The data can be shared.
4. Standards can be enforced.
5. Security restrictions can be applied.
6. Integrity can be maintained.
7. Conflicting requirements can be balanced.

5.2   MANUFACTURING DATA

Any information a manufacturing company requires can in general be referred to as manufacturing data. Carrying out a task requires a certain amount of skill of the person performing the task—for example, the machining of a casting on the turning center. If a familiar machine is used to do the job and the casting is of a conventional material, the operator may already know the appropriate cutting data—speed, feed, tool geometry, and so on—for the casting. But, if the material is unfamiliar, the operator may have to seek help from the supervisor or consult a handbook for the data.

Manufacturing companies need a variety of data. Some examples of typical manufacturing data or information are:

The serial number of the faulty bracket is 235690.

When is the scheduled maintenance of machine # 32 due?

Last year, product ABC contributed 16% to company profit.

How many days was Joe absent last quarter?

The new order of 1,000 units for XYZ bearings is due for shipment on the last day of week # 48.

Could the new composite material being praised highly by the automotive industry be suitable for part # 158?

What is the shear stress of grey cast iron?

What is the appropriate cutting speed for machining the new ceramic part (# 756)?

The last shift production reached a quality level of 120 ppm (parts per million).

These are just a few examples. Manufacturing data is usually quite large and varied even for a small company. In fact, manufacturing "thrives" on appropriate data and their proper use.

Types

Manufacturing data may be static or dynamic. Design data of established products are static; that is, they do not change with time. Product data is usually static, while production data are dynamic, meaning they do change with time. How a change in data affects manufacturing functions is of primary concern. Ideally, relevant information is needed (see Box 5.1) at the right place in the right format at the right time—neither later nor sooner. Delayed information may already have cost the company money, while an early arrival requires temporary storage. For example, for uninterrupted working, a machining center must store the next part program received from the DNC host.

Manufacturing data can broadly be divided into two groups: relatively permanent or temporary in nature. The permanent data will change only when a product is entered into or deleted from a product plan, or if machines change. Examples of such data are structural information about the product, information on process capabilities, operational information on machines, and so on. Temporary data, on the other hand, exists only as long as the part is being processed; it is deleted on completion of the operations. Examples of temporary data include items relating to planning and processing of an order, machine selection and process sequencing, scheduling of shop orders, and similar information.

BOX 5.1 *Making Information Flow*

The people in manufacturing who move information around—and that includes manufacturing engineers—predominate over the people who handle and produce the product. While some of the ratio may still be middle-layer inefficiency awaiting the cut of the knife, the fact is that today direct labor rarely comprises more than 10-15% of payroll. People processing information get the rest.

To cut cost out of manufacturing, companies now need to cut the cost of moving information. Manufacturers have rightly been concentrating on the non-value-added costs of inventory, material movement, and work-in-process, often estimated to be an astounding 50% of the cost of producing a product. The ones who have attacked the problem have discovered that the key to solving it is information: the right information to the right person at the right time.

In a way, information today is material, and companies have to plot and plan its flow as doggedly as they did for material when manufacturing cells and flow lines were set up for production.

Source: From "Making Information Flow" by J. M. Martin, May 1989, *Manufacturing Engineering*, pp. 75–78.

Another classification depends on whether the data is already in the format appropriate for computer systems. Data in handbooks, for example on cutting speeds and feeds, is computer-incompatible; it must be searched for and entered into the computer for storage and subsequent use. Data on floppy disks, paper tapes, or in databases, on the other hand, is computer-compatible.

Appleton (1982) groups CIM data into four classes: (a) product data, (b) production data, (c) operational data, and (d) resource data. According to him, "Product Data is data about the objects to be manufactured. It includes text and geometry data, as well as alphanumeric data. Production Data describes how the objects are to be manufactured. Operational Data is closely related to Production Data, but this data describes the events of production, such as lot size, schedule, sequence of assembly, etc. Resource Data is closely related to Operational Data, but it describes the resources that are involved in operations, e.g., machines, people, and money. The general consensus is that technical data includes all product data, most production data, limited operational data, and no resource data. Nontechnical data is all the rest."

Sources

The various sources of manufacturing data can broadly be divided into two groups: internal and external. Internal sources are company personnel, for both formal and informal data, and various files in hard copy as well as in computer-based soft copy formats. A company can collect internal data or information for the asking, since it is already paid for. External data, on the other hand—especially that not in the public domain—generally costs money. External data may sometimes be available at no cost, especially if the source has some interest in providing it. For example, a company contemplating the use of plastics instead of metal for an existing component may get technical data from the plastics developer or vendor.

Hundreds of external databases are available to manufacturing companies for a moderate fee. In the U.S., Knowledge Publications, with support from the American Society for Information Science, publishes a directory listing all the databases in the world; it includes every technical field.

The sources of both internal and external data are numerous. Some of the common sources and their characteristics are given in Table 5.1.

Table 5.1
Manufacturing data sources and their characteristics

Source	Format	Action	Interface
Book/Encyclopedia/Handbook	written	search and use	literate user
Journal	written	digest and apply	specialist
Drawings	graphic	read and use	trained user
Prototype/part	geometric	measure and use	skilled user
Papertape	coded	use	machinist
Database	digital	search and use	computer
Disk	digital	use	computer

Books and encyclopedias are excellent sources of information on established facts. Handbooks and directories are strong on tables, charts, and lists. Journal articles and research papers cover narrow topics that may be suitable for R&D purposes. All printed materials, however, have three drawbacks: (a) they may lack current information, (b) they can be transferred only one-way, and (c) they generally report on successes, seldom on failures. Moreover, data from such sources must be searched and keyed in (digitized) for computer compatibility. It is then stored in a database or transmitted on-line to its destination. Drawings are the classical source of manufacturing data. Papertapes, disks, and databases are some of the modern sources.

A new approach to generating design data is called **reverse engineering**. This term is used to describe the process in which production development follows a reverse order (hence the name). Rather than the conventional production drawings, the existing product (or its clay model) is the starting point. The part is scanned with a CMM or laser-based digitizing system to capture its geometry for storage in the design database. CMM-based reverse engineering software packages file the data generated by scanning the various part surfaces; they can even account for any wear in the existing part. From there, it follows the conventional CAD/CAM route to manufacture.

The main database in companies producing to their own designs is the design database. Design databases are expected to support four activities: (a) computer-aided drafting, (b) shape description, (c) engineering analysis, and sometimes (d) kinematics. Depending on product complexity, the drafting may be two-dimensional (2D) or three-dimensional (3D), and the shape may be defined using any of the several methods. Some of the methods are parametric equations, sweep repetition, wireframe representation, **constructive solid geometry** (CSG), and boundary representation. The relationships among the various surfaces, edges, and vertices of an object are described by its topology. An object's geometry enables the determination of derived parameters such as surface areas, volume, weight, center of gravity, and more.

5.3   DATABASE TECHNOLOGY

In this section, the technology relating to the creation and maintenance of databases is discussed. Database management is presented in the next section.

Terminology

Computer databases contain data described in bytes of the binary system. A byte is the smallest group of bits that can be addressed individually. A **data item** is the smallest unit of data with a name; data items may be several bits long. A group of data items is called a **data aggregate**; for example, the data aggregate DATE comprises the items MONTH, DAY and YEAR.

A record is a collection of data items or aggregates. It has a unique name. An application program usually fetches one complete record from the database. Since the terms data aggregate and record imply similar meaning, they are used inter-changeably.

A **segment** consists of several data items and is the basic quantum of data han-dled by application programs under the control of database management software.

A file comprises a collection of logical records. In a simple file, records contain the same number of data items, while sophisticated files may have a varying number of data items.

Finally, a **database** is the collection of multiple record types containing relation-ships among records, data items, and data aggregates.

Figure 5.1 illustrates the preceding terms.

Figure 5.1
(a) Database terminology: a file is a collection of records, a record contains several items or fields, and a group of items is known as a data aggregate; (b) A database for customers

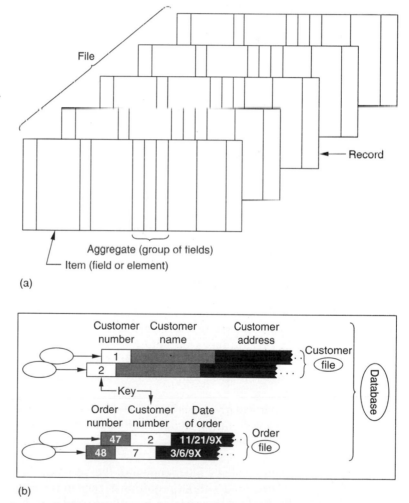

Source: Reprinted with permission from *Using Microcomputers* by D. G. Dologite and R. J. Mockler, 1988, Englewood Cliffs, NJ: Prentice-Hall.

A related set of values is sometimes referred to as a **tuple** (pronounced to rhyme with couple). A tuple with two values is a pair, and one with N values is called an N-tuple. A file consists of a set of tuples; Figure 5.2 shows a nine-tuple (nine values in each row) file. The particular data item used by computers to identify a record (or tuple) is called a **key**; it may be primary (e.g., the first data item in Figure 5.2) or secondary (e.g., the fourth or seventh item).

Basic Concepts

A database is developed to help its users solve problems that involve storing large amounts of data. Fundamental to data storage is the notion of *records*. A record is a set of data items that are logically related. It comprises several **fields** under which the individual items of data are stored. For example,

5394   JOE SMITH   WELDING   INSPECTOR

is a record with four logically related fields: department number (5394), name (JOE SMITH), department (WELDING), and job title (INSPECTOR). To avoid confusion, the fields may be given headings.

A database is essentially a group of "storage bins" within the computer that hold information. A good analogy is a series of filing cabinets, with the difference that the disk or tape medium is magnetic. An important requirement of manufacturing data—for that matter of any data—is that the database be easy to access and maintain.

In a CIM environment, manufacturing data resides predominantly in computer databases. Because a computer's main memory is relatively small, most data are stored on secondary devices connected to the computer via a channel. Such devices include tape and disk units, drums, and devices on which data are stored in demountable cells or cartridges—all these are referred to as **volumes**.

Disks and drums are **direct access storage devices (DASD)**. In a DASD, each physical record has a discrete location at a unique address. DASDs are attractive because data can be read and written directly (randomly) rather than sequentially as in tapes. This saves time, since the recording medium need not be scanned extensively.

A particularly important consideration in database design is the ability to store data so a wide variety of applications can use it and data can be changed quickly and easily. To achieve this:

1. Data should be independent of the programs that use it so data can be added or restructured without requiring change in the programs, and
2. It should be possible to interrogate and search the database without having to write programs in conventional languages. In other words, database query languages could be used.

Figure 5.2
Explanation of terms

Primary key → Employee-number

Secondary keys → Grade, Skill-code

	Name of the attribute:	Employee-number	Name	Sex	Grade	Date	Department	Skill-code	Title	Salary
Form of representation:		N5	AV	B1	N2	N6	N3	N2	AV	N4
Value of the attribute:		53730	JONES BILL W	1	03	100335	044	73	ACCOUNTANT	2000
		28719	BLANAGAN JOE E	1	05	101019	172	43	PLUMBER	1800
Record, segment, tuple		53550	LAWRENCE MARIGOLD	0	07	090938	044	02	CLERK	1100
		79632	ROCKEFELLER FRED	1	11	011132	090	11	CONSULTANT	5000
		15971	ROPLEY ED S	1	13	021242	172	43	PLUMBER	1700
		51883	SMITH TOM P W	1	03	091130	044	73	ACCOUNTANT	2000
		36453	RALNER W LLIAM C	1	08	110941	044	02	CLERK	1200
		41618	HORSERADISH FREDA	0	07	071235	172	07	ENGINEER	2500
		61903	HALL ALBERT JR	1	11	011030	172	21	ARCHITECT	3700
		72921	FAIR CAROLYN	0	03	020442	090	93	PROGRAMMER	2100

Entity identifier

Set of values of one data item (a domain).

Some attributes are themselves entity identifiers of another file.

Source: Reprinted with permission from *Computer Data-base Organization* by J. Martin, 1975, Englewood Cliffs, NJ: Prentice-Hall.

Logical and Physical Views

The description of data and its relationships may be logical or physical. A logical description specifies the data type and associated relationships. A physical description, on the other hand, refers to how data is physically recorded on the hardware. A physical record may contain several logical records to save storage space and access time.

Appleton (1982) differentiates between logical and physical views of data in a simpler way. According to him,

> From the perspective of the manufacturing database, we are mainly concerned about understanding the following:
>
> 1.   Which elements of manufacturing do we need to have data about?
> 2.   What are the relationships among those things?
> 3.   Where do we store the data that we know about those things?
> 4.   How do we keep the data accurate and timely?
>
> The first two issues relate to the logical view of the database, i.e., what the user sees. The second two issues relate to the physical view of the database, i.e., what the computers and technicians see. The distinction between the logical and the physical forms of the CIM database is critical. (Appleton, 1982, p. 5)

If the database only stored data, its organization would be simple. Database management gets complex when we need to show relationships among the data items stored. It is challenging for database designers to develop a user-friendly system; such a system makes accessing logical data simpler.

The logical database description is referred to as a **schema**. It is a chart that remains the same, while the values within may change. With the attribute values removed, Figure 5.2 is a schema. A schema means an overall chart of all the data items and record types stored in the database, while a **subschema** refers to an application programmer's view of the data being used.

A two-dimensional layout of data elements, similar to that in Figure 5.2, is called a **flat file**. A flat file provides the most common technique for associating value with a data item. It also associates data items with the corresponding attributes of relevant entities, allowing storage of items together in a fixed sequence (Figure 5.2). Inside the box, a set of data items is shown along with their values. Each row of data items relates to a particular **entity**, each column to an attribute. Just above the box are the representation forms; for example, in the first column of Figure 5.2, N5 means decimal, fixed point, five digits. The common representation types are:

alphanumeric characters
decimal numbers—fixed or floating point
binary numbers—fixed or floating point
bit strings

Database Requirements

A database is the repository of information for managing the various functions of a company. It should permit both retrieval and modification of data. Searching a database to answer planning and control questions must be user-friendly. A long search time indicates that the database has not been designed efficiently.

An important purpose of databases is minimizing redundancy. Redundancy occurs when the organization stores the same data at several places. This happens because the same data may be needed for different purposes. For example, tolerance data on shaft diameter is required by designers, inspectors, and the operator.

Proper design is critical to the usefulness of databases, since it affects how easily users can retrieve and use stored information. Database design involves the following:

Accessibility. Information should be easy to find and use.

Administration. Information should be organized so the database can evolve smoothly. Standards, procedures, and guidelines are followed for an effective administration.

Cost. The cost of developing and maintaining the database should be kept as low as possible.

Independency. Information should not depend on the type and format of the reports being prepared. Once information has been located, report generation should be straightforward.

Integrity. Consistency, quality, and unambiguity of information are important. Since users' knowledge of a database may be limited, unintentional attempts at contamination are likely. A database should have safeguards against such attempts to ensure integrity and reliability of data.

Interfacing. Being dynamic entities, data change with time. A well-designed database is easier to interface with past data and future expansion.

Migration. Data should be so stored that more commonly used data can be accessed quickly and conveniently.

Performance. Database performance is measured by its effectiveness in achieving intended objectives.

Redundancy. Information should not be stored at more than one location. This saves space and reduces search duration. Remember that database technology has been developed primarily to minimize redundancy.

Reliability. Information should be stored in a format so that reports and forms can easily use it. The database involves two ongoing jobs: creation or updating, and use. To ensure the reliability of data, only authorized personnel should be allowed to create and update it. The creator may be a user, too, but a user need not be the creator or updater.

Response time. The response to a query and the throughput should match or exceed the user dialogue rate.

Search flexibility. The user should be able to ask a variety of questions about the data stored. Even unanticipated types of queries may have to be entertained, requiring faster search.

Security. For security of classified and privileged information, access to unauthorized users must be denied.

Shareability. Users should be allowed fast, uninterrupted access to the same file.

Tunability. Tunability means that the database is adjustable to improve its performance. Database administrators rather than the users are responsible for tuning the database.

Versatility (in representing relationships). To provide users different logical files to work with, these files must be derivable from the same database. Logical files are designed for user convenience and may bear little resemblance to how data is stored physically.

Types of Data Models

There are three types of **data models**: hierarchical, network, and relational or flat file, as shown in Figure 5.3. The relational model uses entity-attribute relationships. For example, a machine part is an entity, while its specifications such as weight, material, and part number are attributes. The flat file format presents information in terms of facts about things. The network model combines aspects of the relational model and the hierarchical model, which is a tree-type structure. The network model has an upper level, called the "root," with one node. The lower levels can have several nodes. Each node is linked to a higher node called the "parent" and several lower nodes called "children." Some nodes may be connected to each other; these are called "leaves."

Consider, for example, an automobile as the product. The root of the network or hierarchical model (Figure 5.4) contains at the uppermost level the vehicle itself. The next lower level contains the engine as a parent node and the camshaft that it comprises as a child node. Bearings may be represented as "leaves" of the camshaft. In the network model, data is identified according to its location as a node on the schema. The other basis of classification may be the task data is associated with. For example, the major components of an automobile—engine, steering, and transmission—can be classified either on the basis of location in the schema or the task. Supporting or interface data, such as nuts and bolts, may serve the needs of more than one node.

The hierarchical model (Figure 5.3) is similar to the network model except that each child can have only one parent. A hierarchical database with a pointer as the guiding search tool is the most common type. The pointer is often a physical address of the sector on the disk that stores the data. Data is stored along with the pointers that identify access paths to the related data. Relationships among data are represented by numerous data structures, which are often complex. Data can be accessed only when an access path is defined.

The major problem with hierarchical database is the difficulty in changing data; for example, adding a field to a record. It may require changes in several paths as well as in programs that use the database. It is considered a good database design practice if the physical details of the database are transparent to (hidden from) the user and only the logical view of the data is made available. This has led to the

Figure 5.3
Three types of data models

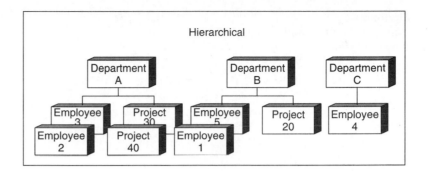

Hierarchical

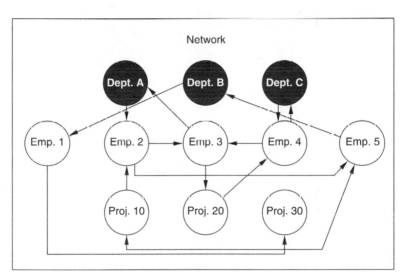

Network

Relational View of Data

Employee Table

Employee Number	Employee Name	Department Number	Project Number
578	Jones	Dept. B	P30
455	Smith	Dept. A	P10
321	Weber	Dept. A	P20
671	Brown	Dept. C	P20
141	Crosby	Dept. B	P10

Department Table

Department Number	Department Name	Location
Dept. A	Finance	New York
Dept. B	Engineering	Los Angeles
Dept. C	Sales	San Francisco

- Simplifies database design
- Easy for end user to understand
- Better end user DP communication

Source: Reprinted with permission from "Data Bases for Manufacturing,"
September 1986, *Manufacturing Systems*, p. 14.

Figure 5.4
Concept of root, parent, children, and leaves

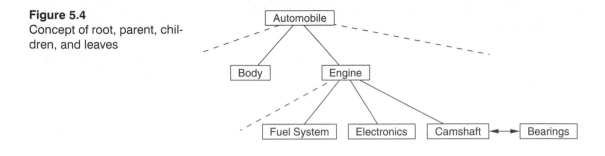

development of relational databases (Box 5.2), which are simply two-dimensional tables of the type normally used to consolidate a given set of data into rows and columns. All the columns or fields have labels, which are the headings of the data under that field. Depending on the system, what constitutes a legal name for the column may be restricted; usually up to 10 characters are allowed.

Rembold, Blume, and Dillmann (1985) classify manufacturing databases into four groups:

1. collection of independent databases
2. centralized or solitary database
3. interfaced database
4. distributed database

Figure 5.5 illustrates these types; Table 5.2 compares them.

Major criteria for database selection include number of users and data files, access speed, flexibility desired, access control and security, type and volume of data to be stored, and maintainability.

File Structures

Files contained in a database may have different structures such as relational, hierarchical, network, free format, or multiuser distributed.

The relational structure is relatively easy to design and modify, and hence is attractive to engineers and CAD/CAM developers. This structure stores data in files containing records; the records in turn are divided in fields. Records can be manipulated (searched, combined, updated, etc.) using the common field. Files are linked through the fields of records.

The hierarchical structure has a fixed connection between the files. This structure also contains records in individual files, but the relationship between records is not through the common fields as in the relational structure. It is rather through a fixed relationship established by the database administrator at the time of system implementation. This becomes a drawback of the hierarchical structure when alterations are required later.

Figure 5.5
Four types of manufacturing databases: (a) Collection of independent databases, (b) Centralized or solitary database, (c) Interfaced database, and (d) Distributed database

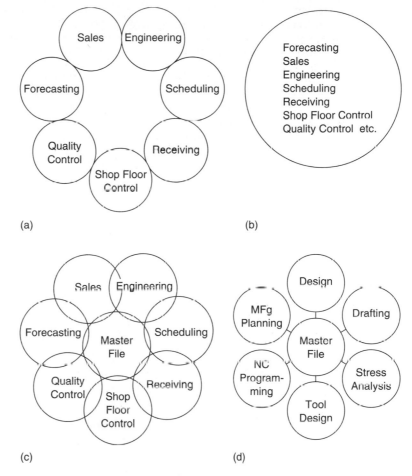

(a)

(b)

(c)

(d)

Source: Reprinted with permission from *Computer-integrated Manufacturing Technology and Systems* by U. Rembold, C. Blume, and R. Dillmann, 1985, New York: Marcel Dekker. Copyright 1985 by Marcel Dekker.

The network structure is an extension of the hierarchical structure; it can relate a file or record to any other file or record in many-to-many rather than one-to-many fashion.

The free format database is a text file, such as the manuscript of this book. The editor performs any manipulation of data. In that respect, a free format database is like the file storage mechanism of a word-processing package.

The multiuser distributed structure allows many users to read the database simultaneously but allows only one of the authorized persons to write at a time. The database is physically distributed but logically integrated. This type of structure is more suited to the needs of CIM.

The task of managing databases is not easy. It requires special computer systems, called database machines, that work on parallel processing principles to

Table 5.2
A comparison of databases illustrated in Figure 5.5

Type	Characteristics	Advantages	Disadvantages
(a)	No consideration of integration. Historically developed for the benefit of the user.	Easier and less expensive to develop. Suits stand-alone applications. High security.	Difficult to combine, expand, and maintain. High data redundancy. Data exchange between files difficult.
(b)	Grew out of the problems with type (a).	All information at one place. Little redundancy.	Administration unwieldy. Access time may be slow. Programming/maintenance difficult due to large size.
(c)	Information entered only once. User's database requests the main database for information.	Suits operations with several common tasks.	Complex. Difficult to control and maintain.
(d)	Common data in master files and specific data in local files.	Suits CIM. Redundancy low.	Complex. Low Security.

achieve efficient data sharing. Such specialized hardware-oriented machines are faster than the software approach using general-purpose computers as platforms.

Relational Database

Database management software that allows several files to be related into one is a relational DBMS. As an illustration, four files have been used in Figure 5.6 (p. 189) to create the invoice by relating one file to another through keys, like a linked chain.

☐ **BOX 5.2**   *Managing Data the Relational Way*

Relational database management systems (RDBMS) are becoming widely used throughout business, including manufacturing. Database management systems (DBMS) are used to accept, organize, store, retrieve and modify computer data. Because one of the principal functions of computers is to manage various types of data, DBMS are fundamental to most computer applications and are used with all classes of computers.

There are three classes of DBMS software:

hierarchical, network and relational. Hierarchical and network DBMS first gained commercial acceptance in the 1960's and 1970's, respectively. Both these require users to know how and where their data is stored, to navigate through the database and access that data step by step, according to pre-defined sequences.

Relational DBMS software, which gained commercial acceptance in decision support applications in the 1980's, provides greater flexibility, ease of use and productivity than hierarchical and network DBMS. Relational DBMS allow users to specify the data they wish to retrieve or modify without knowing how or where the data is physically stored in the computer and without needing to understand how the database works. Because relational systems navigate to the desired data automatically, database information is more readily accessible to users of all experience levels, despite the complexity that underlies the system.

Characteristics of relational DBMS include the following: 1) The data are independent, stored in tables of related information; 2) There are no pre-defined relations/links/paths; 3) A relationship or "view" is established by each user at run time, creating a temporary subset of data; 4) There are no navigational restrictions via path or direction.

A relational database is made up of separate, easily managed tables of related data. A table is made up of rows and columns (set of matrices). A row is one complete set of data (or a record), such as all the information about one project, and a column contains all the data in the same category, for example budgets for all projects. Data, therefore, appear to users as tables and only tables and operations on tables result in new tables (views). All data values are atomic (there are no repeating groups), and each table must have a unique identifier or primary key.

Since relational DBMS stores data in simple two-dimensional tables, data is accessed by specifying the rows and columns desired. The same data can be represented in different ways to different users through multi-table views and joins (an operation that obtains data from two or more tables). These operations allow users to update several files with a single data entry form or generate a report using data from related tables.

The ease of use of relational systems is enhanced because access to the data is "non-procedural," which means that users specify database operations only in terms of what they want done, not how to do it. This enables a wider range of users to perform information queries, create reports, produce informative graphic presentations and update data without having to learn complex programming techniques. It must be noted, however, that while the command language is easy, as the database gets larger and more complex, knowledge of the content of the database is more crucial.

In response to the recent trend that places computing power at the individual workstation or PC, relational DBMS can now also be used in a client-server environment. The workstation, PC or department computer can run applications and control database access against databases set up on the main computer, while the more CPU-intensive work is done by the larger machine. Relational systems can also support a data-sharing model in which multiple workstations access data residing on a central file server.

The relational database approach offers many advantages. The tabular representation is easy to comprehend and implement. Convenient operations (join being an example) make it easy to create new relations. Sensitive data may be secured by placing it in separate relations. Searching is generally much easier than it is within the other schemes, modification is more straightforward, and the clarity and visibility of the database is greatly improved.

Relational systems are now approaching the performance level of other databases for simple queries. They cannot currently match the performance of a structured database, designed for specific transactions, in a complex multi-user environment. However, a full set of maintenance utilities has recently been added to relational feature sets.

Distributed RDBMS, in the near future, will allow users to access data stored on a network of dissimilar computers with the same ease as if all the information were stored on a single computer. Distributed capability will allow users to connect these "islands of information" which frustrate major organization users who know data exist, but are unable to access it and use it where it is needed. Relational systems will do this more easily than non-relational systems.

The advent of relational database management systems, **Structured Query Language (SQL)** and related development tools has ignited a revolution in data processing and application development. Using fourth-generation language (4GL) and Computer Aided Software Engineering (CASE) application development tools in combination with relational technology, developers can create sophisticated transaction processing, reporting and menuing systems—all without conventional programming.

The market for relational DBMS is growing rapidly. Relational systems, by permitting easier end user ad hoc inquiry and analysis, have created a major new category of usage not possible with traditional DBMS technology. Improvement in the performance of relational DBMS, combined with the availability of powerful applications development tools, will increasingly lead users to migrate towards relational databases and tools.

Source: From "Managing Data the Relational Way" by S. Piazzale, (1990, fall), *Askhorizon, 3*(4), 10–13.

The relational type allows changes in data links or relations as the database grows in size and complexity. In other words, data structures can be changed without changing subschemas or application programs. The relational database structure in multiuser distributed environment is more suited to CIM.

A relational structure is a "flat" file, a two-dimensional array of data elements. The file is a table consisting of rows and columns; the rows are called records and the columns are termed fields. An example of a flat file is the following.

CARS		
Make	**Model**	**Cost**
Chrysler	Plymouth Acclaim	$14,500
GM	Geo Metro	$8,250
Ford	Escort	$11,500

The first line is the name of the table, CARS in this case. The second line labels the columns (Make, Model, Cost) and is not a part of the data. This particular flat file contains three records of data (rows) with values assigned to labeled items. The values may be expressed as numeric and/or alphabetic characters.

Figure 5.6
Data from four linked files used to create the customer bill

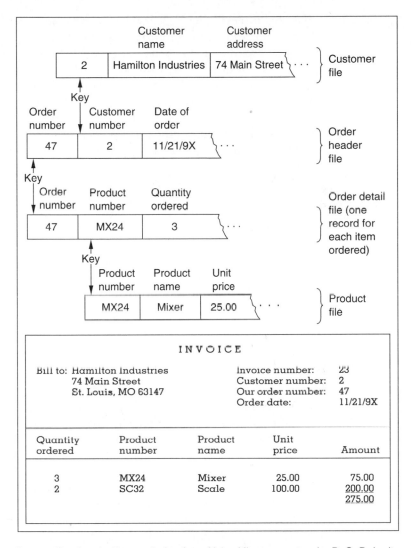

Source: Reprinted with permission from *Using Microcomputers* by D. G. Dologite and R. J. Mockler, 1988, Englewood Cliffs, NJ: Prentice-Hall. Copyright 1988 by Prentice-Hall.

5.4   DATABASE MANAGEMENT

The transition from paper tapes to floppies and hard disks has alleviated a major bottleneck in handling shop-floor data. It has also expanded the role of data management. A single keystroke can wipe out vital data a company may have accumulated over many years. Properly archiving, protecting, and distributing part programs and CAD files, for example, can be a formidable task, partly due to the large size of engineering databases. The size grows faster in a CAD environment due to the temptation associated with "soft" storage. Large companies manage hundreds of thousands of CAD files; Lockheed, for example, is said to maintain over a million files that are active and on-line in its CAD system.

Databases, being complex, must be managed. Most companies employ a **database administrator** (**DBA**) who decides what information should be stored in the database. This individual also decides how data will be represented in the database and coordinates with users to ensure the data they require is available in the desired format. Authorization checks and validation procedures are considered. Also considered are the strategies for backup and recovery in case the systems fails. Lastly, the DBA monitors the database system performance regularly and makes any required changes.

An important tool for the DBA is a **data dictionary**, a catalogue of all the data in the database. The dictionary may include cross-reference information explaining which programs use which data and which department needs which reports. It is like a table of contents, in fact a database itself.

Another tool for the DBA is the DBMS, used to create and maintain databases. DBMS, essentially a collection of programs, is important for other users as well. It is a general-purpose software system that facilitates the tasks of defining, constructing, and manipulating databases for various applications. The database and the software are together called a database system.

User–Database Link

Between the user and the database are three interfaces, shown in Figure 5.7. Note the existence of an **access method**—called the storage subsystem by ANSI—between the DBMS and database. According to Date (1982), "The access method consists of a set of routines whose function is to conceal all device-dependent details from the DBMS and to present the DBMS with a stored record interface."

He further adds, "The stored record interface (Figure 5.7) permits the DBMS to view the storage structure as a collection of stored files, each one consisting of all occurrences of one type of stored record. Specifically, the DBMS knows (a) what stored files exist, and, for each one, (b) the structure of the corresponding stored record, (c) the stored field(s), if any, on which it is sequenced, and (d) the stored field(s), if any, that can be used as search arguments for direct access. This information will all be specified as part of the storage structure definition."

"The DBMS does not know (a) anything about physical records (blocks); (b) how stored fields are associated to form stored records (although in practice this will almost invariably be via physical adjacency); (c) how sequencing is performed; or (d) how direct access is performed. This information is specified to the access method, not to the DBMS," concludes Date (1982, pp. 34–35).

DBMS Versus File Manager

Simple packages called **file managers** that let users work on only one file at a time are insufficient for CIM, where interaction is essential. The DBMS packages allow work with several files at the same time. They can do everything a file manager does,

Figure 5.7
Interfaces between the user
and the database

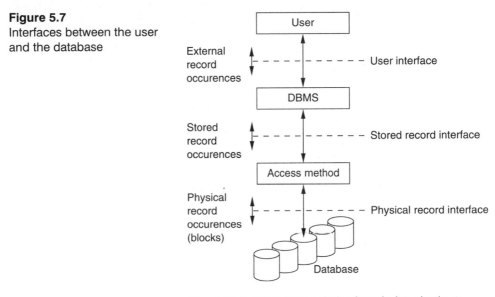

Source: Reprinted with permission from *An Introduction to Database Systems* by C. J. Date, 1982, Menlo Park, CA: Addison-Wesley. Copyright 1982 by Addison-Wesley.

in addition to (a) relating several files to each other, and (b) allowing the development of programs, for special application requirements, using built-in languages.

DBMS software packages are more sophisticated than file managers. Some of them are capable of relating several files to each other on the basis of a "key" field; such packages are called relational DBMS. File manipulation is flexible with relational DBMS. Moreover, a built-in language to program special application requirements may also be available with some DBMSs.

The effectiveness of a DBMS in handling several related files depends on how well the files have been designed. An effective design has the following elements:

1. It does not repeat groups of fields, instead it makes each group a separate record.
2. It eliminates fields that do not depend entirely on the record key.
3. It avoids fields that belong to another subject file.

Another important consideration is to keep only one file for each subject. A well-designed set of files means an error must be corrected only in one file; the DBMS automatically transfers the effects of the correction to other files.

Operation of DBMS

Database management software allows the user to:

create a file
add records to the file
search the file

sort the file
produce reports, if required

Creating a file involves two steps: drawing a layout, then setting up the file based on the layout. For example, a company may be interested in organizing data on the various parts used in one of its assembled products. The data may include the part number, weight, serial number of the subassembly it is fitted to, quantity needed in each subassembly, and unit cost. Also required may be the weights of all other parts and their costs, if the complete assembly weight and cost are to be calculated.

Establishing the file begins with a decision about what the file should contain. This involves identifying and classifying the information into **constant data** and **variable data**. Constant data, such as the headings, does not change from report to report and hence is not included in the file. Variable data, on the other hand, can change each time the report is run. Variable data may be primary or secondary. Primary data items are independent of each other, and hence not computed, while secondary data is computed from other variables. Referring to the example in the last paragraph, the weight and cost of the assembly are secondary data. Once the constants, secondary data elements, and redundant data elements have been be eliminated, what remains on the layout is data for the file.

A file is created by selecting the "create" menu of the DBMS software. Among other likely options in the menu are "set up," "update," "position," "retrieve," and "organize." The screen may also show a dialogue or information box requesting the filename and data. A file contains records of data, which are divided into fields. The field name is another term for the title of a column. The size of the field is set to provide enough space for expressing the data properly and clearly. The field may contain character data such as part name, or numeric data such as cost that may be used in arithmetic calculations. Once fields are established, a data-entry form appears on the screen with the field names. Data entered in response to the corresponding field are stored in the file.

DBMS software also enables users to create custom screen forms for data entry. Once a file is completely loaded, other functions of the database management such as search or sort can be carried out. The software can be used in two different modes. In the menu mode, various screens are displayed and the user chooses one of the options. This mode is slow but helpful to beginners. As experience is gained with use, users prefer to switch to the command mode in which appropriate commands are typed (and the menus are absent). Users who have mastered the commands find this mode faster.

Many packages are capable of checking the accuracy of data as it is entered. The system may provide messages such as "enter data before proceeding," "this field contains only numeric data," or "data entered must be within a set range."

In addition to letting the user create, access, and update the data, database systems address the problems of:

Concurrency—simultaneous updating by more than one programEase of access— very high level interactive language allows simple queries

Efficiency—users can create indexes of their own

Independence—users need not know the intricacies of file processing, and changes in the physical storage of data do not require any change in the programs to access the data

Integrity—only valid updates to the data are allowed

Redundancy—minimum data duplication

Security—the database can be restored in case of a crash, since all transactions are logged

Sharing—data are shared among several users

Some DBMS packages come with built-in programming languages. Such products provide complete application design flexibility. The DBMS programming languages represent the **fourth generation of computer language (4GL)** development and are termed VHLL (very high-level languages). They embody the next development beyond high-level languages (HLL) such as BASIC or COBOL. In comparison with third-generation languages such as BASIC, DBMS languages reduce the number of instructions necessary to carry out a typical task.

Based on 4GL, data management software is an end-user application for a **relational database management system (RDBMS)**. An important characteristic of any RDBMS is that it allow flexible storage of data. A descriptive header for each flat file contained in the directories enables users to store a variety of data about the file, such as drawing and part numbers, date of creation, and product type. Users may search the database for any file or group of files exhibiting a specific combination of data attributes. For example, a user can quickly trace all the files that were created for a particular product.

DBMS packages that also offer graphics capability are more suited to CIM environments. Such packages allow storage of both graphics and text in the database, which is important since engineering drawings contain both graphic and text data.

In the past, DBMSs worked only in the mainframe environment. Recently, however, most established DBMS developers have implemented PC versions of their products. Two such versions are Oracle and PC/FOCUS. The availability of PC-based DBMSs is beneficial to CIM, since they allow manufacturing data residing in company mainframes to be "sliced" for a specific user and downloaded to the user's PC for local analysis and use. Since the same database is used, there is no need to convert the files. This obviously represents a significant step toward CIM. The current trend in database management is changing the role of corporate data processing departments from being the owner of the database to being its trustee.

5.5   AN ILLUSTRATION

Date (1982) offers a valuable illustration of how sample data is stored differently by a manufacturing company having five suppliers; that illustration is reprised here (with permission).

Assume that the company wants to keep four types of data for each supplier: supplier number (S#), name (SNAME), a status value (STATUS), and location (CITY). Assume that each supplier has been allocated a unique value for these items. Each stored file is sequenced by the access method on its primary key—a field whose value identifies all the records in the file.

The first (and simplest) representation consists of a single stored file containing data for each supplier, as shown in Figure 5.8a. Although only five suppliers have been considered here, the actual size may be in the hundreds, even thousands for large companies. Consider that the company has 1,500 suppliers located in 10 different cities. Obviously, repeating the names of these 10 cities 1,500 times in the last column would waste storage space. If each of the 10 cities were allocated a pointer and the amount of space required for the pointer was less than that for the city name, then the resulting representation, illustrated in Figure 5.8b, would save some storage space. The database will need to store two files, however. Note that the pointers are maintained by the DBMS.

Consider now the company need expressed by the query: Find all suppliers in Paris. This will require the alternative representation shown in Figure 5.8c. Two files are needed to answer the query, but the pointers this time are out of the CITY file and into the SUPPLIER file. Note that this representation is better for queries that ask for all suppliers in a city, but worse for queries for all attributes of a given supplier.

The CITY file of Figure 5.8c serves as an index for the other file. In other words, the SUPPLIER file is indexed by the CITY file. Date explains, "The purpose of an index file is to provide an access path to the file it is indexing—that is, a way of getting to the records in that indexed file. A given file may have many associated access paths. One path that is always available is the simple sequential path. An index, then, is a file in which each entry record) consists of a data value together with one or more pointers" p. 38. The advantage of indexing is that it speeds up retrieval, but the disadvantage is that it may slow down the updating of files.

In Figure 5.8c, the CITY file is controlled by the DBMS, not by the access method. In Date's opinion, "It is in fact a dense, secondary index. The term dense denotes that it contains an entry for every stored record occurrence in the indexed file; this means the indexed file need not contain the indexed field—in the example, the supplier file no longer includes a city field. The term secondary means that it is an index on a field other than the primary key" (p. 39).

The representations of Figures 5.8b and 5.8c can be combined to yield the advantages of each, as in Figure 5.8d. Date continues:

> Of course, it also has the update disadvantage mentioned earlier, and in addition it requires slightly more storage. Another disadvantage of secondary index in general is that each stored record occurrence in the index must contain an unpredictable number of pointers. This fact complicates the job of the DBMS in applying changes to the database.

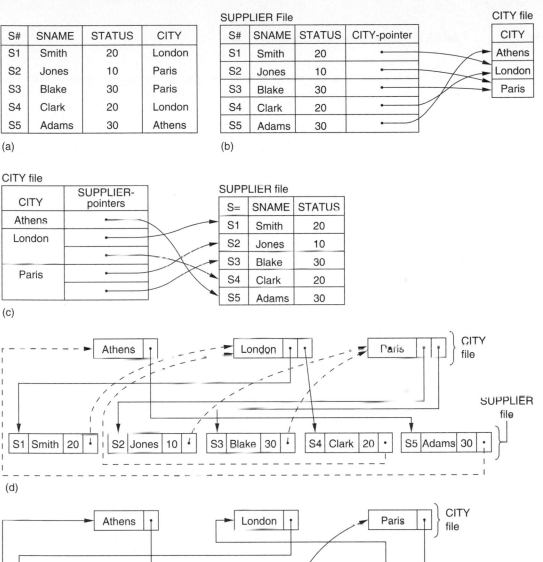

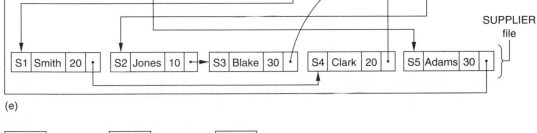

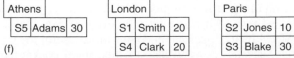

Figure 5.8

Various representations of the data: (a) Sample data, (b) Factoring out the CITY values, (c) Indexing on CITY, (d) Combining the previous two representations, (e) Using pointer chains, and (f) Hierarchical organization. Source: Reprinted with permission from *An Introduction to Database Systems* (pp. 36–41) by C. J. Date, 1982, Reading, MA: Addison-Wesley. Copyright 1982 by Addison-Wesley.

An alternative to the previous representation that avoids this problem is illustrated in Figure 5.8e. In this representation each stored record occurrence (supplier or city) contains just one pointer. Each city points to the first supplier in that city. That supplier then points to the second supplier in the same city, who points to the third, and so on, up to the last, who points back to the city. Thus for each city we have a chain of all suppliers in that city (another example of an access path). The advantage of this representation is that it is easier to apply changes. The disadvantage is that, for a given city, the only way to access the nth supplier is to follow the chain and access the 1st, 2nd, . . ., $(n-1)^{th}$ suppliers, too. If each access involves a seek operation, the time taken to access the nth supplier may be quite considerable.

The representation shown in Figure 5.8e (using pointer chain) is a simple example of multilist organization. In this figure we chained together all suppliers in the same city; in other words, for each city we had a list of corresponding suppliers. In exactly the same way (by means of additional pointers) we could also have a list of suppliers for each distinct status value, for example.

Another representation that should be mentioned is the **hierarchical organization**, illustrated in Figure 5.8f. Here we have one stored file containing three (hierarchical) stored record occurrences, one for each city. Part of each stored record occurrence consists of a variable-length list for supplier entries, one for each supplier in that city, and each supplier entry contains supplier number, name, and status. We have here factored out the CITY values, but we have chosen this time to represent the association between a city and its suppliers by making the city and suppliers all part of one stored record occurrence (instead of using pointers, as in Figure 5.8c, for example). Incidently, a secondary index such as that in Figure 5.8c is in fact a hierarchical file.

The last representation we shall consider is a **hash-addressing** organization. Hash-addressing, or simply hashing, is another example of an access path. The basic idea of hash-addressing is that each stored record occurrence is placed in the database at a location, called a **stored record address (SRA)**, whose address may be computed as some function (hash function) of a value that appears in that occurrence—usually the primary key value. Thus to store the occurrence initially, the DBMS computes the SRA and instructs the access method to place the occurrence at that position; and to retrieve the occurrence subsequently, the DBMS performs the same computation as before and then requests the access method to fetch the occurrence at the computed position. The advantage of this organization is that it provides very fast direct access on the basis of values of the hashed field. (p. 42)

TRENDS

❏ Affordable costs of mass storage devices, such as hard disks, are resulting in local databases on user PCs.

❏ In recent years, hybrid software packages have become more popular. Hybrid software lets the user do several things; for example, a DBMS package may have the added capability of data analysis. Combined with the original data manipulation capability of the DBMS, a hybrid package becomes more useful for CIM tasks. VP Planner is one such package; it looks like a spreadsheet package but creates database files in dBASE format.

❏ Another recent development are DBMS packages that come with "added-on" or "built-in" natural language capability. Such packages allow users to ask search questions in plain English, thus eliminating the need to learn formal commands. This enables the user to import mainframe files into a personal computer and interrogate the files using a natural language accessory package with the personal database package. The natural language capability is based on an internal dictionary that contains words such as job, position, and profit. The words may also be defined; for example, the word cost equals rate multiplied by the time taken to machine the part. Data on rate and time can reside in different files. Users can add more words in the dictionary. The dictionary can learn new information as it is used, thus displaying a touch of artificial intelligence.

❑ Companies use their databases for several purposes. For example, GM has set up electronic bulletin boards (BBs) to serve its customers, clients, and employees, and for advertising purposes.

❑ File systems, hierarchical, network, and relational are considered first-, second-, third-, and fourth-generation DBMS software. Though a few companies are still using first-generation packages, most are working with second- and third-generation tools. Relational database technology is the most recent generation. The next generation of DBMS, currently under development, is called **object-oriented**.

SUMMARY

Basic concepts of database technology and management were presented in this chapter. Database-related terms were explained so that readers can appreciate the possibilities in storing and retrieving data. Manufacturing data has numerous sources, from books and journals to magnetic disks. Data stored in the databases can be conceptualized logically or physically. The requirements of an effective database were discussed and the differences between a file manager and DBMS software were pointed out. The three common data models—hierarchical, network, and relational—were explained. Of the three, the relational database is finding wider acceptance; a distributed approach to RDBMS holds promise for CIM. The fifth generation of object-oriented DBMS, currently under development, is likely to advance the progress of CIM. Finally, database development was illustrated.

KEY TERMS

Access method

Constant data

Constructive solid geometry (CSG)

Data

Data aggregate

Data dictionary

Data item

Data model

Database

Database administrator or administration (DBA)

Database management system (DBMS)

Direct access storage device (DASD)

Entity

Field

File manager

Flat file

Fourth-generation language (4GL)

Hash-addressing organization

Hierarchical organization

Information

Key

Object-oriented DBMS

Redundancy

Relational database management system (RDBMS)

Reverse engineering

Schema

Segment

Stored record address (SRA)

Structured query language (SQL)

Subschema

Tuple

Variable data

Volume

EXERCISES

Note: Exercises marked * are projects.

5.1   Explain the meaning of the terms data, database, DBMS, and RDBMS.

5.2   Discuss the reasons behind the need of database for a manufacturing company.

5.3   The IRS (Internal Revenue Service) maintains all the data of the last three years for 100 million U.S. taxpayers in a database. If each taxpayer files an average of four forms with 2,500 data characters in each, what is the size of the IRS database in bytes? (Refer to Chapter 3, if required.)

5.4   Why is the term *database system* preferred to *database*?

5.5   For a manufactured product of your choice, list 10 important data.

5.6   Of the 10 data in the last exercise, which ones are product data and which ones production data?

5.7   Explain the difference between logical and physical views of data stored in a database.

5.8   List 10 important considerations in the design of a database and discuss them.

5.9   What is the difference between an interfaced and a distributed database?

5.10   How does DBMS software differ from file manager software?

5.11   Write a 300-word essay on relational databases.

5.12*   Visit your organization's central computing facility and learn about its database technology and management.

5.13   Circle T for true or F for false or fill in the blanks.

   a.   File handling and database management are the same thing.   T/F

   b.   Relational databases are generally more flexible than hierarchical
        ones.   T/F

   c.   Because database management in manufacturing deals with
        constantly changing data, special computers implementing
        hardware database processors are commercially not viable.   T/F

   d.   Database management languages represent the fourth generation
        of computer language development.   T/F

   e.   DBMS is basically a hardware.   T/F

   f.   Some common methods for defining shapes are parametric equations,
        sweep representation, and CSG—an acronym for _____

   g.   A flat file is associated with the database model.   T/F

   h.   The database dictionary is a catalogue of _____

SUGGESTED READINGS

Books

Bray, O. H. (1988). *Computer integrated manufacturing—The data management strategy*.
     Bedford, MA: Digital Press.
Date, C. J. (1982). *An introduction to database systems*. Reading, MA: Addison-Wesley. (In its
     third edition, this title presents the basic principles of database technology. The first part
     discusses database system architecture; the second part, the principles of relational
     database. The next two parts explain the hierarchical and network approaches to
     database. The concluding part shows the connection between the three types of data
     models. The book is desirable reading for those interested in the fundamentals of
     database technology.)

Elmasri, R., & Navathe, S. B. (1989). *Fundamentals of database systems*. Redwood City, CA: Benjamin/Cummings.

Hansen, G. W., & Hansen, J. V. (1992). *Database management and design*. Englewood Cliffs, NJ: Prentice-Hall.

Lucas, R. (1988). *Database applications using Prolog*. Chichester, England: Ellis Horwood.

Martin, J. (1975). *Computer database organization*. Englewood Cliffs, NJ: Prentice-Hall.

Ranky, P. G. (1990). *Manufacturing database management and knowledge-based systems*. Guildford, Surrey (England): CIMware Limited.

Rembold, U., et al. (1985). *CIM technology and systems*. New York: Marcel Dekker.

Monographs and Reports

Appleton, D. S. (1982). *The CIM database*. CIM series green book, *1*(4). Dearborn, MI: CASA/SME.

Conkol, G. K. (1991). *The role of CAD/CAM in CIM, Part II, SME blue book series*. Dearborn, MI: CASA/SME.

Modern Machine Shop. A 10-part monthly series on computer-assisted manufacturing (beginning with the October 1989 issue).

Articles

Bata, R. M. (1989, December). Integrated databases for CAD/CAM. *Mechanical Engineering*, p. 20.

Bennett, R. (1990, Spring). Finding engineering information fast. *Journal of Applied Manufacturing Systems*, *3*(1), 43–48.

Business Strategies. (1990, April). Unifying information systems—tough but worth it. *Production*, pp. 48–49.

CIME staff report. (1989). Automating data management. *Mechanical Engineering*, *111*(3), 73–76.

Diesslin, R., & O'Connor, F. (1989, November). Where's the business. *Modern Machine Shop*, pp. 92–100.

Froyd, S. G. (1985, January-February). Relational database: Cornerstone of FMS. *Commline*.

Piazzle, S. (1990, fall). Managing data the relational way. *Askhorizons*. *3*(4), 10–13.

Martin, J. M. (1989, May). Making information flow. *Manufacturing Engineering*, pp. 75–78.

PART III
Technology and Systems

C IM aims at integrating all facets of a manufacturing enterprise using computers. This represents a colossal task, since manufacturing is a complex system composed of many elements that interact closely. In Part III, major elements of CIM technology and their systems are discussed, along with how these systems interact. We deal in this part with topics normally covered as complete courses in themselves.

Discrete-parts production represents a continuum that begins with product design and ends with the shipping of end products. The chapters in Part III follow the usual sequence of this continuum.

In general, production can begin only after the product has been designed. Chapter 6 discusses the design process, use of computers in design, and the importance of product design within the CIM concept. The postdesign function of planning—a preproduction phase—is covered in Chapter 7, the actual production function in Chapter 8, and shop-floor control in Chapter 9. Material handling and transport within (as well as in and out of) the plant and the role of robotics and related technologies on this function are presented in Chapter 10. Chapter 11 deals with the importance of quality assurance and SPC. The third part of the text thus explores what takes place in the design office and at the shop floor to manufacture a product in CIM environments.

CHAPTER 6

Product Design

"CAD/CAM is the granddaddy of CIM."
—Gary K. Conkol in the CASA/SME blue book series,
The Role of CAD/CAM in CIM (1990)

"Historians will recall that the wall between design and manufacturing was crumbling down about the same time as the Berlin Wall."
—S. Kant Vajpayee, The University of Southern Mississippi

"Manufacturing has to give design more standardized information, information that tells designers such things as if a fillet radius is 0.125" (3.17 mm), the tooling will cost five cents, but if it's 0.12773" (3.2443 mm), then the tooling will cost a thousand dollars, and it will be the only tool like it in the company."
—Henry Stoll, Industrial Technology Institute (ITI)

"Studies show a possible reduction in parts of nearly 73% and material cost of nearly 40% for a given component through design for manufacture. Is it any wonder that a product with only one fourth as many parts can be built with higher quality?"
—William E. Scollard, Ford Motor Co.

"Years ago, somebody did both engineering and the business in general a great disservice by fragmenting the product cycle: one designs, one produces . . ."
—Wes Allen, visiting professor, University of Cincinnati

"The U.S. remains highly enamored with technology and is less attuned to making investments to improve upstream design/manufacturing processes. Why didn't automation work? Because major parts of manufacturing start upstream. Since our focus was on manufacturing, (that looked like where the problem was) we were late getting started on doing design processes that lead to manufacturing processes that could be automated."

—H. Barry Bebb, Xerox

"The manufacturing world is viewed as being static. So they do static analysis. But that view has got to be thrown into a cocked hat. The factory floor is dynamic. It's changing by the minute."
—John Layden, vice president and general manager of
Automated Technology Associates, in *Production* (November 1990)

Manufacturing companies fall into two groups. The first group includes those where design is an external activity; such companies manufacture products to clients' designs. The other group carries out design in-house; products are designed, then manufactured for the market. For both the groups, design is an essential prerequisite to manufacturing. This chapter focuses on the design function and its importance in CIM, while Chapters 7 through 11 focus on production.

In successful companies producing to their own in-house designs, the design function is the nucleus of all the technical activities. Until design begins, nothing happens in production, inspection, shipping, or other functions. Design is the most creative task. Design for the sake of creativity, however, does not make much business sense. Management must be convinced that the proposed product will generate a profit. Only then do the R&D and design departments get a green light.

6.1   NEEDS OF THE MARKET

Manufacturing produces both capital and consumer goods. Capital goods, such as machining centers or robots, are sold to other manufacturers who use them to produce other goods. Consumer goods such as VCRs or motorbikes, on the other hand, are sold directly to the consumers for their use. In either case, customers' needs are of paramount importance.

Modern manufacturing is based on the *marketing concept*, which examines the reasons why consumers buy a product. The answers are not easy to come by. In seeking reliable answers, marketing experts draw from subjects as varied as economics, psychology, sociology, and anthropology. Their input is crucial to deciding which products to make, how many to make, and when to make them. Production activities begin with the marketing function, whose input continues through the management and design functions.

An effective marketing strategy takes into account current and future factors, which include

1. rapid introduction of new products
2. new technologies
3. shorter product life
4. public policy, consumerism, and consumer rights
5. emergence of market segmentation

Companies abandoned the hard sell approach in the 1960s in favor of the marketing concept, which stresses consumer analysis and the reasons behind consumer satisfaction. Marketing helps direct a company's resources toward products consumers are likely to buy. The CIM approach further reinforces the market-production tie by including consumers in the decision-making process.

After careful consideration of all the relevant information, especially input from marketing, management decides to proceed with the product design. This sets the ball rolling in the design and engineering departments. Management may sometimes abort a product—especially high-tech products such as computers—even at an advanced stage of design and development.

This chapter discusses design and engineering issues pertaining to CIM. It illustrates how the seeds of computer-integrated manufacturing are planted at the design and engineering phase.

6.2   DESIGN AND ENGINEERING

Design and engineering are the most creative technical tasks in manufacturing. While expert systems (an application of artificial intelligence) and, more recently, neural-network-based computers significantly aid in these tasks, designing remains, and will long remain, a human endeavor.

Design and engineering involve working out all the technical details of the product. Designers and product engineers seek answers to questions such as:

How should the product look?
How will it work?
If it is to be powered, which power source should it use?
What is the best shape and size?
How will the parts be made and assembled?

Designers, product engineers, and production engineers together attempt to solve all possible problems before production begins. They investigate the capabilities of available manufacturing resources to dovetail these with the needs of the consumer, as embodied in the designed product. The century-old practice of "the designer designs and production builds" is ill-suited for the purpose of CIM. Today, design and production need to work very closely.

Designing is the idea-generating phase, with sketches and mock-ups as outputs. Engineering begins with these outputs and aims to translate the designers' ideas into a workable product that will be functionally sound and aesthetically attractive to consumers. This is achieved by applying scientific and mathematical knowledge in engineering the product. The term designing often means both design and product engineering.

Designers' major tasks are:

collecting ideas
sketching ideas resulting in thumbnail and rough sketches
rendering sketches
making mock-ups
making detailed, assembly, and schematic drawings
building and testing prototypes

A design idea may be generated alone, especially if the product is simple. Most products are conceived collectively, however. Designers begin by collecting ideas from colleagues and putting them on paper to see how they look. Their simple, quick sketches are known as **thumbnail sketches**. The more promising ideas are sketched in detail and called **rough sketches**. The final step in sketching is to select the better ones and draw them neatly, showing details; this process is called **rendering**. Next, mock-ups are made to get a feel of the product in three dimensions. A mock-up may be one of three types: paste-up, appearance, or hard. The paste-up type is made by sticking together light materials to get a feel of the design. The appearance type is made to look like the product, but only from the outside. A hard mock-up is generally built using materials the product will actually be made of, but it still is not the prototype. Marketing personnel and other managers contribute their experience at this early phase of design.

Engineering tasks are based on the design team's sketches and mock-ups. Engineers decide about materials for various parts. They do the analysis to ensure that these parts will not fail and to make sure the design is manufacturable. All the results of engineering are eventually conveyed to production through drawings (electronically where CIM exists).

6.3   THE DESIGN PROCESS

The design process is iterative in nature. It can be segmented into six stages as shown by blocks 1 through 6 in Figure 6.1. The process begins with recognition of the need for a new product or major modifications in the existing product. Next, the problem is fully defined to yield a specification. In the third stage, the design undergoes synthesis, joining its various elements. This is followed by analysis and optimization, evaluation, and, finally, presentation to marketing and management.

In Figure 6.1, tasks 3, 4, and 5 are carried out iteratively. The synthesis may start with a crude representation of the idea. The product analysis may reveal weaknesses that are carefully considered for possible improvements. These three tasks are repeated until an acceptable, optimized design has been achieved. The number of iterations depends on the ingenuity and experience of the designers and the tools available to them.

The process of design is repetitive as well as creative. The repetitive tasks lend themselves particularly well to computerization; creative tasks will always rely on humans. CAD contributes to the last four tasks of the design process (blocks A through D in Figure 6.1).

6.4   DESIGN FOR MANUFACTURABILITY (DFM)

Since the maturity of CAD/CAM technology, the term design for manufacturability (design for production in Europe) has been drawing more attention. Even though design should obviously take into account what can be manufactured in the plant, often that practice is not followed. In general, the interaction between design and production functions has been weak. In a non-CIM environment, this interaction is not so critical; in fact, poor interaction gives both the design and production departments some flexibility, and to some extent independence, in achieving their objectives. Recently, CAD/CAM has forced a change in this attitude, primarily because in the long run, improvements in the product can be cost-effective only through design. One might expect that in a CAD/CAM environment the design need not be perfect

Figure 6.1
The design process involves
six tasks

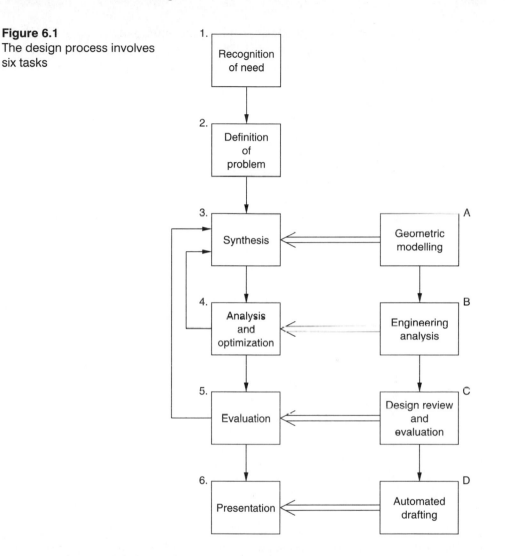

from a manufacturing viewpoint, since computers can accommodate last-minute changes easily. But just the opposite is true. Designers must now pay closer attention to their ideas and drawings from the manufacturability viewpoint. They are expected to design what the available equipment and personnel can cope with.

Designs, therefore, are customized to the production and assembly facilities. The enormous power of CAD/CAM workstations assist in doing this by providing designers with detailed information on the capabilities of existing manufacturing resources.

The term design for manufacturability (DFM) emphasizes design-production interaction, especially in a CAD/CAM environment. The concept of DFM is not new as such, though its popularity in recent years seems to create that impression. DFM simply reinforces the need that, within the functional requirements of the product, designers consider the manufacturability of their design. Recent work by Boothroyd and Dewhurst (1988) challenges designers to apply DFM concepts.

DFM integrates product design, process planning, and production with the objectives of:

1. identifying product concepts that are inherently easy to manufacture
2. focusing on component design for ease of manufacture
3. integrating product design with process design to achieve optimum results

In the broadest sense, the term DFM includes both production and assembly of components. In plants where assembly is the major activity—and there are many such facilities—DFM primarily considers the ease with which components will fit together. Thus, DFM does not end with the production of components, but extends well beyond it, to include assembling (see Box 6.1) and other downstream functions.

With CIM, designers are expected to have significant knowledge of manufacturing processes and of the service department's requirements. Effective communications among marketing, manufacturing, and service personnel is important as well. For example, designers must be proficient in cost estimation, raw materials and their characteristics, and the processes by which parts are shaped, machined, and assembled. Only then can designers standardize the parts across the models, and across the products, to minimize tooling and other fixed costs to achieve optimum design.

To highlight the importance of design in manufacturing, the so-called Rule of Ten is often cited. According to this, if it costs $1 to correct a product defect at the design stage, and that correction is not made, then it will cost $10 (10 times the amount) to correct it at the production stage and $100 (another 10 times) at the cus-

BOX 6.1 *Brown and Sharpe's MicroVal CMM*

The MicroVal is a compact CMM with a measuring envelope of 14 × 16 × 12 inches. For just over $10,000 (1989 price) it surprised many that a CMM of sufficient capability can be available at low cost. The MicroVal's success is attributed to the application of 'design for manufacture and assembly' (DFMA) concepts developed by Boothroyd and Dewhurst. The product has won the 'Excellence in Design' award by Design News.

The development of MicroVal started with the goal of: (a) high-quality product, (b) easy to use, and (c) affordable. The development process concentrated on a total examination and rationalization of all aspects of CIM including material selection. Considerations were also made for having near-net-shape parts. Enough coordination between design and manufacturing personnel took place. Every aspect of design was questioned; for example, why six cap screws and not four? The objective approach of DFMA, rather than the conventional subjective review of design, was found to be of much help, according to Don Herman—the manager of value analysis at Brown and Sharpe.

Source: "Quality is a function of design at Brown and Sharpe" by G. S. Vasilash, April 1989, *Production*, pp. 48–50.

tomer point—when the product has been installed. The message is loud and clear: the time to contain product cost is at the design stage.

Component Design

Designers follow several rules to achieve DFM. The rules depend on the type of production process. These processes can be grouped into three: (a) manual, (b) soft-automated, and (c) hard-automated. DFM rules vary widely among the process groups.

Some typical guidelines for implementing DFM to component production are:

1. Design with as few parts as possible.
2. Minimize part variations.
3. Attempt multifunctionality of parts.
4. Design parts for multiple uses.
5. Design for ease of fabrication.

Design for Assembly

Several manufactured products are just one piece, such as envelope openers or disposable razors. If the product is simple enough for one-piece design, then assembling cost is eliminated. In general, a product can be one piece if:

1. It does not need to be made of different materials.
2. There is no relative movement during its operation.
3. It does not require disassembly during service.
4. Any combination of the above three.

Most products, however, comprise more than one component. These components are first produced or purchased, and then assembled. Obviously, assembling takes time and costs money. A trend to minimize assembly is occurring in the design of newer products. The concept of **design for assembly (DFA)** denotes all the efforts at the product design stage to ensure ease of assembling. Boothroyd and Dewhurst (1988) have made significant contributions in this area by cataloging the principles to follow at the design stage for minimizing assembly cost. While it may simplify production, designers must ensure that DFA does not cause an inordinate increase in the cost of producing the individual parts. Minimizing product cost thus requires a balance between component and assembly costs.

One of the primary rules of DFA is to opt for uniaxial assembly, meaning components can be placed on the base subassembly from the same direction. In automated assembly, this practice requires less sophisticated robots, since they would need one less degree of freedom. Another useful DFA rule is to keep the number of assembly steps to a minimum. This rule may not apply, however, if the steps are complicated and can be broken down into several simpler ones.

Some typical guidelines to achieve DFA are:

1. Make design modular.
2. Minimize the number of directions along which assembly takes place.
3. Design for top-down assembly.
4. Design for ease of assembly to maximize compliance.
5. Orient the parts to be assembled to minimize handling.
6. Eliminate or simplify adjustments.
7. Consider alternate assembly methods.

Design for manufacturing and assembly (DFMA) emphasizes the assembly component as well. DFMA concepts are very effective in reducing costs, since more than two-thirds of product development, assembly, and production costs get built in during the design stage. A variety of proprietary DFMA software packages are available to help designers. With similar objectives, **group technology (GT)** is considered an element of DFMA.

Taguchi has also contributed significantly to improving manufacturing design. While Boothroyd and Dewhurst suggest using time-study techniques to minimize the cost of manufacture and assembly within the constraints of design's functional requirements, the **Taguchi method** (Morgan, 1991) aims at optimal design by applying the theories of statistical design of experiments. Taguchi introduced the concept of **loss function**, which is the cost to the customer and the company—and eventually to the society—because of deviations in the product from what was intended.

In recent years, a trend to design and manufacture disposable products has gained momentum. This trend is slowing down, however, as we become aware of the harm some of these products do to the environment when discarded. Today's products are designed to be recyclable ("green" design) within the constraints of cost and technology (Ashley, 1993).

6.5   COMPUTER-AIDED DESIGN (CAD)

Design is one of the few areas of manufacturing that used computers as early as the 1960s. The initial use was limited to drafting tasks, which justified investment in CAD systems with payback periods of two years or less. Today, computers are used in other design tasks as well, and in quite sophisticated ways, such as engineering analyses and unambiguous part description.

CAD may be defined simply as the use of computers in various facets of design and its presentation. Computers increase the productivity of designers, and hence of the design process, by making it easier to draw, modify, copy, test, save, and simulate design ideas and concepts. In the 1960s and early 1970s, CAD meant computer-aided drafting; today, it is an acronym for computer-aided design. Some people use CADD for computer-aided design and drafting to differentiate it from CAD (computer-aided design). In general, CAD includes drafting.

Most early uses of computers in design began in the United States, primarily because of resources of the U.S. space program and, to some extent, the computer and automobile industries.

Areas of Application

A CAD system helps the designer in various ways, the main ones being:

1. Invites and promotes interaction through various I/O (input/output) devices such as light pen, digitizer, mouse, and function keys
2. Allows manipulation of soft drawings, using graphics, through translation, scaling, and rotation
3. Enables the designer to carry out engineering analyses for stress, vibration, noise, thermal distortions, and more. Analyses are usually based on the finite element technique, in which the component is modeled as being composed of small elements. The resulting computer model is analyzed under expected loads or environments.
4. Encourages design optimization through simulation and animation. The soft model is subjected to various what-if questions to achieve the best design.

    CAD use in manufacturing, in terms of time, is divided as:

    32% in engineering activities
    37% in drafting
    15% in BOM (bill of materials) generation
    16% in routine work

This distribution is changing, however. Drafting is less important as newer CIM systems work directly with the design database.

Benefits of CAD

All CAD activities are based on a model of the object's geometric characteristics. Once created and stored in the computer memory, the model can be manipulated rapidly in an error-free manner.

The use of computers in design results in the following benefits:

1. New products are designed faster.
2. Errors during design changes are less likely.
3. Documentation and drawings are of better quality, so there is less ambiguity and improved clarity.
4. Detailed or assembly drawings can be automatically generated for the production function, technical illustrations for marketing, a BOM file for materials requirement planning (MRP), parts inventory for the service department, and so on.
5. On-line modal analysis is possible.
6. Prototyping efforts are reduced.
7. Unnecessary components, tools, or fixtures can be controlled through a part-family approach and GT.

8. Reduced lead times for quotation and design help manufacturing across the board, especially the marketing function.
9. Design can be linked with production and other functions within a CIM framework. For example, on-line inspection using a CMM, or probes off the automatic tool changer (ATC) of machining centers, can compare the actual part geometry with what it should have been, by referring to the design database.
10. DFM can be effective.

Computer Graphics

CAD applications are not limited only to manufacturing industries. They have many other applications, such as architectural design, art, and scientific research. Why is CAD so versatile? Undoubtedly, because of its graphics capability.

All CAD systems are based on interactive graphics, a user-oriented system that uses the computer to create, transform, and display data in the form of graphics or symbols. The results of user interactions are displayed on the screen, which is based on a **cathode ray tube (CRT)**. Most CRTs display graphics as well as text, based on vector or raster technique, as illustrated in Figure 6.2.

In the vector technique shown in Figure 6.2a, an electron beam "paints" the data on the screen by drawing lines on its phosphor surface. It also paints a grid directly behind the screen to energize the image continuously. This results in high resolution, generating clear, bright lines on the display. The vector technique is also used in oscilloscopes and radar displays. The beam moves simultaneously in arbitrary X and Y directions along a straight or curved path. It is essentially an electronic pencil. The vector technique is also called stroke writing, line drawing, random position, directed beam, or cursive or calligraphic writing. The term vector is used to describe how the computer locates two specified points and connects them. The resulting line is a vector, since the direction is as important as the distance between the points. Vector screens are of two types: storage and refresh. In the storage type, selective changes in the display cannot be made; the whole screen must be

Figure 6.2
Two common methods of generating images in computer graphics: (a) Stroke writing; (b) Raster scan

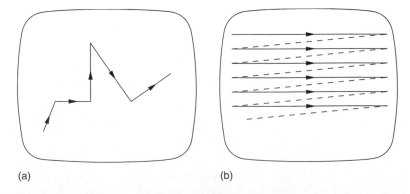

(a)   (b)

Source: Reprinted with permission from *CAD/CAM: Computer-aided Design and Manufacturing* by M. P. Groover and E. W. Zimmers, Jr., 1984, Englewood Cliffs, NJ: Prentice-Hall.

repainted. This redrawing of the entire screen is a drawback since it takes time, a few minutes if the drawing is complex. In the refresh vector screens, any change by the designer is implemented instantly, since the screen is refreshed continuously. Thus, selective erasing of the display graphics is possible. If refreshing is slow, however, the resulting flicker can strain the user's eyes.

The raster technique, illustrated in Figure 6.2b, offers a screen that combines the advantages of vector refresh and storage CRTs, while eliminating their limitations. Thus, the display is flicker-free and can be erased selectively. In the raster technique, a beam is sent continuously across the screen following left-to-right movement from top to bottom. The electron beam moves in a fixed pattern. Movement is rapid when scanning horizontal lines—about 67 microseconds per line. The beam moves back to the left end of the next line and repeats the process. It takes about 16 milliseconds (corresponding to 60 Hz of the power frequency) to scan the entire screen once. The image is created by activating the data-specific dots as the beam scans the screen. This process is called **rasterization**. Television screens work this way; the display is basically a grid pattern of dots lighted selectively to create the picture. Each dot is called a pixel, a contraction for picture elements. The sharpness of the display depends on the closeness between pixels. Denser dots (higher resolution) yield a sharper picture. Raster screens are also refreshed, since the beam continues to scan through the screen, generally at 30–60 Hz.

Another technique creates a third type of screen. Called storage tube display, it allows drawing of pictures as in vector writing. But it eliminates the need for refreshing by having the screen surface coated with phosphor, which holds the picture for a longer time.

Many times, text is displayed along with graphics. For larger and complex drawings, this may make the display look congested or cluttered. In such applications, it is desirable to have two screens—one for graphics and the other for text—connected to the same workstation.

Box 6.2 describes recent developments in graphics cards that enhance the capabilities of PCs.

CAD Hardware and Software

CAD hardware and software are simply tailored versions of computer systems for use in drafting and design. The principles behind their operation and use are similar to those of general-purpose computer systems, discussed in Chapter 3.

With an installed base of 300,000 users, AutoCAD is the most popular CAD software, followed by VersaCAD, with a base of 85,000. The latest version of AutoCAD sells for about $3,500, while VersaCAD sells for $2,500.

CAD/CAM Workstations

In the early 1980s, CAD was host-based, so that work carried out on terminals was connected to a time-shared computer. This has changed with the emergence of PCs and high-performance workstations. Table 6.1 illustrates the growth in popularity of PC-based systems and workstations over host-based systems.

BOX 6.2   *Getting a Clear Picture*

The right graphics board can add workstation muscle to a 32-bit PC. But with hundreds of products to choose from, selection is difficult. To compare, look at the controller chip, resolution, number of colors, type of monitor supported, and the price.

Selecting a graphics board is considered by many engineers to be the most important part of turning a 32-bit personal computer into a full-fledged PC-based engineering workstation.

In the high-resolution category (defined as the ability to display at least 1024-by-768 pixels) alone, engineers can choose from more than 250 boards from over 100 vendors.

Picking a Chip.   Graphics cards can be compared on the basis of the following criteria: type of graphics controller chip, resolution, number of colors, type of monitor supported, and price. The graphics controller chip is the most important element of the hardware. It is the primary factor in determining how quickly a graphics card generates pixels. The fastest of the group can draw 3300 pixels per second.

Different Resolutions.   Graphics cards are available in a wide variety of resolutions. The EGA-, Super EGA-, and Hercules-compatible devices commonly found on IBM PC/AT-class compatible computers are in the medium-resolution category. Also included in this category are the VGA (640-by-480) and the Super VGA (800-by-600) boards, which are appropriate for PCs built around the Intel 32-bit 80386 microprocessor.

For professional CAD users, high-resolution cards such as the Extended VGA graphics boards with 1024-by-768 pixels are recommended. This category also includes a variety of special purpose cards marketed specifically to users of CAD, image processing, and presentation graphics.

Cards for these applications will provide resolutions as high as 2048-by-1530 pixels in the next few years, predicts Peddie.

Color Capability.   For 2-D line drawings, boards that support 8 or 16 colors are probably sufficient. For 3-D applications, 256 or more colors are required. Boards with very high resolution, defined as those capable of generating at least 1024-by-1024 pixels, may provide as many as 25,000 colors.

Card price increases with the resolution and the number of displayable colors. Prices vary largely because additional memory is required to generate higher resolutions and more colors. A minimum of 1.25 MB is normally required for a board to produce a resolution of 1280-by-1024 pixels and simultaneously display up to 256 colors.

Highly complicated line drawings may require as much as 4 MB of display-list memory. The amount of display-list memory required by an application is a function of the number of vectors used. Engineers often overlook the fact that alphanumeric characters are also formed from vectors. "There are between 8 and 16 KB of information stored per vector, and just the letter r, which might have 10 vectors, could require between 80 and 160 KB," said Peddie. Consequently, drawings that are not particularly complex yet heavily annotated can require substantial display list memory.

Users of 3-D applications such as solids modeling and rendering generally require the most powerful graphics cards, partly because cards for these applications must be able to calculate the areas of polygonal surfaces for shading.

Source: From "Getting a Clear Picture," March 1990, *Mechanical Engineering*, pp. 30–34.

Table 6.1
Distribution of CAD systems

Year	PC-based	Workstations	Host-based
1986	29%	12%	59%
1990	44%	21%	35%
1995 (estimate)	55%	30%	15%

A PC-based CAD system on a network is now a viable alternative. At a price of $5,000, it is very attractive in comparison to workstations at $20,000 and host-based systems that can cost as much as $100,000. Moreover, if the past is any guide, the cost of PC-based systems continues to decrease without any sacrifice in performance; if anything, performance continues to improve.

6.6   THREE-DIMENSIONAL CAPABILITIES

In a true three-dimensional (3D) system, the part is usually designed as a 3D entity, with 2D views extracted for dimensional details. In 3D systems, the user makes the change only at one location within the solid; the effects of this change are automatically made at all appropriate locations. The features to look for in 3D software are automatic hidden-line suppression, calculation capabilities for mass properties such as moment of inertia or weight, and interfaceability to **FEA (finite element analysis)** packages and computer-controlled equipment of the plant. It should be noted, however, that 3D programs take longer to learn and run slowly due to the large amounts of data crunching. From a CIM viewpoint, however, a 3D program facilitates integration.

With respect to 3D programs, the two major considerations are: (a) the ease with which 3D objects can be constructed, and (b) the associativity of 2D and 3D views. Associativity refers to the degree to which changes in one view of a 3D object are reflected in the other views. Ideally, any change in one view should automatically be effected in all other views along with the dimensions.

Principles of Curve Generation

Methods for defining curves and surfaces have proliferated as a result of CAD developers seeking flexible ways to deal with geometrically complex shapes, such as car bodies, as well as offering users the friendliness of shape manipulation (Beard, 1990).

Let us look at the mathematical building blocks CAD/CAM systems use to create and manipulate part geometry. Consider first the 2D case. The most basic entities CAD/CAM systems use are line segments, circles, and arcs. While these entities can be grouped together to generate simple 2D shapes, they are inadequate to define more complex or free-flowing surfaces. Such surfaces require more sophisticated methods of defining shapes. The three common methods are

splines
Bezier curves
B-splines

Figure 6.3

Techniques of curve generation: (a) A contoured surface
represented by a series of
chords; (b) B-splines in which
the curve is controlled by
points outside the curve; (c)
Curve approximated by arcs

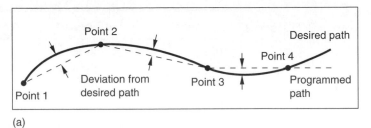

(a)

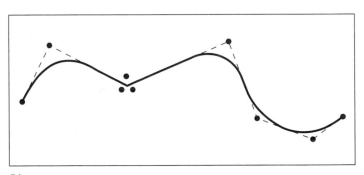

(b)

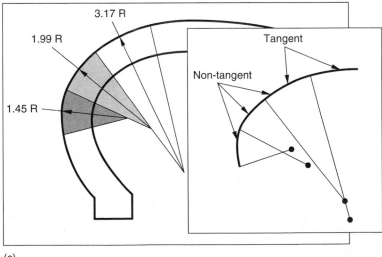

(c)

The spline principle is the computer equivalent of drafters' French curve—a flexible rule used for drawing free-flowing curves. In the **spline technique**, the control points, which are used to control the shape of the curve, are located on the curve itself, as shown in Figure 6.3a. The designer manipulates the shape of the curve by indicating the necessary number of points. The computer then connects these points or dots. Complex shapes are created by joining several splines end to end, as in the Bezier curves method. The spline technique works well when the input data are in digital form, as in reverse engineering. It is the least sophisticated of the three techniques.

A **Bezier curve** is defined with four control points: two as endpoints and two as handle points. The handle points, which are not on the curve, are meant to simulate the pull of gravity on the curve. The Bezier mathematical formula behind the curve-fitting can cope with several points. Most CAD systems enable users to create the desired shape by blending several simple curves, end to end. Bezier curve-based systems allow very smooth curves that are relatively easy to manipulate and edit. Until recently, only large systems could offer Bezier curve capability; today it is available even on PC-based CAD systems.

Like the Bezier curve, **B-splines** are fixed with end points; the shape of the curve depends on control points, which are not on the curve, as shown in Figure 6.3b. The control points are used to define a series of line segments, joined end to end. The designer effects the pulling of the curve near or through the line segments at their center points. The **rational B-spline** adds one more degree of control by varying the force of the pull.

CAD systems based on B-splines are sophisticated, since B-splines represent a complex form as one curve, unlike the Bezier curve, which is composed of several blended segments. This offers designers more control over the curve, since dragging a control point affects the curve only in the point's vicinity. Sometimes a curve is approximated by arcs, as shown in Figure 6.3c, to enhance the representation.

These basic principles of curve generation can be extended to define 3D sculptured surfaces as well. In some CAM systems, a 3D shape is visualized merely as a grid of X, Y, Z coordinate points similar to the way splines are constructed. In others, a 3D surface is generated as a series of parallel curves spaced at user-defined intervals. Still another method to represent a 3D surface is a mesh—sometimes called line mesh—of perpendicular intersecting curves. While these techniques can generate three-axis data for contouring, they do not really define the true surface. For example, in the line mesh method, the available information corresponds only to the coordinate points, which are finitely distributed. No information is available about the rest of the surface; in other words, the model has holes in it.

Although CAD systems keep the associated mathematical complexities transparent from the user, the principles of curve generation differ from system to system. The difference may become obvious, and sometimes critical, when the two systems try to exchange geometric data. When the two systems generate curves differently, exchange is imprecise and slow—an undesirable situation for CIM.

Representation of 3D Surfaces

True representation of a 3D surface must be defined exactly by mathematical formulas. Such an approach defines each point on the surface, resulting in an exact shape, not an approximation or representation.

Several techniques use mathematical formulas to define surfaces in CAD/CAM systems. As in the case of curve generation, the formulas use control factors to manipulate the 3D shape in such a way that what is described mathematically is as close as possible to what is desired. In order of complexity, the techniques are:

1. coons patches
2. Bezier surfaces
3. B-splines
4. nonuniform B-splines (**NUBS**) and their derivatives
5. nonuniform rational B-splines (**NURBS**)

These techniques increase in complexity from coons patches to NURBS by offering more control variables, thus handling shapes that are geometrically more complex. Although the coons patch technique represents the simplest formula-based generation of 3D surfaces, it is implemented on several CAD/CAM systems. This is so because it constructs complex forms by piecing together many patches, similar to the way several Bezier curves are blended together to represent a sophisticated curve.

CAD vendors favor more complex formulas, since the formulas provide greater flexibility in constructing and manipulating 3D shapes. As pointed out by Beard (1990), "Because each higher-level formula uses more variables to express a single surface, it has the mathematical capability to exactly express any surface created with a lower form." For example, a Bezier formula used to express a coons patch yields the same results, but the reverse is not true. Thus, coons patch capability can be considered a subset of Bezier surface, which in turn is a subset of the NUBS. This fact is of significant importance to CIM. It means that the surfaces constructed on a CAM system based on a low-level formula are transportable to all other systems that support a higher-level formula. But the reverse is not true. For example, the geometric model of a complex surface created with NUBS cannot be imported to a CAM system that supports only coons patches. In such a case, the model must be reconstructed, which may not yield a true replica of the NUBS-based model.

Since the NUBS technique represents a higher level of sophistication in generating true 3D surfaces, a CAM system supporting NUBS can import geometry intact from any CAD system, irrespective of the construction technique employed in developing the model. It is so powerful that NUBS can be used to import or construct multiple surfaces independently and merge them into one composite surface expressed as a single entity. NUBS has become viable for most CAM systems, even those that are PC-based, now that microprocessors' performance-to-price ratio has improved. The latest method is based on the NURBS technique, a further development of NUBS. Briare (1992) called NURBS the technique of the future in 3D surface generation.

From CAD to CAM

For CIM, an accurate and complete description of the surface is essential in CAD, since it ultimately becomes the roadmap by which CAM systems generate tool path codes that the machine tool executes as a series of point-to-point axis commands. The CAM system refers to the surface, or the curve, and reconstructs or translates the shape along the tool path as a series of chords linked end to end so that the endpoints are on the original surface or curve. Any deviation between the desired surface and that actually produced depends on the resolution of the translation process. In sophisticated CAM systems, the user simply specifies the deviation limit, and the system takes care of the associated decision making relating to the translation. Obviously, the shorter the chords (i.e., the larger their numbers), the more accurate the shape representation. But shorter chords generate more data for analysis. It is necessary, therefore, to strike an appropriate balance between the desired accuracy of shape representation and the amount of resulting data analysis.

In 3D-sculptured surfaces, the tool must be moved in all three directions. CAM systems account for the **gouging** that can occur due to abrupt changes in the topography, such as when two or three surface areas converge. Models based on true surface techniques are superior in controlling gouging to those based on representation, such as the line mesh method. Only by increasing the resolution of surface definition can gouging be reduced in line mesh methods, but more chords mean more data for analysis, which slows the process

A true surface approach also yields consistency. In some CAM systems, the user merely specifies the maximum kerf height and surface error, and the system automatically generates cutter paths to the specified tolerances. In exacting jobs such as mold making, such capabilities allow accurate curvature at the boundaries or edges of an interrupted surface—at the parting lines of the mold, for example.

Solid modelers usually conduct Boolean algebra operations on geometric primitives, e.g., blocks, cones, or spheres, to define complex machining operations. For example, to model a drilled hole, the designer creates the workpiece on screen, locates the hole, creates a cylinder of diameter equal to the hole size, and then—using Boolean algebra—subtracts the cylinder from the workpiece. This leaves a drilled hole at the specified location. A hole location is changed by filling the previous hole and starting again. The process can be expedited by using a parametric or feature-based design system that stores objects rather than geometric images in the real-time knowledge base. The user only has to specify the hole with one command. Following built-in rules, the system aids the designer with useful information; for example, the system may respond by displaying: "It is impossible to drill at the specified spot." It can even suggest avoiding that particular hole for reducing manufacturing cost. Such capabilities encourage what-if interaction from the designer, leading to better designs.

Most CAM systems follow a similar approach, using formulas and algorithms to construct the part geometry. The performance, flexibility, speed, and user friendliness of these systems depend on how equations are manipulated and data is processed, stored, and moved in and out of memory.

6.7   COMPUTER-AIDED ENGINEERING (CAE)

Computer-aided engineering (CAE) is a generic term that denotes the use of computers in tasks essential to engineering a product. It involves material selection and analysis for stress, vibration, noise, thermal distortion, and so forth. To ensure safe operation of the product, stress analysis is widely practiced. In the 1990s, knowledge-based systems are being used in CAE, as discussed in Box 6.3.

☐ **BOX 6.3**   *Using Knowledge-Based Engineering*

In less than one day, Eastman Kodak Co. (Rochester, NY) can perform the equivalent of 12 to 18 weeks of engineering for the design and configuration of plastic injection mold bases.

They are using knowledge-based engineering software, a means of reducing engineering effort, time, or both. A fundamental reason behind this interest is the ability to computerize the repetitive and the routine in mechanical engineering, which often exceeds 80 percent of the total engineering effort.

What It Is.   Knowledge-based engineering is software technology that provides a means of storing product or process information as a set of engineering attributes, rules, and requirements. The rules and requirements can generate designs or tooling, or process plans automatically. The systems typically reside on powerful general engineering workstations.

Unlike traditional computer-assisted drafting (CAD) programs that capture geometric information only, knowledge-based engineering systems capture the intent behind the product design— the how and why, in addition to the what of the design.

One important aspect of knowledge-based engineering is the ability to generate data instead of just storing it. The technology encourages development of a generic model. For example, plastic injection molds are made up of standard components. Mold base configurations are based on constraints relating to resin temperatures and

pressures, molding requirements, and the type of mold and molding machine to be used. Instead of storing thousands of complete mold designs or hundreds of pre-assembled macros in a CAD library, knowledge-based engineering software stores the methodology, or rules, by which such mold base designs are created.

Using It.   To deploy a knowledge-based engineering system, engineers use a design language to build a smart model of a product or process. The smart model typically includes: (i) the product structure (virtually equivalent to a bill of materials); (ii) rules for reconfiguring or changing the product structure when there are new inputs; (iii) dependencies and relationships among features and parts of the product so changes to one part or feature automatically change those that depend on it; (iv) functional, physical and geometric attributes; (v) engineering rules for contributory engineering disciplines (e.g., manufacturability, structural analysis); (vi) engineering rules for design optimization.

The Benefits.   Knowledge-based engineering software offers three basic benefits: reduction in time to market, capture of engineering knowledge, and facilitation of concurrent engineering.

Source: From "Using Knowledge-Based Engineering" by H. W. Rosenfeld, November 1989, *Production*, pp. 74–76.

Finite Element Technique

A manufactured product should be designed to operate safely. Structural analysis, a major task of design, ensures that the components and their assemblies will not strain beyond limits. The finite element method is a structural analysis technique that models the product in a computer. The computer views the structure as a finite number of small elements. With appropriate values of material properties for these elements and their end conditions, the structure is analyzed mathematically to predict stress field, thermal distortion, vibration severity, noise emission, and so on.

The finite element analysis (FEA) is also used to analyze the workpiece. For example, in the case of metal forming, FEA can answer the question, "What is the likely residual stress in the turbine blade after forging?" Such information can help improve product quality and reduce cost.

FEA was once an analysis tool requiring mainframes. Not any more. The 80386 microprocessor with its 32-bit data path and up to 4 GBytes of RAM allows even a PC to solve an FEA problem with more than 50,000 nodes almost as fast as a mainframe. Although a mainframe can solve the finite simultaneous equations for the model faster, PCs excel in facilitating the development of the model, which usually is a time-consuming task.

FEA requires representation of the workpiece geometry in computer memory— a task usually carried out early in the drafting stage. This work may have to be duplicated if integration within CAD functions is weak. CIM is eliminating such duplications. Integration is achieved via neutral-file formats for transferring geometry data from CAD packages to FEA packages. A de facto standard for neutral format in PC environments was set in 1982 by Autodesk in the form of **DXF (data exchange format)** files. However, DXF is somewhat limited, since it is only a 2D specification (at most, 2½D). Approaches similar to Autodesk's are available in other packages such as CADKEY and VersaCAD.

6.8   TRANSPORTABILITY

Transportability refers to the ease with which two computer systems can exchange design data. Since most CAD systems, whether micro-, mini-, or mainframe-based, use proprietary formats for storing drawing data in files and databases, transferring data from one system to another requires some intermediate format legible to both systems. Some translation is generally required.

Mini- and mainframe-based systems transfer data through magnetic tapes while PC-based systems do so via floppy disks. On-line transfer between the PC and larger system is over a wire requiring asynchronous communications via telephone lines or a hard-wired connection between the two. CAD data transfer between a PC and larger system involves the following five steps:

1. Choose a data exchange format common to both systems.
2. Select a common data transfer method.
3. Convert the source data in the common data exchange format.

4. Transfer the converted data to the receiving system.
5. Convert the transferred data to the receiver's format.

According to Carpenter (1988), "Problems can arise at any of these steps. There is almost always some loss of associativity, intelligence, or organization in transferring drawings between systems. The more complex the drawing, in terms of symbols, layers, text styles, and geometry (two-dimensional or three-dimensional, types of entities used, etc.), the more likely some loss will be experienced. The drawings may look the same on a superficial level, but attempts to modify or add to them will reveal the changes" (p. 1828).

The complete process of data transfer between two CAD systems is illustrated in Figure 6.4.

Proprietary Formats

Several vendors have developed proprietary data exchange formats. Proprietary formats can become de facto standards especially if the CAD system is popular. Examples are the CALCOMP 925 plot file format and, more recently, AutoCAD's DXF as a drawing interchange file format. AutoCAD's popularity has dictated that other PC-based systems also provide the capability of creating and reading the DXF format. In many cases, proprietary formats such as DXF are found to work better than the standard formats, such as **IGES (initial graphics exchange specification)**.

Whatever the data exchange format, the first step in transferring data is to convert it. In AutoCAD, the DXFOUT command converts the source data into exchange format. In both the IGES and DXF formats, the converted data is in an ASCII format file that a text editor can view and modify. Normally the converted file is longer than the original.

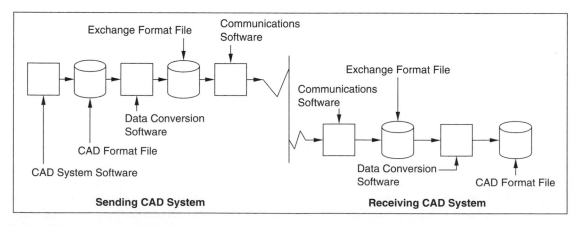

Figure 6.4
Process of data transfer between two CAD systems
Source: Reprinted with permission from "Making the Connection: Micro to Mainframe CAD System Communication" by S. Carpenter, *1988 ASEE Conference Proceedings*, 1988, p. 1827. Copyright 1988 The American Society for Engineering Education.

The next step involves transferring the converted data, which can be done in one of several ways. The simplest method for micro-to-micro transfer is via floppy disk. For micro-to-mainframe transfer, data can be sent over the asynchronous ports under the command of communications software. The software used can simply be a terminal emulation package having file upload and download capability. At the other extreme, it may be a file transfer system such as **Kermit** or a proprietary system. An important consideration in transmitting data is the time elapsed. For example, at a transmission rate of 9600 bps, a 200-Kbyte drawing file will take 21 seconds ($200 \times 1{,}024/9600 = 20.8$ seconds) to transfer. The actual transfer time is usually longer than the theoretical value due to the need for sending handshaking and error correction bits.

After the CAD data has been transferred to its destination, the last step is converting the data exchange format into the receiver's format. This is done either as a CAD system command, such as AutoCAD's DXFIN or IGES's IGESIN, or through a separate program run before the CAD system is invoked.

Plot File Formats

The plot file format is the earliest method of exchanging data. It is based on **CALCOMP (California Computer Products)** plotters, which most CAD systems can support. Vendors usually develop the software that creates the tape for transferring data to match the proprietary CAD system. The plot file format is used primarily with minis and mainframes. It is limited because the drawing consists entirely of vectors with little organization into symbols, layers, entities, and so on. In addition, plot file drawings are large and difficult to edit once converted into a CAD format.

Kermit

Developed at Columbia University as a simple, reliable method for transferring data between computer systems, Kermit's original intent was to allow students to save mainframe-resident files on microcomputers. Over the years Kermit has become popular for data transfer between PCs and mainframes. Kermit provides both terminal emulation for micros and file transfer capabilities. While transferring the file, Kermit also checks for any transmission errors. See Box 4.6 for a further discussion of Kermit.

Standard Formats—IGES

For commercial reasons, CAD system vendors prefer to develop proprietary data exchange formats that, for users, represent the antithesis of the CIM concept. Customers of CAD systems have recently been pressing for the development of standardized data exchange formats.

The evolution of a market for CAD data transfer prompted many vendors to develop special file formats for data exchange. This created considerable confusion in the beginning, but that ended in 1981 when the National Bureau of Standards (**NBS**) introduced the IGES (Initial Graphics Exchange Specification) format.

NBS's clout as a government agency encouraged acceptance of IGES. As the most universal standard, IGES facilitates data exchange among CAD systems of different makes. It is based on the principle that each type of CAD data assumes a standard format. Numerous subsets within IGES represent various mathematical entities to implement translation. For example, IGES entity 100 is a circle or an arc, while 128 is a B-spline surface.

Originally published in 1980 by NBS (now the National Institute of Standards and Technology, **NIST**)—IGES specifications as a means of representing geometry and other design data—such as dimensions, layers, and annotation—in a file format have been created by a consortium of government and private-sector representatives.

According to Kwok and Eagle (1991),

> The IGES exchange format is often a convenient means of exchanging data between CAD systems because it does not require a CAD system vendor to reveal proprietary information regarding the internal representation of geometry, database structure, or algorithmic techniques. An existing drawing on a CAD system can be preprocessed using a software program provided by the CAD system vendor to translate the drawing into an IGES format file. The IGES file can be post-processed by a software program on a completely different CAD system to convert it into a drawing. Although there are many cases where conversion problems exist, the neutral file format is useful as a practical means of exchanging design data.
>
> An example of an IGES file is shown in Figure 6.5. The IGES representation is an ASCII text file composed of five major sections:
>
> The START section, containing a human-readable header
> The GLOBAL section, containing information about the CAD system where the drawing originated
> The DIRECTORY ENTRY section, which lists all entities (such as lines, circles, arc, and dimensions) and relationships that exist among them
> The PARAMETER DATA section, which gives specific information about entities such as the starting and ending points of a line
> The TERMINATE section, containing a count of all lines that should be in each preceding section for data integrity purposes.
>
> These five sections are used concurrently to describe all aspects of a drawing in a CAD system. In addition, strong internal relationships exist between the contents of the various sections in order to assure the integrity of the data in the file.

CAD/CAM systems use different entities to represent the same shape, and often they handle and store the entities differently. The use of mathematical formulations, as in IGES, may not achieve point-to-point duplication of complex models between the two systems; the user may still need to manipulate information. What standards such as IGES do is to help the user in this manipulation, as an alternative to completely reconstructing the model.

IGES is a system-independent standard format that allows transfer of data from one CAD system to another with minimum information loss. CAD vendors write

IGES file generated from an AutoCAD drawing by the IGES   S0000001
translator from Autodesk, Inc., translator version IGESOUT-2.0.   S0000002
,,6HCMMRES, 10HCMMRES. IGS, 13HAutoCAD-10 c2, 11HIGESOUT-2.0,   G0000001
16, 38, 6, 99, 15, 6HCMMRES, 1.0, 1, 4HINCH, 32767, 3.2767D1, 13H901205.   G0000002
132040, 1.0D-8, 7.2071067811865D0, 6HThroop, 14HAutodesk, Inc., 4.0;   G0000003

110	1	1	1		00000000D0000001
110			1		D0000002
110	2	1	1		00000000D0000003
110			1		D0000004
110	3	1	1		00000000D0000005
110			1		D0000006
110	4	1	1		00000000D0000007
110			1		D0000008
100	5	1	1	0	00000000D0000009
100			1		D0000010
100	6	1	1	0	00000000D0000011
100			1		D0000012
100	7	1	1	0	00000000D0000013
100			1		D0000014
100	8	1	1	0	00000000D0000015
100			1		D0000016

110, 2.0, 2.0, 0.0, 5.0, 2.0, 0.0;   1P0000001
110, 6.0, 3.0, 0.0, 6.0, 6.0, 0.0;   3P0000002
110, 5.0, 7.0, 0.0, 2.0, 7.0, 0.0;   5P0000003
110, 1.0, 6.0, 0.0, 1.0, 3.0, 0.0;   7P0000004
100, 0.0, 5.0, 3.0, 5.0, 2.0, 6.0, 3.0;   9P0000005
100, 0.0, 5.0, 6.0, 6.0, 6.0, 5.0, 7.0;   11P0000006
100, 0.0, 2.0, 6.0, 2.0, 7.0, 1.0, 6.0;   13P0000007
100, 0.0, 2.0, 3.0, 1.0, 3.0, 2.0, 2.0;   15P0000008
S0000002G0000003D0000016P0000008   T0000001

Figure 6.5
Data representation in IGES exchange protocol as a file (for a simple 2D object) during
transfer between two CAD systems
Source: Reprinted with permission from "Reverse Engineering: Extracting CAD Data from Existing
Parts" by W. Kwok and P. J. Eagle, March 1991, *Mechanical Engineering*, pp. 52–55. Copyright 1991
The American Society of Mechanical Engineers.

two programs—a preprocessor and a postprocessor. The preprocessor converts the
sending vendor's proprietary data format into standard IGES format for transfer to
other vendors' systems. The postprocessor, on the other hand, converts from the
IGES format to the specific vendor's system format. Each type of data has an entity
number and a specific definition. The entities are defined for (a) geometry elements
such as lines, planes, and circles, (b) annotations such as dimensions, centerlines,
and notes, and (c) structures such as views, subfigures, and properties. Even then,
IGES does not guarantee error-free data transfer if no corollary exists for each
IGES entity or if the vendor does not implement all entity types in the processors.

　　IGES is still evolving. It has suffered from inconsistent and incomplete vendor
implementation, periodic release of newer IGES versions (currently in the third version), large data files, and the requirement of extensive technical labor for transla-
tions.

Automobile manufacturers have established their own standard for tape formatting; suppliers use this format to communicate with them or each other. Suppliers working with more than one company probably need to use more than one system.

Product Definition Exchange Specification

The translation of CAD-generated geometric models in formats suitable for other tasks of manufacturing remains problematic, especially if the devices are not from the same vendor. A new standard called Product Data Exchange Specification (**PDES**) is being developed to solve this problem. PDES is intended to facilitate a less ambiguous exchange of geometric data. It includes a wide range of product attributes and manufacturing information, such as surface finish or assembly instructions. PDES is still on the drawing board and CIM enthusiasts eagerly await its commercialization. In the meantime, users are making the best use of existing standards, notably the IGES and DXF formats.

6.9   NEEDS OF CIM

In CIM, the critical need for CAD (and CAM) is the ability to use design data throughout the enterprise without having to reenter the data. Not only does this save time and money, it also eliminates any risk of data corruption. The rapid proliferation of microcomputer-based CAD systems in recent years, however, has created data incompatibility problems that need to be addressed.

CAD-CAM Continuum

In a CIM environment, CAD is linked with various other functions of CAM. This is illustrated in Figure 6.6, for a typical manufacturing company that produces to order, by the CAD blocks within the dark area on the left. Note the entry (sales order processing) and exit points (shipping) shown by the wider arrows.

Major CAM tasks directly influenced by CAD are:

data processing
NC/CNC part programming
robot programming
CMM programming
FMS operation
tool design
MRP system
product development

As a terminology, CAD/CAM describes the monolithism between design and manufacture, especially when they are computer-assisted. The concept of CAD/CAM as a seamless continuum (see Box 6.4) is becoming a reality now. The difficulties arising from the proliferation of nonstandardized CAD systems speaking different languages are being addressed through standardization.

Figure 6.6
A CIM model showing CAD and its interactions with other subsystems

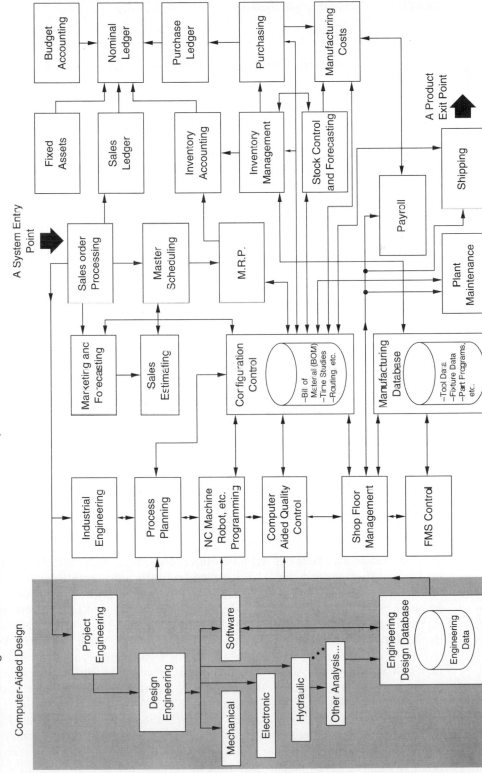

Source: Reprinted with permission from *Computer Integrated Manufacturing* (p. 92) by Paul G. Ranky, Copyright 1986, Englewood Cliffs, NJ: Prentice-Hall.

▢ **BOX 6.4**	*From Start to Finish with CAD/CAM*

We were the traditional model shop that made master patterns for the automobile and furniture industries. From our masters someone else made the tracing pattern to get to the final mold or die.

Our march to becoming a full-service house that could take a project from a submitted design to the final mold/die began five years ago with the purchase of a simple two and one-half axis CAD/CAM system that produced wire-frame designs. We then started to generate NC part programs. At the time, we did not fully appreciate the power of integration that allowed us to help customers come up with the final detailed designs from their concepts.

Expanding Capabilities. The next step was the purchase of a Battelle ToolChest CAD/CAM system from Battelle Columbus Division. This 3D system, which operates on small computers, includes surface modeling, drafting, CNC programming, finite element analysis, database conversion, documentation, direct NC, shaded image modeling, and CMM programming capabilities. Each part of the system is completely integrated with the others, allowing seamless transfer of data between the various functions.

The system has three different curve-fitting options: spline curves, curves-fitted-to-points, and Bezier curves. In building surfaces, we depended heavily on the numerous surfaces available in the system: fillet, spline, polynomial and Bezier. The fillet option was used to generate constant or blended radius surfaces; the spline option to create complex contoured surfaces; and the polynomial option where the smoothness of the surface created was particularly important.

The final step was the development of checking fixtures and CMM inspection routines for use by the arm rest (the part as the subject of the article). Again, we used the same design geometry database. Programming a CMM is very similar to programming a CNC machine tool. The user selects critical points for inspection and defines the inspection path. The resulting program is output into a DMIS file—the format used by most popular CMMs.

Source: From "From Start to Finish with CAD/CAM" by D. Spaulding, July 1990, *Modern Machine Shop*, pp. 82–88.

The **Automotive Industry Action Group** (**AIAG**) and other industry groups realized this difficulty early on. AIAG, which includes representatives from the Big Three automakers (Ford, GM, Chrysler) and their suppliers, has been heavily involved in developing standards—such as for bar coding and EDI (electronic data interchange)—and has recently turned its attention to CAD-to-CAM incompatibility. AIAG's white paper "CAD Data Exchange within the North American Automotive Industry" describes the state-of-the-art in translation technology and related issues. According to AIAG, "IGES is the most recognized, but not necessarily the best translation standard. . . . While IGES is a suitable standard for transferring CAD data, it is deficient in interfacing CAD to CAM." Still, IGES will be widely used until PDES is fully developed. The difference between IGES and PDES is obvious from their names; also, IGES has been designed to swap geometry data between CAD systems, whereas PDES also includes textual information such as features, tolerances, material specifications, and surface finish.

CAD-CAM Link

In the past, NC programmers or software were concerned primarily with creating or recreating part geometry in CAM, since technical limitations precluded importing geometric data files intact from different CAD systems. This is no longer true. New technology allows translation and transfer of data from the CAD database directly into the CAM system. Still, file transfer remains problematic in many systems because geometric part definition gets distorted during the translation process. Furthermore, it is not only this translation but also the consideration of how precisely an NC program ultimately approximates the desired shape. The CAM system's ability to geometrically define the part surface is solely responsible for achieving precision on machined surfaces, and all CAM systems do not do it the same way. The ability to import or create complex geometry on a CAM system is very important for shops that machine sculptured surfaces.

6.10   REVERSE ENGINEERING

Reverse engineering is a new concept that denotes the process of generating engineering design data from existing components. These data may be physical dimensions, coordinate values, surfaces, orthographic drawings, and so forth. Reverse engineering has been made possible by the digital measurement capabilities of CMMs and the graphic communications features of CAD systems. Its application areas fall into two groups:

1. *Aesthetics*. The primary design characteristic of some products, such as automobile dashboards or shampoo bottles, is a pleasing appearance. The CAD models in such cases contain surface data that can be generated following the concept of reverse engineering. Using a CMM system, the existing surface is digitized and brought back into a CAD package with the help of IGES. The CAD system then generates drawings or NC programs.
2. *Drawings*. Sometimes, a company may have a part but no production drawings or design. Reverse engineering can help obtain relevant data and drawings from the existing part by digitizing it with a CMM.

IGES plays an important role in reverse engineering. Data transfer between CMM and the CAD units takes place via IGES, which is a nonproprietary neutral format for exchanging design data between different CAD systems.

6.11   SIMULTANEOUS ENGINEERING

Merging the efforts of product designers and manufacturing engineers to improve manufacturing processes and products has recently been termed **simultaneous engineering (SE)**. According to W. David Lee of Arthur D. Little, "Simultaneous engineering is the process in which key design engineering and manufacturing professionals provide input during the early phase to reduce the downstream difficulties and build in quality, cost reduction, and reliability at the outset." Thus, SE represents the embodiment of the design-production continuum. It is a procedure, not a specific product or software. SE reinforces the message that engineers should optimize product design for ease of manufacture, high quality and reliability, low cost, and shortest possible time from concept to delivery. It has allowed automobile manufacturers to reduce the design cycle—from concept approval to volume production—from five to three years. In aerospace manufacturing, SE has reduced development time by almost half, and cost by 30%.

Henry W. Stool of the Industrial Technology Institute suggests the following "four C's" of simultaneous engineering:

1. *Concurrence*. To emphasize the fact that product and process design go hand-in-hand.
2. *Constraints*. To emphasize that process constraints be considered a part of the product design. The term *design for manufacturability* has a similar connotation.
3. *Coordination*. To remind that the product and processes need to be coordinated closely to achieve the best quality, cost, and delivery.
4. *Consensus*. To stress the desirability of decision making based on a consensus among the members of the simultaneous engineering team.

Simultaneous engineering results in: (a) products that have been designed to reduce their manufacturing cost, (b) planning and implementing those advanced technologies that are appropriate to the company, and (c) integrating new technologies into existing ones.

What role does simultaneous engineering play in CIM? At the least, the SE approach brings design and manufacturing functions closer; at best, SE forces integration.

TRENDS

❏ The importance of CAD data in achieving CIM is increasingly being realized. Vendors are developing software products that enable exchange of data among CAD systems, as well as direct transfer to CAM systems and other manufacturing functions.

❏ Some recent advances in defining surfaces, such as Bezier and B-splines, render CAD systems more compatible.

❏ The automation of prototyping using stereolithography—a new technology—is gaining industry acceptance (see Box 6.5).

❏ The use of knowledge-based technologies in CAE continues to expand.

❏ The DFM concept is being implemented under various names, such as simultaneous engineering.

❏ There is a trend to computerize as many subfunctions of design as possible.

❏ The demand of the market for product variety and rapid response to customer orders continues to bring sophistication in CAD systems.

BOX 6.5   *Stereolithography Automates Prototyping*

Stereolithography apparatus is a new equipment that can transform a 3-D design into 3-D output, cutting the time of prototyping from weeks/months into hours. It does so through the process of photopolymerization in which a liquid plastic monomer under ultraviolet light gets transformed into a solid polymer. The process is similar to the processing of photographic film in which silver changes state when exposed to visible light. Automotive and aerospace manufacturers whose work involves complex parts and assemblies can justify the SLA's cost of $200,000 (1990 dollars).

The prototyping process begins when the file containing a 3-D surface or solid CAD model is loaded into the SLA's control unit. Based upon the desired accuracy, the control unit, which works similarly to an NC machine tool "slices"

the model into a series of small cross sections in 0.005–0.030 inch thickness range. The mirrors as directed by the control unit focus the light beam to solidify a 2-D cross section of specified thickness on the surface of the liquid. The elevator table then drops for a solid polymer coating with another layer of liquid that is solidified by the laser. The process continues working from the inside of the part to its exterior until the prototype is completely built up. The "green" prototype is finally cured by exposing it to an intense long-wave ultraviolet light. It may next be sandblasted and painted as desired.

Source: "Stereolithography automates prototyping" by D. Deitz, May 1990, *Modern Machine Shop*, p. 36.

SUMMARY

Under the CIM concept, the design function has become more important in manufacturing. It is now the hub of activities involved in producing a tangible product for the market. The concept of DFM helps integration and, thus, CIM. An analysis of the design function was presented in this chapter to highlight the activities that are computer-aidable. The role of graphics in CAD is enormously important. Recent developments in curve generation—such as B-splines and NURBS—that enhance the compatibility of design data have been discussed. The roles of reverse engineering and simultaneous engineering in design were described. The needs of CIM as far as CAD is concerned were also discussed.

KEY TERMS

Automotive Industry Action Group (AIAG)

B-splines

Bezier curve

California Computer Products (CALCOMP)

Cathode ray tube (CRT)

Data exchange format (DXF)

Design for assembly (DFA)

Design for manufacture and assembly (DFMA)

Finite element analysis (FEA)

Gouging

Group technology (GT)

Initial graphics exchange specification (IGES)

Kermit

Loss function

National Bureau of Standards (NBS)

National Institute of Standards and Technology (NIST)

NUBS

NURBS

Product data exchange specification (PDES)

Rasterization

Rational B-spline

Rendering

Rough sketches

Simultaneous engineering (SE)

Spline technique

Taguchi method

Thumbnail sketches

EXERCISES

Note: Exercises marked * are projects.

6.1   Explain in about 300 words the role of DFM in modern manufacturing.

6.2   What is the Rule of Ten? What is its message?

6.3   What are B-splines? How do they differ from coons patches as far as surface generation is concerned?

6.4   Compare IGES and PDES on the basis of six important criteria.

6.5   Write a 400-word essay on CAE. (A literature search in the library is recommended.)

6.6   What is simultaneous engineering and how does it help achieve CIM?

6.7   What is reverse engineering and when is it practiced?

6.8*   It is claimed that DFM principles were fully exploited in the development of MicroVal CMM. Investigate this product as a case study and present your findings to the class. (One of the students may be given this project. The student should refer to Box 6.1, carry out further literature research, and also write to Brown and Sharpe for additional information on this product's development, design, and manufacture.)

6.9*   Refer to *Design for Assembly: A Designer's Handbook*, by Boothroyd and Dewhurst (in the library, if required), and list 10 major guidelines for DFA. If this title is unavailable in the library, ask the librarian to procure a copy.

6.10*   Visit a local manufacturing company to study its efforts for achieving DFMA.

6.11*   Prepare an inventory of the hardware and software available at your college or university in the area of CAD. Which academic programs use these resources and for which tasks?

6.12   Circle T for true or F for false or fill in the blanks.

   a.   In the case of CAD, three-dimensional modeling is just another name for solid modeling.   T/F

   b.   3D drawings are impossible without solid modelers.   T/F

   c.   Finite element technique is associated with CAD.   T/F

   d.   PDES, not IGES, better serves the objective of CIM.   T/F

   e.   CIM and simultaneous engineering are synonymous.   T/F

   f.   IGES stands for _____ and PDES for
   _____

   g.   The main difference between B-splines and rational B-splines is that the latter _____

   h.   NURBS is an acronym for _____

SUGGESTED READINGS

Books

Besant, C. B. (1983). *Computer-aided design and manufacture*. Chichester, England: Ellis Horwood.

Boothroyd, G., & Dewhurst, P. (1988). *Design for assembly: A designer's handbook*. Wakefield, RI: Boothroyd Dewhurst.

Groover, M. P., & Zimmers, E. W., Jr. (1984). *CAD/CAM: Computer-aided design and manufacturing*. New York: Prentice-Hall.

Hyde, W. F. *Improving productivity by classification, coding, and data base standardization: The key to maximizing CAD/CAM and group technology*. New York: Marcel Dekker.

Lamit, L. G., & Paige, V. (1987). *Computer-aided design and drafting*. Columbus: Merrill.

Mellott, D. W., Jr. (1983). *Fundamentals of consumer behavior*. Tulsa, OK: PennWell.

Onwubiko, C. (1989). *Foundations of computer-aided design*. St. Paul, MN: West.(This book explains the mathematical principles behind CAD.)

Stover, R. (1984). *An analysis of CAD/CAM applications with an introduction to CIM*. New York: Prentice-Hall.

Monographs/Reports

American National Standards Institute ANSI Y14.5M-82. (This standard describes the basics of graphics essential for DFM.)

Conkol, G. K. (Ed.) (1990). *The role of CAD/CAM in CIM*. SME Blue Book series. Dearborn, MI: SME. (The preface of this monograph states, "What better place to go for a real review of how CIM should work and what issues are central to the technology? By bringing representatives of these groups together, we were able to tap the opinions of over 100,000 end-users.")

Articles

Adams, J. A. (1988). Descriptive geometry and geometric modeling. *1988 ASEE Annual Conference Proceedings*, pp. 938–46.

Ashley, S. (1993, March). Designing for the environment. *Mechanical Engineering*, pp. 52–55.

Bancroft, C. E. (1988). Design for manufacturability: Half-speed ahead. *Mechanical Engineering, 101*(3), 67–69.

Beard, T. (1990, July). Coping with complex surfaces. *Modern Machine Shop*, pp. 52–62.

Beard, T. (1991, July). In search of the perfect curve. *Modern Machine Shop*, pp. 58–68.

Briare, K. (1992, June). In CAD/CAM, the future is NURBS. *Modern Machine Shop*, pp. 78–84.

Carpenter, S. (1988). Making the connection: Micro to mainframe CAD system communication. *1988 ASEE Annual Conference Proceedings*, pp. 1827–31.

Constance, J. (1992, May). DFMA: Learning to design for manufacture and assembly. *Mechanical Engineering*, pp. 70–74.

Deitz, D. (1990). Stereolithography automates prototyping. *Mechanical Engineering, 112*(2), 34–39.

Dixon, J. R. (1991, March). New goals for engineering education. *Mechanical Engineering*, pp. 56–62. (The first part of this article appeared in the February 1991 issue.)

Kwok, Wai-Lun, & Eagle, P. J. (1991, March). Reverse engineering: Extracting CAD data from exiting parts. *Mechanical Engineering*, pp. 52–55.

Morgan, R. E. (1991). The challenge of teaching the Taguchi method in industrial engineering. *1991 ASEE Annual Conference Proceedings*, pp. 1815–22.

Noaker, P. M. (1992, June). Manufacturing by design. *Manufacturing Engineering*, pp. 57–59.

Rosenfeld, L. W. (1989, November). Using knowledge-based engineering. *Production*, pp. 74–76.

Stoll, H. W. (1988, January). Design for manufacture. *Manufacturing Engineering*, pp. 67–73.

CHAPTER 7
Production Planning

"Computers enhance the design process, but the competitive edge still goes to companies that create data, not merely manipulate it."
—Douglas E. Booth (president, SME) in *Manufacturing Engineering,* June 1992

"In 25 years of working with North American and European manufacturing firms, though, I rarely encountered one that had machine and process capability data for their operations. Even if they had gathered such information, it was usually for machine acceptance, not for product design."
—Myron J. Schmenk in *Manufacturing Engineering,* June 1992

"Wouldn't it be great if you could test your ideas on increasing production efficiency, reducing cost, saving floor space, and decreasing labor content without betting your company's bankroll or your career? You can with a new generation of PC-based software designed to simulate shop-floor processes."
—John R. Coleman, editor-in-chief, *Manufacturing Engineering*

"Many of the best-known corporations still have design empires, production enclaves, and precious little cooperation. After years of specialization, it's tough to abandon the concept of efficient division of labor and embrace a concept like simultaneous engineering."
—Myron J. Schmenk in *Manufacturing Engineering,* June 1992

This chapter discusses the functions that are prerequisites to actual production. These functions are postdesign but preproduction. In the case of manufacture-to-order companies, they represent the beginning, since the design function is absent. The functions described in this chapter are commonly grouped together under one heading, computer-aided planning (CAP). CAP is a module of CAM in its broadest sense.

7.1   INTRODUCTION

As discussed in Chapter 2, manufacturing companies fall into two categories: (a) those producing to someone else's design, and (b) those producing from an in-house design. Companies in the first group ship to customers or distributors immediately after production. The second group stocks products in finished-goods inventory for later distribution or sale by the marketing department. In both cases, the production is planned and controlled for timely shipping or storage. But preplanning tasks are very different for the two groups. Companies producing to a customer's design can begin planning only if they have a final order. This requires quoting prices and delivery dates, which must be acceptable to the client. Thus, cost estimating is an important task within such companies. The other type of companies, conversely, forecast the demand for their products and prepare a master production schedule for them. Such companies also must estimate costs, albeit for internal use in predicting profit.

The job of production planning usually occurs in a department called production and inventory control (PIC). In the United States, the American Production and Inventory Control Society (APICS) disseminates knowledge in this area.

7.2   COMPUTER-AIDED COST ESTIMATING

Before a product is manufactured—even before its design is approved—management requires an estimate of its cost to decide whether it is marketable. Job shops that make to order are first asked by the client, "Can you produce to our drawing? If yes, how much will it cost and when can you deliver?" The answers to these three seemingly simple questions are critical to the produce-to-order companies, since the estimation results either yield the order or they don't. These questions in effect require a dry run of the planning and production tasks. The answers are based on what the plant is processing at that time and how the order would affect machine loading, MRP II (manufacturing resources planning), and so forth.

Quoting is so important that a special group of personnel called cost-estimators are entrusted with this task to ensure profitability. To project an accurate estimate, estimators obtain input from personnel in a number of departments, ranging from procurement of raw materials to packaging and shipping, to determine process times and related costs for manufacturing and delivering the product.

The computer has proved to be an effective and increasingly popular tool in estimating costs. Modern **computer-aided cost estimating** (CACE) systems not only produce complete estimates, they also formulate and print detailed shop routings and process plans. The work plans are printed out for use by programmers and operators in planning, programming, and set-up. These systems also serve as a quick-access library of past bids and routings, which is useful to other departments. CACE systems help companies hone their competitive edge better than manual estimates. They reportedly offer payback periods of less than a year.

Several software packages can estimate costs accurately and rapidly. They incorporate a database of common workpiece materials, which can be modified and expanded. Some packages allow for up to 500 different materials. The cost can be estimated for machining parts from standard stock, forgings, or castings. The user inputs data on both raw materials and finished parts, and the software estimates the machining cost using these and other data in the database.

Some packages come with information on as many as 1,500 metals, alloys, and plastics, together with data on machine tooling, noncutting elements, and shop labor standards that aid in complying with various standards such as those of the U.S. Department of Defense. The user can tailor generic packages to suit individual needs by editing them with the right figures and formulas; for example, adding special information on machining speeds and feeds. The user can also adapt the supplier's database.

CACE software contains the values and formulas needed to make an accurate estimate. It prompts the user to select the materials and the processes. An important benefit of computerized cost estimating is the accuracy of estimates, since chances of errors are minimized and personal factors do not creep into the calculations. What-if capabilities of CACE software also allow better make-or-buy decisions, since all possible alternatives can be considered.

In some CACE systems, automation in transferring information from the drawing is also possible using a measuring device called a sonic digitizer. The user simply traces over the part drawing with an electronic pen to obtain tool travel and part dimensions. The pen emits sound waves, and its movement is trapped by two sound-activated clocks for the x and y coordinates. Using this data, the software calculates the distance the pen moved, which ultimately reveals the part's surface area, volume, and weight. The digitizer can also measure part models or CAD screen images, taking into account any scaling of the drawing. An override function allows flexibility in fine-tuning quotes for unusual jobs that do not conform to the standard information stored in the software.

7.3   PRODUCTION PLANNING AND CONTROL

The function of production planning and control is to extrapolate from sales predictions or customer orders to human resources and facility requirements. The resulting schedules and capacity plans set the tempo for the factory. They direct the purchasing function to coordinate vendor deliveries with shop-floor activities. The use of CAD/CAM forces closer interactions among engineering, production control, and shop-floor functions. This chapter focuses on planning and long-term control, while Chapter 9 highlights short-term production control.

MRP II

MRP II is a plant-wide, long-range monitoring and control software system. It represents the natural evolution of closed-loop MRP (materials requirements planning). MRP II encompasses not just manufacturing but finance, marketing, engineering, purchasing, distribution, and more. Thus, we can view MRP II as the upper stratum of CIM, the lower one being the **plant-floor monitoring and control system (PFM&C)**. Except for the **shop-floor control (SFC)** module—the module concerned with execution—all other modules of MRP II are planning tools and have little to do with execution of plans at the shop floor. SFC systems track costs at the shop floor and provide place markers to determine where a specific order is. They assume no need for data in other areas, such as quality, operator, or equipment. In a full-blown CIM environment, however, decision makers need such data.

MRP II has conventionally been a mainframe system. Through the **management information system** (**MIS**) department, such a system might distribute computer reports to each production planner on Monday mornings. By the Tuesday afternoon, the information contained in the report would be out of date. Some companies distribute MRP II results on-line from the mainframe to high-end networked PCs, thus providing current information. Another advantage of PC-based MRP II systems is that users tend to be more familiar with them. Moreover, PC-based systems are more cost-effective to implement. But this trend is resisted by managers of centralized systems installed when PCs did not exist or were not fully developed.

CAPP

Once a part has been designed, it is produced using various types of machines and other resources, including materials and personnel. Planning these activities so that production can begin is called **process planning**. This step determines the sequence of individual operations necessary to produce the parts, and eventually the product. All the related information is documented on a **route sheet**. Such information includes cutting conditions—such as feed and spindle speed—that determine the total time of production, and hence the product cost.

Process planning is challenging, since it involves selecting the best plan out of several possible ways for manufacturing a component. A relatively simple product, such as an expander sleeve, may have dozens of possible plans. This is where the application of expert systems in process planning holds promise. The ingenuity of the process planner becomes evident in preparing a plan that achieves the management objective, such as minimum product cost.

Traditionally, route sheets have been prepared manually. Since the advent of computers, however, companies have attempted to computerize process planning tasks. The use of computers in process planning is termed computer-aided process planning (CAPP). CAPP systems are either retrieval or generative type. The retrieval type, also called variant type, is suitable for a family of parts. This system draws a standard process plan and stores it in the database. Whenever a different part from the family is to be processed, the standard plan is retrieved and appropriately modified—hence the name retrieval or variant. The retrieval method relies on the concept of GT for part coding and classification. It is also compatible with the concept of **cellular manufacturing**, in which cells are designed and laid out for family-of-parts production. MIPLAN is an example of retrieval-type CAPPs (see Box 7.1).

The generative method of developing process plans involves starting from scratch every time a different part is to be processed; no plans are available as a baseline. Users enter the input data, and the CAPP system develops the plan following algorithms by which process sequences are synthesized, based on part geometry, material, and other related factors. Obviously, generative systems are more sophisticated and costlier than the retrieval ones. GENPLAN, developed by Lockheed-

---

◻	**BOX 7.1** *Capitalizing on CAPP*

Technological advances are reshaping the face of manufacturing, creating paperless manufacturing environments in which computer-automated process planning (CAPP) will play a preeminent role. There are two reasons for this effect:

Costs are declining, which encourages partnerships between CAD and CAPP developers.

Access to manufacturing data (graphics, text, or machine language) is becoming easier to accomplish in multivendor environments. This is primarily due to three reasons: (i) increasing use of LANs, (ii) IGES and the like are facilitating the transfer of data from one point to another on the network, and (iii) relational databases (RDBs) and the associated structured query language (SQL) allow distributed data processing and data access.

The developments in RDBs and SQL are popularizing CAPP. First, SQL provides unprecedented access to data, especially when more than one application, for example MRP schedules to CAPP process plans, is involved. Second, the ANSI's SQL encourages data access even with different hardware platforms. Third, over the long term, RDB technology promises to play an important role in both knowledge acquisition and representation in computer-readable form. The expert systems based on CAPP rely on RDB technology.

Forty to 50 years ago, process definition as we understand it was almost totally absent. Process steps were left to the judgement of those in manufacturing, as were some aspects of the design. As factories grew in size and complexity, the need to document manufacturing processes emerged, and process plans were created.

In the 1960s, the capacity requirements planning module of MRP required process routings to compute machine tool loads. Ten years later, closed-loop MRP systems required routings to monitor production and report progress against the plan.

The foundations for modern process planning were laid in the '70s. Manufacturing engineers embraced the first systems designed to automate process planning. MIPLAN, developed by the Organization for Industrial Research (OIR), and TNO's Metl-instutuut, a manufacturing research arm of the Dutch government, was a commercial success.

MIPLAN contained several innovative features:

1. Retrieval used TNO's MICLASS group technology (GT) classification for metal parts. Users no longer had to memorize part numbers to retrieve similar process plans. Flexible retrieval enabled users to identify parts on the basis of any feature or combination of features in any sequence.
2. Process flow analysis (PFA), the ability to identify families of parts on the basis of process similarities, was another capability.
3. Key words and formats embedded in standard text could be used to trigger standard calculations and activate searches of machinability databases automatically.
4. The CAPP database now had greater utility. GT classification and other methods of data access made every part of the database available to the user.
5. Tool management modules linked tooling specified in process instructions to tool availability.
6. IGES allowed graphics interfaces, making it possible to show in-process views of completion within the process plan.

RDBs eliminate these problems. They store data in tables, not in hierarchical structures or networks. Using SQL, a nonprocedural fourth-generation language, the user specifies what the computer is to do, not how it is to be done. In this environment, data are not associated with an application program that makes use of them. Moreover, the user need not be concerned with the location of the data, the device on which they are stored, or the manufacturer of that device. Indeed, data need not be stored in one location at all; they may be retrieved from anywhere in the network.

How does one make the transition from hierarchical to RDB environments? At first, a mixture of relational and nonrelational applications are present. For example, most CAPP systems have migrated to RDBs, while MRP systems have not. To obtain data from nonrelational applications, utilities called loaders are employed to extract data and map it into a table in an RDB. Loaders make data available to relational and nonrelational applications.

Source: From "Capitalizing on CAPP" by J. A. Nolen, May 1989, *Manufacturing Engineering*, pp. 70–73.

Georgia Co., is an example of the generative type. For CIM, the generative technique is superior since it can be integrated with the product design function where required data are available. Thus, this technique eliminates the need for human intervention.

7.4   GROUP TECHNOLOGY

In group technology, similar parts are grouped together to improve manufacturing effectiveness. The family of parts can be processed by a well-configured set of machines, tools, and fixtures. The technique is inappropriate where mass production is feasible due to large product volume. GT facilitates production in committed small cells of machines and associated accessories such as tools, pallets, and fixtures.

GT implementation begins by classifying the various parts the company produces. The parts are next coded into groups based on design and/or manufacturing similarities. Manufacturing-based grouping enables the use of similar process plans for all the parts in the family. The family concept maximizes the use of available resources.

The use of GT has been reported to significantly improve the flow of material through the plant. As an example, compare Figure 7.1 closely with Figure 7.2 to appreciate the improvement in the production flow involving 150 similar parts. Figure 7.1 corresponds to the current practice, while Figure 7.2 is the result of GT implementation.

While Box 7.2 described a case study, Box 7.3 discusses why a large number of companies have not yet adopted GT.

7.5   SIMULATION

Development of new products or major modifications in the existing ones involve building and testing of prototypes, usually a time-consuming and costly step. In the past prototypes were physical models. Today, with the help of graphics, computers

---

| **BOX 7.2** | *Grappling with Group Technology* |

We make gas turbine engines for helicopters, commercial aircraft, air-cushioned marine craft, and Abrams M1 main battle tanks. In the early '80s, production demands accelerated beyond job-shop capabilities, making change necessary to stay competitive.

Group technology became the organizational backbone of an eight-year, $60-million makeover. In three basic stages, production planners set up a GT database; grouped parts into families with common shape, function, or manufacturing processes and tooling; and redesigned the shop floor into 14 manufacturing centers.

GT also closed the gap between design and manufacturing. One computerized database of part families formed the basis for automated part-specific design systems. This reduced part proliferation, standardized designs, and provided a manufacturing producibility interface. As a result there are fewer design changes.

Demystifying Classifying. We found a major stumbling block when moving to GT: traditional classification and coding tools work best for a limited number of parts and processes. Using them on an inventory such as ours would take years. Here's how we built a database with approximate part families in 90 days.

Production planners first identified parts that could be bought cost-effectively or for which the plant didn't have processing capacity. This left around 1000 parts for in-house manufacture. They then defined production levels for all engine models to find machine requirements.

The next step relied on production flow analysis (PFA), which uses matrix clustering to group elements with similar processing requirements.

The technique works best with a small number of parts and processes. We expanded its function by combining it with a decision support system. The result was a computer-based analytical tool called OPSNET which allowed production planners to interactively query discrete data on part geometry and processing. Parts then trickled through the decision support system into sets with common attributes. The tool also provided a conceptual presentation of design processes, using part information and routings, time standards, production levels, and data on existing machines.

Facts on the Facelift. Design teams divided final shop-floor design into 28 steps and three design reviews. First, they merged mainframe-based files of current and interim routings, manufacturing standards, and production requirements into several databases. This information formed the basis of a PFA matrix, with part numbers compared to machine codes that included operating conditions and equipment capabilities.

The plant's military business tripled and the commercial engine program increased fivefold. The GT program helped to eliminate 700 of the 1400 machine tools in the factory. Overall, production efficiency is up 50%, and rework and scrap are down 85%.

It may look like our production problems have disappeared. Far from it. You must stay on top of GT, or workers will drift back to a job-shop mentality, losing track of the total manufacturing concept.

Source: From "Grappling with Group Technology" by M. Propen, July 1990, *Manufacturing Engineering*, pp. 80–82.

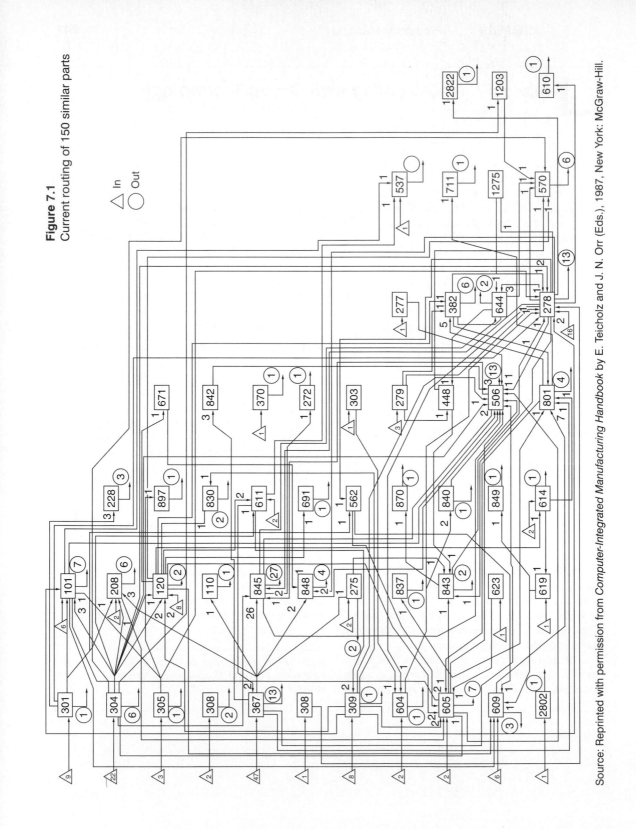

Figure 7.1
Current routing of 150 similar parts

△ In
○ Out

Source: Reprinted with permission from *Computer-Integrated Manufacturing Handbook* by E. Teicholz and J. N. Orr (Eds.), 1987, New York: McGraw-Hill.

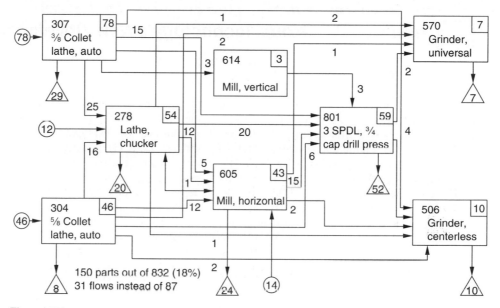

Figure 7.2

Improvement in routing following GT implementation

Source: Reprinted with permission from *Computer-Integrated Manufacturing Handbook* by E. Teicholz and J. N. Orr (Eds.), 1987, New York: McGraw-Hill.

past prototypes were physical models. Today, with the help of graphics, computers provide an alternative in which the prototype is a simulated "soft" model.

Simulation is a software tool that models physical systems to study their behavior. It can be applied in a variety of areas, including manufacturing. According to Keith Wheeler of Litton Industries (Coleman, 1990), "Simulation is a descriptive modeling technique that is most valuable when subsystem interaction makes evaluating total system performance unreliable. Simulation can provide more information about system performance than static engineering calculations."

A common use of simulation is in optimizing the design of a product before building it. This approach avoids costly design and safety errors. The technique can also help improve the performance of existing products or systems. Besides modeling the products, simulation can also model production processes or facilities. It can also verify tool paths to ensure that the part program will do what is expected of it. A solid 3D model of the raw stock is displayed on the screen along with a simulation of programmed cutting motions, and finally the finished workpiece is shown. Besides identifying the program mistakes, the technique can also be used as an excellent training tool.

Major benefits of simulating processes are that it:

1. provides a realistic model of the process
2. accounts for random events such as downtime
3. can analyze effects of proposed changes in the process

BOX 7.3   *Group Technology's New Developments*

Group Technology (GT) has been around for more than 30 years. It was first documented by a Russian engineer, S. P. Mitrofanov in 1958. Early studies predicted that 50 to 75 percent of all American plants would be using GT by 1990. This has not happened. Why?

For one thing very few universities teach the subject. Secondly, many different GT classification and coding systems are on the market, but not one of them has emerged as a clear leader. As a result some major companies have developed their own coding system and they use it to develop CAPP (computer-aided process planning) for their own shop floor.

Within CAPP there is the variant philosophy that is defined as altering an existing standard process plan to make a new part or assembly. On the other hand, the generative process planning systems start by developing a code that describes the part using GT. This can be done by hand or with computer assist. Generative systems select machines as well as calculate speeds, feeds, depths of cut, time values, sequences, tooling, and all other steps needed to produce a part. The volume of data that such a system needs often defeats its application. Some plants have spent many man years developing such a data base. Every time a new machine or process is added or removed, the data base must be updated.

Source: From *Modern Machine Shop*, May 1990, p. 154.

---

4. predicts dynamic interaction among system components
5. works as an off-line training platform
6. determines system operating procedures.

Two major limitations of simulation are: (a) perfecting the model requires a time investment, and (b) model verification requires data from the actual process or system, which may also be time-consuming.

PC-based simulation systems have recently become powerful enough to handle moderately large, detailed models. The software is either language-based, requiring some program coding such as SLAMSYSTEM or SIMAN, or data-table driven, such as WITNESS or SIMFACTORY. Most packages offer a graphics interface with a mouse and pull-down menus that obviate the tedious coding associated with simulation. Over the years, their user-friendliness has drastically reduced the time of developing and "what-iffing" the model.

7.6   PART PROGRAMMING

NC **part programming** involves writing coded programs that make machine slides and tools move relative to each other for machining a part. This step falls between design and production and is an important function within CAD/CAM. In fact, CAM's earlier meaning was limited to the development of part programs only.

Part programs can be stored on papertapes, magnetic tapes, or disks. The input medium to the MCU is usually punched tape or floppy diskette. The alternative is to input the data directly via the keyboard of the MCU panel. Some controllers can speak two languages—the conventional G-codes and simple English prompts. These controllers can translate one into the other, thus providing increased versatility. Most are very powerful on graphics, showing on screen the part's likely final shape in comparison to the solid blank begun with. A 3D shaded view that can be rotated in any axis with the possibility of magnifying certain areas is generated. The levels of shading are used to define different surface depths.

In general, a part program can exist for the processing of any part and its related assemblies. In the limited sense, a part program is prepared for parts that are machined or formed on machine tools. The procedure of part programming comprises three basic tasks:

1. describing the part geometry
2. relating manufacturing operations and tools to the geometry
3. specifying and formatting the information to suit the machine tool on which the part will be processed

The first two tasks are handled together and the resulting output is called a **cutter location (CL) file** or data. The CL file is modified to suit the requirements of the machine that will process the part (the third task). Modification translates the CL file to render it machine-specific. The translation is known as postprocessing, and the translating software is called **postprocessor**. The output of the entire part programming process is ultimately a set of information called **machine control data (MCD)** or "G" codes. **G codes** are created by postprocessors, which in a computerized part programming environment are really the link between the CL file and machine tool.

Postprocessing

Postprocessing converts CL data into a format understandable to the specific machine tool. Postprocessors are software, either customized or generic. Customized postprocessors are designed specifically for a given environment of the machine tool and the associated CNC. Such a processor is called a **C-post**. A generalized postprocessor, called a **G-post**, is written so that with minor modifications it can be used in any machine–CNC combination.

The early processors of the 1960s were C-posts. They were three-axis contouring versions designed to run on APT systems and conformed to the guidelines laid down by the Illinois Institute of Technology Research Institute (IITRI). Each C-post was designed for a specific machine, a specific NC, and a specific computer. If any of these were different, the postprocessor supplier would implement the changes, usually at a significant cost. As the variety of NC units and machine configurations proliferated, the number of C-posts also increased, creating difficulties for large machine shops. As a result, in the mid-1970s Rockwell developed an alternative course of BCL INPUT (see Chapter 4). The concept of part program input had two effects: (a) it forced the postprocessing task out of computer rooms and into the CNC, and (b) it laid the groundwork for G-posts.

G-posts comprise a set of separate software modules that allow modifications for the specific machining environment only to the particular module without affecting the functions in other modules. They are also provided with a "user configuration file" so that the end user can effect the modification at the shop floor. Some typical parameters that might need modifying are: maximum spindle speed and feed rate, axis limits, add/delete axis, modification of words, and addition of special characters. G-posts have also helped in development of flexible C-posts, which involves combining existing G-post modules rather than starting from scratch.

Which of the two, C- or G-post, is better? The answer depends on the machining environment. For sophisticated machines with motion capabilities along more than five axes, the C-post is preferred; a good example is a 10-axis contouring machine for laying composite tapes. G-posts, on the other hand, are suitable for machines of up to four axes—the most common environment in discrete manufacturing. G-posts cost under $10,000; C-posts can be twice as expensive. G-posts have been gaining in popularity recently for the obvious reason that users can tailor them to their needs.

Programming Methods

There are four different methods of part programming:

manual
digitizing
language-based computer assist
graphics-based computer assist

Manual Programming.   Manual programming is carried out at the MCD level in the language of the machine control unit (MCU)—a microprocessor serving as an interface between operator and machine tool. It requires the programmer, who may also be the operator, to calculate, code, and format all the data necessary to machine the part. In other words, all three tasks of part programming, discussed earlier in this section, are carried out as one. This suits stand-alone environments, but could become time-consuming for complicated parts. In industry, about one-third of part programs are developed manually. With the proliferation of PCs, however, computer-assisted methods have nearly replaced manual programming.

Digitizing.   Digitizing relies on a template and a computer-interfaced stylus or scanner to enter the part geometry and the sequence for the tool to follow. It may generate either a CL file or MCD. The principle behind digitizing is simple. A measuring probe—contact or noncontact type—is positioned manually or by a special part program at a series of XY coordinate points over the part/model. At each point the Z-axis dimension is recorded with the aid of the probe and stored along with the coordinate values. This can be done on a CNC machine or a CMM.

Digitizing is also called **scanning**. Scanning traces a model of the part while recording the coordinate surface data. These data are merged with the necessary programming commands and stored or output as a complete part program. The part program is then used to machine duplicates of the model on the same machine or another compatible one.

On advanced 3D tracing systems, special features such as box cycle or stylus deflection correction are available. The digitizer also allows operator or programmer inputs such as programmed feed rate, spindle speed, part scaling factors, and special M and G codes. The coordinate data when combined with user input information generates the part program.

In the early 1960s, digitizing was considered an effective solution for creating part programs for automotive body panel dies. It provided a means of gathering data directly from the original wood or plaster concept model of the car. As CAD systems developed in the meantime, digitizing did not advance very far. Today, a part can be designed on a CAD system and its geometry converted to a part program. Alternatively, design data can be transferred to another system for developing the program. Where a proven design already exists for a master part, digitizing can be used to duplicate its geometry.

Language-Based Computer Assist.   This method uses a high-level language such as APT (automatically programmed tools) or COMPACT II for creating the source program (also called manuscript). In processing, the source program generates a CL file that is postprocessed to output the MCD.

Graphics-Based Computer Assist.   This method relies on graphic systems that describe the geometry and tool paths. Both language- and graphics-based programming methods allow communication with the machine through punched tape, magnetic tape, or DNC interface.

Advances in NC programming are making program verification easier. This is being achieved through: (a) visualization, (b) animation, and (c) solid modeling. Box 7.4 describes how the machining process can itself be modeled to generate NC part programs.

Modern computers enable the bunching of more and more downstream preproduction tasks into the design and process planning phase. Following this trend, NC part programming is being pushed further upstream to merge with design work. Part program generation at the product design stage is a major step in CAD-CAM integration.

Program Verification

NC program verification is an integral task of generating the data required for NC equipment operation. The term used at the shop floor to describe this task is "**tty**" for tape tryout or "**tto**" for tape tryout. Tape tryout is usually the last step prior to program data's release to the production department.

Program verification is necessary because NC programmers might make mistakes in areas such as the following:

Have realistic feeds and speeds been programmed?
Could tools and part collide when the tool turret rotates?
Could there be typing mistakes in the punched tape?
Have the correct formats and codes been used?

⬜ **BOX 7.4**   *Model the Machining Process to Generate NC Part Programs*

Traditional NC part programming includes defining geometry to describe a part, then determining the appropriate tools, feeds, and speeds. The next step is programming tool path in reference to part geometry. Finally, a CL (cutter location) file is postprocessed to format the program for the particular machine tool/control unit that will execute it. CAD/CAM systems that include graphics interfaces or CAD front ends have made this process easier over the years. But multiple steps are often required.

A modeling approach to creating a part program lets the user create, change, and interact with a graphic model of machine tool motion on a computer screen and get the program directly from the model for the appropriate machine tool.

The dynamic model is comprised of three integrated components:

Properties: The machining parameters associated with tool motion;
Sequence: The order of machining operations or events; and
Geometry: The elements defining tool path.

The term "properties" refers to the information that describes how a part is to be machined, including such things as tool type, shape, and size; Z level; and tool offset for each motion.

"Sequence" specifies when events will occur during the machine process. For example, it might indicate motion, a feed or speed change, or a switch from milling to turning mode.

"Geometry" defines where machine motion will occur. With the modeling concept, geometry implies tool path. The user defines tool path at the same time geometry is defined. The difference between geometry that represents tool motion and geometry that represents clamps, fixture, or rough stock is in their properties. Each point, line, arc, curve or surface that has a tool property assigned to it represents some type of motion event on the machine tool. A line is a linear move or cut; an arc is a circular motion. Geometry that doesn't have tool properties, but instead has layer properties for example, is treated as clamps, fixtures, and so on.

Properties can be assigned to any geometric element as a group or individually, at any time. As the user builds his graphic model, he creates geometry, assigns properties and manipulates sequences dynamically. Interaction and changes can be made as the process continues. The part program is being constructed, changed, and modified as the process takes place.

Source: From "Model the Machining Process to Generate NC Part Programs," July 1990, *Modern Machine Shop*, pp. 166–168.

The number of times a tape needs verification depends on the programmer's experience. On average, each first run of NC data requires three tryouts before a tape is ready for production. Obviously, tape tryouts represent unnecessary cost. U.S. industries spend billions of dollars each year to prove (verify) NC data. Software-based verification attempts to reduce this cost. Modern NC program verification software works on both CAD/CAM workstations and PC-based systems.

In a recent study by Wean Limited, Flavell found 19 types of common errors in an NC program, eight of them due to the part programmer. These are: dimension error, prep code error, setup (programming, shop), function error, excess stock, routing change, speed and feed rates, and link error (postprocessor). The dimension errors and programming technique problems are more common.

An NC program free of programming error will not necessarily be able to machine the part; the program may have nonprogramming errors. Such errors in part programs can have a detrimental effect on the machine and/or fixture. NC programmers and machine operators are always apprehensive when trying a new part program, since programming errors can easily slip into the program.

The three ways of verifying a part program are:

1. manual verification
2. verification on the machine itself
3. simulation

Manual verification involves proofreading the source program and the post-processed listing of MCD. This method is more effective when done by someone other than the programmer. It is inefficient and should be backed up with tape try-outs on the machine.

Verification on the machine involves cutting in air or cutting softer inexpensive material such as wax, wood, or proofboard. As a dry run, this method can reveal programming errors provided few tool and workpiece motions occur simultaneously. A major drawback of this method is the loss of machine time during verification. Other negative factors are the risk to tools and machines during verification, and the need to retry if the program undergoes major changes. Sometimes a table-top model of the machine is used for tryouts. In such cases, however, the proofing material may not reveal the problems associated with actual machining involving higher feed and speed, chatter, cutting forces, and so on. Nevertheless, the scaled model approach attempts to imitate real-world conditions.

Simulation-based program verification is becoming popular. It relies on a model of the machine, part, tool, and associated motions on a computer screen as an image. Its advantages are efficiency, freeing the machine for production, and the possibility of program optimization. The use of color, shading, and other image controls such as zoom and rotation for better viewing are possible. On the other hand, it is more difficult to comprehend, especially if the program is sophisticated, and dimensional integrity may not be as accurate.

The graphic representation for simulation may be a wireframe model, shaded image, or true solid model. The computer model is attractive for CIM since it can be linked with the design phase. In other words, designers can develop the part program and its verification as an integral task of design. Interactive simulation of tool paths, especially with solid models, is becoming a standard feature. New software allows users to interrupt the simulation and request dimensional and other information. Simulation of material loading and unloading using robotic arms and inspection with touch probes can also be combined with the verification of relative tool-workpiece motions. Use of expert systems is also possible in this area.

Parametric Programming

Parametric programming is a technique that automatically modifies existing part programs to develop programs for similar parts. Depending on the control manufacturer, the feature may be called **custom macro**, **user macro** or **Qroutine programming**. Some typical applications of parametric programming are in manufacturing various sizes of piston rings, circular cams, or sockets. This technique is best when parts belong to a family so that changes required in developing the new program are manageable. Thus, parametric programming represents an extension of the GT concept to the CAM area.

Parametric programming applications fall into three categories.

Family of Parts.   Parametric programming is most effective when the parts belong to the same machining family. Changes from one part to another are simple with the parametric programming technique. Features such as bolt holes, pocketing, or repetitive machining patterns are most compatible with the parametric concept.

Complex Part Shapes.   Even when powerful CAM systems are available as programming aids, the parametric approach saves time in complex geometrical shapes. Wherever programmers can identify similarity within the geometry, the technique can help yield shorter and more efficient programs.

Driving Machine Options.   The application of parametric programming to influence the driving motions using probes, in-process or post-process gaging, and so on, is complex and is probably best handled by the machine tool builders or the manufacturer of option devices.

Parametric programming is possible with modern MCUs having built-in computer language capability such as BASIC or C. This allows handling of statement labels, looping (the DO statement), branching—both unconditional (GOTO) and conditional (IF statement)—arithmetic calculations, and other features normally associated with a language. According to Lynch (1991), "Parametric programming may be among the best kept secrets of CNC programming, for only a small percentage of CNC users know of its existence" (p. 156).

TRENDS

❑   An increasing number of PC-based application programs for various preproduction tasks such as process planning are becoming available in the manufacturing software market.

❑   Most application programs for the planning phase of production are suitable for the DOS environment. Programs for UNIX platforms are being developed slowly.

❑   Cost-estimating packages have almost matured as a product.

❑ A strong trend to computer-integrate the various tasks of production planning and control using distributed PCs has set in.

❑ Recent developments in CAPP, especially the use of expert systems, have begun to build an effective bridge between CAD and CAM.

SUMMARY

Following the product design, an important activity in discrete parts manufacturing is production planning and control. This task also includes process planning and development of route sheets. Chapter 7 discussed production planning and control issues relating to CIM. The importance of computer-aided cost-estimating has been emphasized. NC part programming was also covered in this chapter, since it is basically a preproduction activity. The contributions of computer-assisted part programming to CIM were highlighted. The roles of simulation and group technology (GT) in the planning phase also were discussed.

KEY TERMS

C-post
Cellular manufacturing
Computer-aided cost-estimating (CACE)
Custom macro
Cutter location file (CL file)
G codes
G-post
Machine control data (also called G code) (MCD)
Management information system (MIS)
Parametric programming
Part programming
Plant-floor monitoring and control (PFM&C)
Postprocessor
Process planning
Qroutine programming
Route sheet
Scanning
Shop-floor control (SFC)
Tape tryout (tto)
Tape tryout (tty)
User macro

EXERCISES

Note: Exercises marked * are projects.

7.1   Explain why the use of computer systems in cost-estimating is especially useful for companies that produce to order.

7.2   Discuss the statement: MRP II is a major building block of CIM.

7.3   Name the methods for developing part programs. Compare the advantages and limitations of each in a table. Which one is the best for CIM and why?

7.4   Discuss the difficulties that may arise when product designers are entrusted with the task of developing part programs. How will such an approach affect CIM?

7.5   Compare C-posts with G-posts on the basis of five criteria selected in terms of importance.

7.6*   List the application programs available in your department, school, or organization in the area of production planning and control.

7.7*   Compare five of the major programs (all, if less than five) in the above exercise on the basis of hardware requirements, cost, vendor, and their strengths from a CIM viewpoint.

7.8*   Elect someone from within the class to learn more about one of the programs in the above exercise, and ask that individual to give a 30-minute talk to the class. Question the speaker on how well the program can integrate with other application programs of a discrete manufacturing enterprise.

7.9*   Visit a local manufacturing facility to study the extent of production planning and control computerization there. Discuss with appropriate personnel how the company might be planning to broaden the current practices to encompass CIM.

7.10   Circle T for true or F for false or fill in the blanks.

    a.   Cost-estimating is necessary only in companies that produce to order.   T/F

    b.   The usual timescale for the master production schedule is one year.   T/F

    c.   MRP II denotes the second-generation software for materials requirement planning.   T/F

    d.   The group technology concept evolved in the 1950s.   T/F

    e.   "Cutting in air" is a simulation technique.   T/F

    f.   CAPP packages are of two types: variant and _____

    g.   Name five common errors in NC part programs that are due to the part programmer.

    h.   Name three distinct advantages of simulation-based part program verification.

    i.   Name two major limitations of simulation as a tool for solving manufacturing problems.

SUGGESTED READINGS

Books

Aft, L. S. (1987). *Production and inventory control*. New York: Harcourt Brace Jovanovich.

Bedworth, D. D., & Bailey, J. E. (1982). *Integrated production control systems: Management, analysis, design*. New York: Wiley.

Buffa, E. S., & Miller, J. G. (1979). *Production-inventory systems: Planning and control*. Richard D. Irwin.

Chang, T., & Wyst, R. (1985). *An introduction to automated process planning systems*. New York: Prentice-Hall.

Greene, J. H. (Ed.). (1987). *Production and inventory control handbook*. New York: McGraw-Hill.

Nolen, J. A. (1989). *Computer-automated process planning for world-class manufacturing*. New York: Marcel Dekker.

Riggs, J. L. (1992). *Production systems: Planning, analysis, and control*. Prospect Heights, IL: Waveland.

Monographs/Reports

Klein, L. R. (Project Leader). (1989). *Integrating process planning and production*. CASA/SME Technical Council. SME Blue Book series. Dearborn, MI: SME.

Articles

Coleman, J. R. (1990, April). PCs make it with MEs. *Manufacturing Engineering*, pp. 52–57.

Hawkins, R. C. (1989, Spring). The future of computer simulation in the year 2000. *Journal of Industrial Technology*, 5(2), 24–25.

Martin, J. M. (1989, February). A strategy for NC programming. *Manufacturing Engineering*, 102(2), 82–84.

Noaker, P. M. (1989, July). The demassification solution. *Production*, pp. 72–74.

Nolen, J. A. (1989, May). Capitalizing on CAPP. *Manufacturing Engineering*, pp. 70–73.

Propen, M. (1990, July). Grappling with group technology. *Manufacturing Engineering*, pp. 80–82.

CHAPTER 8
Production

"The modern turning center is to the engine lathe what the F-16 is to the camel."
—*Manufacturing Engineering*, June 1989

"No precise definitions exist for robots or FMSs."
—A careful reader

"As long as shops are willing to tolerate a proliferation of postprocessors and put up with NC programs that can't be exchanged, they haven't fully embraced the key concepts of CIM, at least not on the shop floor where it really counts."
—Mark Albert, *Modern Machine Shop*, July 1990

"When does a cell become a system? Ask 10 people, get 10 definitions."
—*Production*, February 1990

"With 1 million in Japan and 3 million in the U.S., per capita population of machine tools in the U.S. is fifty per cent more than in Japan."
—S. Kant Vajpayee, University of Southern Mississippi

"BCL is a 'CIM glue' that bonds machine tools, NC programming, DNC (distributed numerical control), CAD and CAM into a tight, cohesive, and comprehensive system. BCL makes everything stick together as part of the overall CIM strategy."
—Mark Albert, *Modern Machine Shop*, July 1990

"But now 32-bit control technology has thrown the door wide open. We've got machines today that are miles ahead of the machines of just a few years ago."
—Jim Berger, Turning Products Manager, Mazak Corp., *Production*, November 1989

"Almost half of the 50,000 robots being used in Japan are in the automotive industry. Japan also has about 300 FMSs."
—A 1987 survey by Japan's MITI (Ministry of International Trade and Industry)

This chapter discusses CIM-related issues of production, such as forming, machining, assembly, and inspection. Short-term problems arising from a lack of coordination among production functions are covered in the next chapter, which deals with shop-floor control.

8.1   BASIC PROCESSES

Depending on the sophistication of products, a variety of processes manufacture them. The level of sophistication depends on factors such as product volume, lot size, workpiece material, physical shape and size, and desired tolerances and surface finish. Newer materials, too large or too small dimensions, closer tolerances, and finer finishes may require unconventional processes such as electric discharge machining (EDM), laser machining, or water-jet machining. In general, discrete-manufacturing processes fall into four categories: forming, machining, assembly, and auxiliary operations such as welding, washing, or heat treatment.

Forming

Forming is any process that shapes raw material into the product. Shaping may be under mechanical force, as in presses, or under gravity, as in drop hammers. The term forming pertains to both metallic and nonmetallic materials such as plastics. With plastics widely replacing metal, plastics forming has become a specialized discipline. Most of the principles discussed in this section apply both to metals and nonmetals.

Metal forming involves transformation of the raw stock, usually called a blank or workpiece, into a useful part by deforming it plastically. A workpiece can also be deformed with mechanical pressure between it and the production tool (often called a die).

New developments in forming usually lag behind those in machining, primarily because the forming machine tools market is smaller. In recent years, though, flexible automation technologies, proven effective with metal cutting machine tools, have been immediately adapted to forming systems. Several techniques targeted at improving forming productivity have also been developed. One such development is the **quick die change** (**QDC**), which reduces setup time. Box 8.1 describes how die change time is reduced through a set of QDC methods.

In modern plants, forming machines are arranged as cells or systems for a family of parts, based on the GT technique. A metal-forming cell is defined as two or more metal-forming machines—or one machine combining the functions of several automatic material handling units—and a control system. The control system consists of individual machine or equipment control devices and a cell controller or computer. Cell-based forming facilities are more suited to CIM's integration needs.

Machining

Component machining is often a major activity in discrete-parts manufacturing. Prior to assembling the end product, formed or raw parts are machined at strategic locations according to the functional requirements of tolerance and finish.

☐ **BOX 8.1**	*Changing Dies Quickly*

Die-change aids are as old as presses and dies. Over time these aids became increasingly sophisticated and made real advances toward shorter changeover times for presses and presslines. Despite die-change aids, changeover times—even recently—often took hours. Only with development of large-panel transfer presses and modern control technology was it possible to reduce them to less than 10 minutes.

Quick die change (QDC) is more than the right hardware. Properly done, it is a finely tuned combination of hardware, personnel, and timing. With hydraulic or mechanical clamps, die lifters, and other devices to simplify die change, setup

time is shorter, and productivity will increase.

Once the die is positioned, you need power for clamping. Using clamps allows boltless die changing. With no studs, T-bolts, shims, backup blocks, or nuts, time and labor savings are substantial.

QDC in Automaking.   Changing large dies used to take the auto companies an entire shift, and sometimes more. "The aim at Wayne (a stamping plant) was a five-minute die set versus a current average of four to eight hours," notes Jerry Berendt of Ford. QDC is the trend also in Europe and Japan.

Source: From "Changing Dies Quickly" by K. H. Miska, April 1990, *Manufacturing Engineering*, pp. 45–49.

Machine Tools and Machining Centers.   **Machine tools** are the basic equipment in discrete-parts production. A machine tool is a machine that can produce itself with the help of an operator. Using a lathe, it is possible to make another lathe; hence lathes are machine tools. A sewing machine, on the other hand, cannot produce another sewing machine; it is therefore a machine, not a machine tool.

The milling machine, grinder, shaper, and gear hobbing machine are some of the other machine tools. Some machine tools, such as lathes, are designed to be general-purpose while others are geared for specific uses, such as gear hobbing machines for machining gears. Conventional machine tools such as lathes and milling machines that carry out specific machining operations are becoming obsolete. The trend is toward machining and turning centers that combine several operations in one machine tool.

A **machining center** is a multifunction machine tool that performs milling, drilling, tapping, boring, and other similar operations. It also offers automation options such as an automatic tool changer (ATC) and pallet changer. It is capable of performing, in one setup, several operations on a variety of parts. This saves the time and cost of setting up on different machines and of transporting workpieces from one machine to another. **Turning centers** are machining centers with turning as the primary operation. Boxes 8.2 and 8.3 describe recent advances in turning centers.

Depending on the orientation of the main axis of rotation, a machining center can be vertical or horizontal. In the United States, vertical machining centers are more popular since their setup and the associated tool and part handlings are usu-

---

BOX 8.2   *What Can Possibly Be New Here?*

If you're talking about lathes—or turning—this is probably a valid question. Let's face it, as technology goes, this one's about as old as they get. But don't let that fool you. Today, a lathe by any other name, indeed, may be something entirely different.

Yet since the early 1980s, the lathe's modern descendent, the turning center, has been appearing in a baffling array of new guises. One view of what's been happening centers around two such innovations: multi-processing and 32-bit control.

Multi-processing means the ability to turn, mill, drill, and perform other operations in a single turning center in one setup. The benefits of such flexibility are significant. Parts are produced more quickly; errors resulting from repeated setups are eliminated; material handling, with its inherent expense, consumption of time, risk of part damage, and necessity for part orientation, is reduced; in-process inventory and the related floor space requirements are lessened; labor content diminishes; and scheduling is simplified.

The 32-bit CNC entered the U.S. market in 1987 when Mazak Corp. (Florence, KY) introduced its M-32 control for machining centers. The first 32-bit controller specifically for turning applications was introduced the following year. Why 32-bit? Don't the old CNCs work just fine? Yes, of course. But 32-bit CNC technology produces a greater amount of information faster, and this opens the door to increased machine speed and precision, the ability to monitor and to respond to cutting conditions and other functions, communication with cell or system controllers during metal cutting, simplified programming, built-in databases that decrease the machining know-how required of an operator, as well as other benefits.

Source: From "What Can Possibly Be New Here?" November 1989, *Production*, pp. 50–54.

BOX 8.3   *Turning Centers: A Revolution in Manufacturing*

In what has been called the second industrial revolution, producers of CNC turning centers are responding rapidly to industry's incessant calls for ever greater flexibility, productivity, and quality of product. As a result the classic distinction between milling machines and the turning machines is becoming blurred.

Today, many CNC turning centers are equipped with a programmable C axis that can stop the rotation of the spindle at any time and allow power-driven tools to approach the work-piece from all directions to mill, drill, tap, bore, and perform other milling-type operations.

In addition, turning centers now come equipped with automatic parts loading, tool changing, and chuck changing. They are also able to inspect work in progress and can automatically examine tools for wear and signs of breakage.

The ultimate dream: all parts made in a single setup, all secondary operations eliminated.

What triggered this significant change in manufacturing? For one thing, the all-pervasive demand for quality. For another, the impact of just-in-time (JIT) delivery, as well as the disappearance of skilled machinists who used to know how to coax high performance out of antiquated machinery.

There is another factor. Many companies are now outsourcing their parts. They don't want the overhead or the labor and supervision problems, nor do they want to stock raw material and finished inventory.

In choosing a parts supplier, the first thing they ask for is a facilities list. That tells them what kind of machines the vendor has and what kind of quality and production can be expected.

Bob Siewert predicts major changes in the turning machine business, and indeed, in the entire machine tool industry. For one thing, he sees a drastic reduction in the amount of metal being used, with a swing instead toward plastics and composites. He points out that there are many ways other than machining to shape parts made of these materials. Moreover, says Siewert, near net-shape technology is becoming an important factor in manufacturing. He points to investment casting as an example, noting that they are being held to tolerances unheard of five years ago.

"We're now removing about 10% of the metal we used to," he says. "It all adds up to the fact that we're going to see fewer metal parts in the shop, and the ones that are there will need less metal removed."

Siewert sees tomorrow's turning machines operating at speeds up to 50% faster than today's, the result of the demand for reduced cycle times and finer finishes. In addition, he says, turning machines will be using more automation, following the basic designs that are already in place.

Machines will be equipped with automatic parts loading, in the form of gantries and robots, and with a large number of tools as a means of reducing setup time and providing increased machining flexibility. Automatic tool changers and interchangeable turrets will be commonplace in shops everywhere.

[The setup] has a central lubrication system which automatically lubricates the slides after 60 feet of X and Y-axis travel. It also features a 20-HP A-C spindle and can achieve 0–6000 RPM in only three seconds.

Tool data are measured by simply bringing a tool tip into contact with the tool eye sensor. Measurement and compensation of tool wear and inspection for broken inserts are automatically performed, permitting accurate machining during periods of unmanned operation. The tool eye system is also used for tool setting. Only a few seconds are required to measure and register data for each tool.

Horizontal or vertical, high speed or low, the revolution in turning continues, with all players headed for these common goals: unattended machining, zero defects, total quality—in short, perfection in manufacturing.

Source: From "Turning Centers: A Revolution in Manufacturing" by R. Eade, June 1989, *Manufacturing Engineering*, pp. 75–79.

ally easier. Horizontal machines, on the other hand, facilitate the fall and easier removal of chips, leading to a faster metal removal rate, better surface finish, and improved tool life. A vertical machining center of comparable power, capacity, and accuracy is less expensive than a horizontal or universal machine. Normally, workpiece shapes dictate the type of preferred axis orientation. If the workpiece is chunky or prismatic, for example, a horizontal machine is preferable. For workpieces that are relatively flat and long compared to their height, a vertical machine is suitable.

A machining center is called universal if it can provide spindle rotation along both vertical and horizontal axes. In one recent design, **universality** was achieved by combining the horizontal spindle with the vertical spindle in such a way that both operate from the same common drive.

Modern machining centers are the "keystones" of FMCs and FMSs. As stand-alone equipment, machining centers are used about half the time. In a cell, the same machine is used more than 80% of the available time, by automating pallet changing and tool management functions.

Recent Advances in Machine Tools.   All modern machine tools are incorporated with a built-in computer called a machine control unit (MCU). The MCU is a machine tool's interface with the users as well as with other equipment. The decreasing cost of microprocessors enables enhanced computing and communications capabilities as an integral part of the machine tool. This is obviously desirable for CIM.

Newer machine tools offer the following features:

1. Known as machining centers or turning centers, they are multifunctional. The trend is to equip a single machine with capabilities for as many different operations as possible (see Boxes 8.2 and 8.3).
2. To facilitate material handling, their working heights are about the same even if their "footprints" differ.
3. The design is continually improving to exploit the higher cutting speeds possible with newer cutting tool materials.
4. Both horsepower and maximum speed available at the spindle have been increasing over the years. Some machines provide as high as 60 HP and 50,000 RPM.
5. Structures are made of materials such as ceramics to minimize frame distortion due to heat generated by machining.
6. Dynamic characteristics of the machine tools can be stored in the design database so that process planning can be based on a machine's three-dimensional accuracy, dynamic rigidities, and chatter proneness.
7. Machine control units are based on 32-bit processors (see Box 8.2).
8. **Mean-time-between-failures** (**MTBF**) is as high as 14,000 hours. Based on an eight-hour shift and 250 working days per year, this means seven years of life with no failure!
9. Newer machine tools come "geared" with a host of sensors that watch their "health" and performance.
10. Their design takes into account integration potential, an important feature for CIM.
11. Some new machine tools are mill-cum-turn centers (see Box 8.4) so that most of the processing can be completed in one setup.
12. Their smaller footprints save floor space.
13. Machining accuracies are as high as ±12 microinch per full stroke axis with repeatability as high as ±40 microinches. The average NC machine tool has a resolution of one micron (0.001 mm), about 0.00004 inch.
14. Part tolerances under 100 microinches are possible.

Better accuracy on modern machine tools is possible using sensors and the associated electronics for thermal correction and feedback control, which compensate

---

BOX 8.4   *Driven Tools Turn on Turning Centers*

Long runs, once the rule, are becoming the exception. Orders can be spread over 10 to 20 partial deliveries on short notice. Economical production of such batch sizes requires complete machining in a single setup. Sounds like machining center work, doesn't it?

An alternative may be a new generation of turning machines equipped with powered tool spindles in their turrets. These combination turning/milling machines are ideal for small to medium-sized workpieces that require cylindrical as well as prismatic machining.

"You can now get a small machining center and a medium-size turning center in one package," says James Bushong of Hitachi Seiki USA. "Instead of transferring parts from a turning machine to a machining center, with possible queuing of parts and tolerance buildup, a manufacturing engineer can do all the processing on one machine. You can also save on fixturing. Often irregularly shaped castings or forgings can be held in a three-jaw chuck. The part is chucked once and not released until the job is finished. The point is that once you release a casting, it's hard to align it again."

Advantages and Limitations. When a turning center equipped with driven tools engages the C axis, a part can be rotated to maintain prismatic

feature-to-feature accuracies not possible when refixturing parts for subsequent milling and drilling operations. This reduces the risk of rejects, and production quality goes up at no additional cost.

There's also economics here. Usually the cycle times for drilling and milling operations after turning are so short that the cost of reloading the parts onto a machining center is prohibitive. When milling exceeds 40 to 50% of lathe processing time, it may call for a secondary machine tool.

There is a tendency to associate CNC turning centers with relatively small, large-volume parts. US auto-makers, however, are using turning centers with driven tools to make aluminum alloy wheels ranging from 13 to 17 inches in diameter.

According to Bill Dunsmore, "With JIT production, quantities are decreasing. Stockpiling is out. If you're not going to do large runs, cell systems really aren't efficient because you must set up three or four machines to make 50 to 1000 parts, which may vary depending on the part's complexity. That's why we are using a turning center with driven tools."

Source: From "Driven Tools Turn on Turning Centers" by K. H. Miska, May 1990, *Manufacturing Engineering*, pp. 63–66.

---

for the machine's mechanical errors. Some describe this error compensation capability as "electronic pills for mechanical ills."

Recent improvements in the design of machining centers have helped speed up machining. For example, polymer composite materials are used for the machine base to achieve higher rigidity and superior vibration-damping characteristics.

Cutting Data.   In any machining operation, cutting parameters must be set at appropriate values to achieve optimum results on cost, production rate, profit, and more. Over the years, these values have been determined by conducting actual tests on a variety of workpiece materials and tool materials.

Machining Data Handbook from Metcut Research Associates is a useful resource for data on machining parameters such as speed, feed, and tool geometry

---

BOX 8.5   *PC-Based Machining Recommendations Software Package*

CutData is a PC-based software package for planning machining operations, determining machine tool requirements, and estimating part production time. The software is based on Metcut's *Machining Data Handbook* and includes over 84,000 recommendations for 40 machining and grinding operations on more than 1500 different materials. Custom data and experience from the user's own shop also can be captured and added to the database.

The software calculates cutting time per pass, material removal rates, and horsepower requirements. It also interpolates data for various depths of cut and converts units. The platform required is a PC with at least 512K of memory.

Source: From "PC-Based Machining Recommendations Software Package," May 1989, *Modern Machine Shop*, p. 198.

---

for a given workpiece material. To eliminate the task of finding such data and keyboarding it into a computer, data for 40 common operations on more than 1,500 workpiece materials is available on a floppy disk (see Box 8.5). This makes cutting-data selection a matter of keystrokes rather than a search through the pages of the source. Database-resident cutting data obviously are more suited to CIM.

8.2   NC, CNC, AND DNC

The term NC (numerical control) denotes the technology originally developed at the Massachusetts Institute of Technology for building the first prototype of an NC machine tool. Modern machine tools are CNC (computer numerical control), which can operate in **DNC** (**direct or distributed numerical control**) mode. A CNC machine tool has a built-in computer to support its various functions, such as determining the angular orientation of a hole when x and y coordinates are given. It behaves as a DNC system when it can communicate with other systems inside and outside of the plant. While NC as a basic technology continues to prevail both in CNC and DNC systems, "plain" NC machine tools are almost obsolete.

New CNC machine tools are generally expensive. As an alternative, sometimes it is possible to retrofit a good quality, used standard machine tool with CNC capabilities (see Box 8.6).

During the last decade, CNC technology has improved tremendously. The clearer graphics, for example (see Box 8.7), enable tool path and process simulation prior to actual machining. With the incorporation of 32-bit controls, modern CNC systems handle large amounts of information faster. Some of the benefits of 32-bit controls in CNC are in the areas of sculptured shapes, sophisticated internal commands, faster editing and conversational programming, new machine features, and more (see Box 8.8).

BOX 8.6   *Take Control with CNC*

Brain transplants are science fiction for humans but engineering reality for machine tools. The elective surgery—called CNC retrofitting—can transform a brain-dead iron hulk into one smart investment.

It's worth considering when a machine tool is in fairly good mechanical condition but the control is old and obsolete. Even a five-year old controller may inhibit a machine from performing at full potential.

The Three Rs. [There are] three ways to breathe life back into an aging machine tool: remanufacture, rebuild, or retrofit. Remanufacturing costs the most. It involves making an old machine better than new by adding capabilities such as probing and torque-controlled machining, safety features not available when the machine was built, and a tool changer. Rebuilding entails repairing or replacing worn or damaged components or remachining them to bring the machine back to original specs. Retrofitting, the focus here, is the least expensive. It involves adding a new control system and possibly replacing existing hydraulics with a new electric drive system.

A CNC retrofit virtually ensures increased single-run time, primarily because the new controller is more reliable. In addition, new controls can support faster spindle and axis drives with quicker acceleration/deceleration, tool changing, table indexing, and pallet shutting. No time is lost waiting for the control to catch up to the process.

"Today's 16- and 32-bit CNCs run rings around their predecessors, especially in part programming, program manipulation, and machine opera-

tion," says Dobson of DynaPath Systems Inc. "For example, new controls are almost transparent to the programmer. Years ago there was a clear distinction between manual and computer-assisted part programming. Today that's not so. When a CNC programmer manually inputs part print information to define pockets or angle intersections, that programmer is relying on the controller to do the complex trigonometric functions and translate the solutions into machine operation instructional codes."

Dobson notes that new CNCs also have more memory. This, coupled with features such as canned and customizable macros, allows pulling together blocks of information from existing programs to form new ones, which is ideal in family-of-parts situations where you repeat certain operations such as bolt hole patterns. Armed with RS-232 and RS-422 ports, modern CNCs also can link a machine tool to hierarchical controls in LANs or full-scale CAM systems.

No Electronic Pills for Mechanical Ills. Although CNC retrofitting can rejuvenate aging equipment—possibly even compensating for some machine sloppiness such as backlash and lead-screw pitch error—it can't do much for badly abused machines.

Better machine utilization should result from using RS-494 (also called Binary Cutter Location, or BCL), which provides program transportability among similar machines.

Source: From "Take Control with CNC" by J. R. Coleman, May 1990, *Manufacturing Engineering*, pp. 35–40.

Selecting an appropriate CNC system is no easy task; it involves a number of factors. Even if one buys the same brand of CNC, it may not be program-compatible with an older model in the plant. In such a situation, according to Herrin (1990), users have three options:

	BOX 8.7	*Today's CNC Graphics*

Much of the graphics capability that has made personal computers so desirable is now being provided in many of today's CNCs. Graphics is important because so much information flows to us through our eyes in the form of images. There is little doubt that the human mind has the ability to extract certain information from visual images easier than from text; an example is the dimensioned drawing of a part to be manufactured.

Graphics images in a computer system are created by defining an array of dots on the face of the CRT (screen). Each dot has an addressable X and Y location and may be defined by color as well as shade in most systems. The number of X and Y definable positions determines the resolution of the system. The proper array of dots specified by the computer form an image. Each dot is called a picture element (pixel for short). The more pixels the better the image definition.

The graphics capability of PCs is specified by the names of the various graphics adapters such as CGA, EGA, VGA and so on. The type of graphics adapter does not necessarily indicate what capability a specific graphics software package may provide like animate, rotate, or scale, but it does indicate what the display capability is. The following is a definition of the more common graphics adapters used in PCs.

	Resolution	Colors
CGA (color graphics array)	320 × 200	4
EGA (enhanced graphics array)	640 × 350	16
VGA (video graphics array)	640 × 480	16
SVGA (super VGA)	800 × 600	256
8514	1024 × 768	256

Even though the widespread use of PC graphics has made the technology affordable in CNCs, it is difficult to make a direct comparison to one of the basic PC graphics arrays since most CNCs use proprietary solutions.

Source: From "Today's CNC Graphics" by G. E. Herrin, July 1990, *Modern Machine Shop*, pp. 124–128.

---

1. Prepare and run only new part programs on the new machine.
2. Reprogram or repostprocess existing part programs.
3. Purchase an emulation package to convert existing part programs.

Herrin further adds, "Emulation, if performed in the CNC, may circumvent the costly part recertification process. It is important to note here that in addition to emulation, BCL and ACL inputs are alternative solutions that allow a common part program to be run across a variety of machine/control configurations."

Thus, the wrong CNC can deter integration if the buyer is not careful about its compatibility with the existing systems. Box 8.9 highlights the major issues relating to CNC selection; note the questions that must be answered to avoid a poor selection.

BOX 8.8   *The Advantages of 32-bit Controls*

The newest generation of CNC controls being manufactured today are standardizing on a 32-bit-based microprocessor. While there is a dramatic processing advantage to the 32-bit control (over the older 16-bit control), there also have been over-enthusiastic claims as to the benefits this new control type will reap for the end user. The speed of the CPU in the control, of course, will directly affect the internal processing time for all commands being executed.

If you have experience with personal computers, you can easily draw an analogy to the processing time savings from a 286-based computer to a 386-based computer. The 386-based computer will outperform the 286 in all processing and have no need for the faster processing speed. The same thing goes for the new 32-bit CNC controls over the 16-bit control.

The areas where the 32-bit control will decrease processing time and increase productivity are the following.

Sculptured Shapes.   The most beneficial feature of the 32-bit control is realized when running programs (possibly generated from a CAM system) that are made up of a series of very small axis movements. An example of this would be a program to machine a 3D shape in the form of an aircraft airfoil. When actually running the part, a 16-bit control would be limited to a given feed rate since it would "bog down" and not be able to process the program commands quickly enough. This feed rate limitation would also be directly related to the size of the axis movements. The 32-bit control would easily allow the machining feed rate to be doubled in comparison to the 16-bit control, and in some instances, could actually perform as much as 3 to 5 times as fast.

Complex "Internal" Commands.   Depending on the machine tool builder's interface, certain commands have always been a thorn in the side of control execution time. Things like tool changes, spindle range change, table indexes, and pallet changes have always taken more time to execute than machine tools builders would like. The 32-bit control's faster execution time will reduce the processing time to less than half of the 16-bit predecessor's. This is NOT to imply that these commands will be reduced to less than half of what they were. We are saying only that the processing time will be reduced. Generally speaking, this will reduce the execution time of any one "miscellaneous" function by about 10 percent.

Faster Editing.   Editing programs on a 32-bit control will be much faster. Searching functions especially will be improved. Generally speaking, all editing functions will execute about 30 percent faster on a 32-bit control.

Faster Conversational Programming.   For those of you who are interested in conversational controls, there is a definite benefit to a 32-bit processor for this type of control. Conversational controls require a great deal of "executing time," especially while the control is creating the CNC program used to machine the workpiece. On a 16-bit conversational control, it can be quite frustrating to wait while the control is "thinking" about what it is going to do next. The 32-bit control will decrease this waiting time by about 40 percent.

New Machine Features.   Due to the faster "sweep time" of the 32-bit control, many machine-tool builders are incorporating features not feasible on a 16-bit control. Adaptive control, artificial intelligence, and improved in-process gaging systems are among the new features available with the 32-bit control.

As discussed, there are distinct advantages to the new 32-bit controls. However, for general purpose machining, its advantages are somewhat limited.

Source: From "The Advantages of 32-bit Controls" by M. Lynch, June 1990, *Modern Machine Shop*, pp. 126–30.

◻ **BOX 8.9**   *Picking a CNC*

A CNC (computerized numerical control) is a combination of hardware and software that comprises the electronic brain of the machine tool. It performs a number of functions to accomplish this task, including the classical one of motion control and the relatively recent function of controlling the input/output (I/O) ports. The CNC also functions as the man/machine interface for control of the machine tool as well as a communications partner to other host computers within the factory network.

CNCs have changed in the last 10 years primarily because of the reduced cost and increased power of the available hardware. These developments have provided software developers with a "playground" to implement more and more functions into the controls. Their two main development thrusts have been to harness and direct this increased power and the software capability into the areas of part programming at the CNC screen and to increase distributed numerical control (DNC) links with a host computer.

The selection of a CNC begins with the specification of the NC machine tool. When you decide to purchase a piece of NC equipment, the machine tool features will typically be more important than the CNC product. Workpiece parameters—size, weight, number of machining operations, amount of material to be removed, and so forth—will dictate your choice of machine tool. Once a selection is made, users will then typically have a choice of two or three controls that the machine tool builder has standardized in its product line. CNCs outside the standard range can be generally accommodated as well, albeit at a premium price to reflect the unfamiliar engineering interface task facing the builder.

"Everything that belongs to the machine tool itself—such as motion control, tool changers, and adaptive control—should be controlled by the local CNC." It is important that the CNC have an open architecture and a reasonably sophisticated communication link. The CNC has to have all the data available for a host—such as a cell controller—on a cost-effective interface. Schaffer (Robert Bosch Corp.) has an equally simple formula for determining whether or not the CNC has, in fact, an open architecture suitable for further integration within a manufacturing network. The question to ask, he says, is this: "Can the DNC link-access bidirectionally all the data stored in the RAM (random access memory) of the CNC and the PLC? If the answer is yes, you are in good shape."

A number of specific questions must be asked and answered to determine network compatibility of the CNC:

Can the host or cell controller access diagnostic and machine status data in the CNC?

In addition to part program and tool data transmission, can the host transmit to the CNC commands such as Select Program B and Station Operation?

Can an actively running program, calling a gaging cycle, communicate with the host in on-line mode and transmit the measuring results to the host?

Can the host, via the DNC link, access PLC data?

Can the operator panel of the CNC be used in a terminal emulation mode?

Is the CNC capable of sending back vital statistical information for efficiency evaluation and report purposes?

Today's reality is that the DNC is no longer just a tool for downloading part programs to the CNC. Rather, it is a vehicle for implementing various degrees of remote control over the CNC to facilitate unattended operation of a machine tool or manufacturing cell. With that new functional responsibility, the CNC's ability to go beyond its original charge—automatically controlling the machine tool—to collect, store, send, and receive data is critical. Bosch, in fact, calls its CNCs "computer controls" (CCs) because they control much more than motion, according to Schaffer. Indeed, with this enhanced functionality, the CNC itself operates as a mininetwork of its own. For example, it divides some of its responsibility with a resident PLC, first combined with the CNC at the beginning of this decade to replace the relay logic control. In general, the PLC's responsibilities will include everything that is controlled in a digital manner through the digital I/Os. The CNC will handle the servo controls and, of course, the user interface. Since the PLC may be handling things like tool changes and the pallet shuttle moves, which require information resident in the CNC, the window between the two must be wide open.

Source: From "Picking a CNC" by J. M. Martin, May 1989, *Manufacturing Engineering*, pp. 59–62.

DNC

Over the years, the term DNC has been an acronym for direct numerical control as well as distributed numerical control.

Direct Numerical Control.   As an acronym for direct numerical control, DNC originally meant direct control of NC machines or processes by a mainframe (host) computer. The host computer is external to the machine tool and commands several machine tools as well as other NC equipment and devices. In this mode of operation, the instructions (e.g., part programs) are stored in the host's RAM/ROM or in its secondary memory and downloaded electronically as and when required by the machine's NC controller. Their interpretation and resulting actions are the responsibility of the NC device itself.

Distributed Numerical Control.   With the advent of microprocessors and advances in this technology, DNC's meaning has changed over the years. In modern usage, DNC is an acronym for distributed numerical control in which control is exercised by distributing the NC machining files and other data files to the networked devices. This context of DNC is more meaningful in a CIM environment. In distributed numerical control, the DNC is a conduit for collecting and transmitting data. In this broad sense, the term NC (last two letters of DNC) implies control using not only numeric data but also nonnumeric data, such as the letters of the alphabet.

Usually, the DNC operates in two modes, download and upload. In download or downstream mode, DNC sends a variety of manufacturing support documentation; process planning related data such as setup sheets, tool and gauge lists, inspection instructions, router; and other manufacturing information to the computer systems hierarchically below. In upload or upstream mode, DNC keeps the host system informed of work status at the subordinate levels. Shop-floor data can be collected three ways: (a) manually via keyboard entry, (b) semiautomatically via bar code and other devices, or (c) fully automated via discrete signal acquisition in a closed-loop system. A DNC configuration supports any of these or their combinations. In the broad sense, DNC is the on-site monitor, transmitter, and formatter of essential information for CIM.

Distributed numerical control is not just a mechanism for transporting part programs from the central program's library to the shop floor. It has several other uses. It is in fact a highly flexible factory floor network for communicating and managing a broad range of information both to and from the shop. Besides the part programs, shop-floor personnel can have instant access to setup instructions, current scheduling, routers, inspection instructions, and more. For upstream needs, the DNC network functions as an electronic highway for gathering shop-floor data. Some examples of such data are machine status, part run times, downtime, and quality problems. Best of all, DNC facilitates a modular approach to integration by enabling CIM development in stages.

Consider a typical need of manufacturing, that of labor reporting. In the early stages of computerization, this task was handled through punched cards handed to the time clerks. Later, magnetic strip badges replaced the cards. A DNC system can handle this task easily and thoroughly. For example, data entered into the DNC loop can reveal the exact time the operator was waiting to begin machining. Further analysis of data can show the cause of waiting, such as wrong fixturing, absence of an NC file, delay in delivery of work material, incomplete tooling, insufficient process-plan documentation, or a revised program with an outmoded fixture.

Automated discrete signal acquisition under DNC can reveal useful information such as power on, auto-mode, feedhold, or speed or feed **overrides** that indicate whether the feed is too high for the desired finish or the speed is too high for an economic tool life. Power-on data shows how often the machine was actually on, thus pointing out any bottleneck problem or capacity planning needs. Auto-mode denotes that the CNC machine was cutting material. Nonauto time may be due to manual cycling of tape prove-out. Thus, data collected under DNC can effectively be used as a tool for production management. Moreover, the (distributed) DNC can connect to a LAN or WAN (wide area network) to move data from the CAM domain to other functions within CIM.

Implementing a DNC involves the following three requirements:

1. The first requirement is a machine interface for connecting the CNC machine with the DNC network. This is achieved via standard CNC protocols such as RS-491, RS-494, or proprietary communication protocols compatible with the controller. The newer MCUs have communication capabilities requiring less sophisticated interfaces.
2. The second requirement is a terminal for each shop-floor operator or, where feasible, for a group of operators. In DNC environments, an active terminal allowing two-way data transfer is desirable. The terminal should be capable of collecting data at the station or stations if connected with sensors. In the download mode it would receive part programs, process planning data, and so forth. If necessary, a resident text editor to modify part programs is also desirable.
3. The third requirement is a link between the terminal and the host computer as an interface between corporate and shop-floor data. Through a LAN, the link can benefit other business functions—such as order entry, marketing and sales, cost and finance—that may need actual shop-floor data.

CNC Software

Various CNC software can be grouped into the following three categories: executive, application, and front-end.

Executifve Software.   Executive software, embedded in the CNC, provides the basic architecture and control features. The CNC builder must supply and maintain the executive software as a component of the CNC unit. Because this type of software is usually proprietary, the source code may not be available to the user for effecting changes. The CNC supplier usually makes changes as an upgrade or to tailor the program at the user's request. Depending on the number of other customers who may have a need for similar tailoring, the cost of customizing varies widely.

Application Software.   Application programs are provided by machine tool builders to match the machine with the CNC. They reside in a special partition of the CNC that acts like a program controller. Machine tool builders can also provide custom subroutines to carry out specific programming tasks. Basic functions commonly available are homing sequence, toolchange cycles, pallet shuttle, logic for various machine pushbuttons and light or limit switches, and so on. Recent additions allow generation of M codes, G codes, and mnemonic codes.

Front-end Software.   This software manages data transfer into and out of the CNC. Some typical tasks of front-end software are transferring part programs, collecting and analyzing status information, and locating incompatible communication protocols. Front-end software is usually outside the CNC; it may, for example, be in a connected PC. Front-end processors can be developed in-house by the user or customized by independent or third-party software houses or the machine tool builder. Powerful front-end software is essential for achieving CIM in the plant.

Most manufacturing companies, especially smaller ones, find the available off-the-shelf CNC software sufficient for their needs. When the available software cannot provide the desired performance, however, the alternatives are to develop one from scratch or to modify the bought software if legally allowed and technically cost-effective. Users with limited in-house computing expertise hire independent software firms to write customized CNC software. Sometimes, machine tool builders carry out the customization for a fee.

Binary Cutter Location (BCL)

EIA RS-274 has outlived its purpose in light of modern CNC control features. The practice of adding digits and decimal points to M and G words by CNC builders outside of RS-274 is not conducive to CIM. There is a need to achieve universal data exchange among NC machines in the form of a **neutral language**.

A practical approach to fulfilling this need for **part program compatibility** is through **binary cutter location** (BCL) input, which several manufacturers support. A strong supporter of BCL is Rock Island Arsenal, a defense manufacturer. Box 8.10 describes the efforts at this company and the benefits of BCL in interfacing most CIM requirements at the plant level.

▢ BOX 8.10   *Interfacing NC with CIM*

The Rock Island Arsenal presents a strong case for using BCL as an "NC interface language" that firmly bonds programming and machining into a CIM network. With BCL, NC programs truly become generic production process descriptions.

Most shops think of an NC program as the data required to make a certain machine tool cut a particular workpiece. Planners at the U.S. Army's Rock Island Arsenal Operations Directorate used to think that way too. But when they began to implement a CIM strategy a few years ago, that thinking clearly had to go. Now they think of each NC program as a "production process description that allows identical parts to be machined at any available piece of appropriate equipment."

They can think this way because that's what their NC programs have, in fact, become, at least for over 30 CNC machine tools now in operation. The Arsenal has been able to make this CIM-inspired transition by adopting a standard for NC programs that has been largely overlooked by the metalworking industry.

This standard is EIA 494, which is usually referred to as BCL. It became a national standard in August 1983, after years of effort by a handful of persistent user advocates. BCL is short for Binary Cutter Location, which comes from the EIA standard's title, "32-Bit Binary CL Exchange (BCL) Input Format for Numerically Controlled Machines."

BCL was originally conceived over 15 years ago as a way to bypass the traditional postprocessors required for NC programming and thus create exchangeable NC programs, that is, programs that could be used to machine a workpiece regardless of the make of machine tool or control unit. But planners at the Arsenal recognized that BCL offered more than that.

Planners at the Arsenal define CIM as the application of computer networks that share a single database to unify and streamline the manufacturing process from design engineering to final shipping.

Productivity (at Arsenal) hinges on flexible, efficient NC programming. This is why looking at fresh approaches to NC programming was a top priority when planning for CIM began in earnest in the mid 1980s. It was clear that BCL might very well provide the missing link for integrating NC in a CIM strategy. In October 1985, the Arsenal officially adopted BCL for new machines. The following February, the first specifications for machine tools that included BCL as a required program input format were issued. The new machines make Rock Island one of the largest users of this input format in the country to date, and the largest user among DoD (Department of Defense) operated plants.

Rethinking NC. "At the outset, we did not see BCL as simply a means of eliminating postprocessors," recalls Steve Harris. "Problems with postprocessors were just symptoms of a larger issue facing us—data communication across our whole organization was not as smooth-flowing or complete as it needed to be. This is what CIM is all about—communication and flexibility."

In keeping with this understanding of CIM, the project focuses on manufacturing integration through open systems interfaces and neutral or standard data file specifications. *Open systems* essentially means nonproprietary—not restricted to certain vendors or subject to influences dictated by the commercial interests of suppliers.

Most people think of an NC program as simply the data you need to run a machine tool. And that's essentially all you have with a conventionally processed, nonexchangeable program—a set of process instructions to run one and only one machine tool. But a BCL program is different. It is also a set of process instructions, but it is stated in a neutral manufacturing language. The make or model of the available process equipment doesn't matter. And yet, the response of the equipment is clearly defined. A BCL part program is a production process description allowing identical machined parts to be manufactured at any location.

BCL in Action.   In most shops, moving a workpiece from one machine to another necessarily involves reprocessing, or at least some editing of the NC program, especially if the machines come from different builders and have different control units. In Arsenal's case, the program was copied at the K & T machine as a disk file onto a 3.5-inch diskette, which was carried to the T-30 (a different machine tool) and loaded into the disk drive at the Acramatic (the receiving machine's control). Of course, tooling and fixturing were duplicated so that the setup matched, but no consideration was given to the geometry of the new machine; that is, whether it had the same clearance planes for tool changing and axis travel, whether home position was the same, whether origin of the coordinate system was the same, whether maximum traverse rates matched, and so on. The program was not modified in any way. It was executed exactly "as is." The program ran perfectly—the workpiece was machined without a hitch.

How BCL Works.   On the shop floor you need machine tools that have BCL-compatible controllers. Such machines may have special executive software that not only performs the usual processing functions of a CNC, but also interprets the BCL format and incorporates the data transformations and processing routines usually handled by the traditional postprocessor.

This capability of handling the postprocessor function internally is what distinguishes a BCL control unit from other CNCs. The machine itself is no different from standard numerically controlled models. Shifting the postprocessor function from the NC programming computer to the machine tool control unit is the key to the BCL concept. With the postprocessing function embedded in the control unit, tool path information in the NC program remains oriented only to the geometry of the workpiece itself. BCL advocates often refer to this orientation as "part space" as opposed to "machine space," the orientation required for the specific machine tool. The re-orientation that makes a program specific to one machine tool/control unit combination happens only in the control unit itself. This is illustrated in Fig. 8.1. As seen in the right hand column of this figure, the program written in APT is interpreted by the APT processor to generate a CL file. The CL file contains information about where and how the tip of the cutting tool is going to move as it machines the workpiece. This information defines the path of the tool and is the heart of any NC program. For APT users, this information is usually referred to as cutter location data or CL data.

The BCL concept essentially says, take the CL data as output of the APT processor, use them as input to the control unit, and let the control unit do its own postprocessing. So, for BCL, a programming department must be generating CL data. It can do this with an APT processor, as the Arsenal does, or create it with some other programming system whose output is identical to APT CL data. Many CAD/CAM systems with graphical NC programming modules generate APT-like CL data, for example. No postprocessing is required by the programming department (postprocessing is now a shop floor function).

BCL eliminates these traditional postprocessors. However, the BCL standard does specify a 32-bit binary code for NC programs. To meet this specification, BCL requires a CL converter, a rather simple software program that formats the CL data and translates it into 32-bit binary code.

Standardized CAD/CAM Output.   To carry exchangeability of BCL one step further, the Arsenal now specifies BCL as the NC programming output of its current and future CAD/CAM systems. Thus, BCL not only defines input to the machine tool, but it also defines what the output of any NC programming system must look like to ensure that the system can be integrated easily into the Arsenal's CIM network.

By standardizing the output of NC programming systems, the company expects its programming department to enjoy many of the same advantages that BCL offers on the shop floor. For example, any change in computer hardware or operating system has no effect on existing programs. Likewise, programmers have increased flexibility in processing NC programs. Simple programs can be processed on systems running on PCs, while mainframe or the more powerful workstations can be reserved for complex, multi-axis programs. Availability of appropriate postprocessors for each system is no longer an issue.

Other benefits are:

1. The ease with which programming staff turnover can be faced. New programmers can be trained more quickly because there is only one part-programming manual, the EIA 494 document.
2. Machine functionality is uniform and clearly defined.
3. Programming errors are less likely and programming productivity is improved.
4. Saving of time or resources acquiring, debugging, storing, and maintaining postprocessors.
5. Enhancement of scheduling since the part can be machined on any machine having the right technical specifications.

BCL and ASCII.   Binary was originally proposed when the concept of using CL data as input to the machine tool first emerged in the mid 1970s. At the time, the processing units in CNCs had limited speed and power, whereas postprocessing in real time puts considerable demand on the memory and logic sections of a CNC.

Today's CNCs do not face this limitation. Their processing speed and power is many times what it was when the first BCL control went on-line in 1975. However, binary poses an apparent obstacle to integrating NC with CIM. The main objection to binary format is that it is not human-readable. This problem, fortunately, suggested its own solution. By virtue of its simplicity, binary part program can readily be translated into other character codes that are readable as the English-like command words and symbols of the programming language. [Tables 3.2 and 3.3 (Chapter 3)

compare binary code with an alphanumeric code.]

ASCII, one of the most widely used character codes in data processing, was chosen for this translation. A symbolic equivalent of BCL in ASCII can be produced and displayed quite easily. Following this thinking, ASCII display of BCL programs became part of the original specifications for the machine tools ordered by the Arsenal.

An ASCII display of a BCL program is very useful on the shop floor. It allows manual editing of programs and even allows an operator to write BCL programs for simple workpieces using the manual data input mode. Furthermore, the ASCII equivalent of BCL data is nearly identical to APT; therefore, program statements are recognizable and familiar. Operators who recognize and are used to working with M codes and G codes and dynamic axis position readouts on the Arsenal's conventional CNCs had no difficulty adapting to BCL.

The advantages of ASCII have not been lost on other users or potential users of BCL. The BCL Users Group and the standards forming committee it works with are actively considering the adoption of an ASCII equivalent of BCL as part of the standard.

Enhancing Exchangeability.   An ASCII display of BCL is not the only instance where BCL is shaping CIM and CIM is shaping BCL. To maximize exchangeabililty and flexibility, the Arsenal specified "real-time" BCL record processing of data about tooling and workpiece fixture location. Real-time record processing means that values for tool dimensions and workpiece location are applied to axis motion coordinates as the program is being executed on the shop floor.

Tool management has always been a snag in schemes for exchangeability. Tool path data in an NC program must obviously reflect certain characteristics of the cutting tool involved. Length and diameter of the cutter, for example, determine offset values and the corresponding tool path coordinate.

Typically, BCL part programs are developed with zero tool length specifications, that is, as if the tool had no length. A BCL LOAD/TOOL,n command points to a "logical" tool number memory location in the BCL control system where specific cutting information has been entered. This information includes such items as tool length, tool location in the machine tool's tool changer, tool life data, or redundant tool locations.

This approach allows the programmer to program tools by inventory numbers—not specific tools. In other words, tool data represent "generic" tools, not unique pieces of hardware. Actual tool dimensions are entered on the shop floor from data received from the tool preset areas. It is not until the last second that tooling-dependent data is applied to any axis motion data of the program. Thus, the exchangeability of the program is preserved yet precise cutting tool requirements are definable as part of the process description that a BCL program represents in a CIM environment. Various tool management and workpiece location features have been incorporated into the latest version of the BCL standard. For example, LOAD/TOOL, xxxx program line means load tool from tool inventory number xxxx. The tool is supposed to be one inch in diameter and four inches long. If the operator finds that such a tool can be accommodated in station 8 of the tool storage and that the actual tool is 4.002 inches long and 0.996 inches in diameter due to sharpening, the operator would enter these data when prompted by the screen cursor for logical tool xxxx. Similarly, ORIGIN/n command points to a BCL control memory location where the actual origin values that locate part space geometry to machine space geometry are entered.

CIM Means Risks and Rewards. Adopting BCL at Arsenal was a bold move. Because CIM is a concept (and a radical one at that), not a highly defined and well proven technology, the prospect of implementing a CIM strategy naturally arouses doubts and anxieties. It's risky. Specifying BCL, a standard that has only a handful of installations to judge it by, seems to add to the apparent risk of CIM.

BCL is not without its tradeoffs. Nor do all machine tool builders actively support it, and their enthusiasm for it is lukewarm at best.

The first major revision to the BCL standard is just now in the final stages of review and approval. Nevertheless, this revision contains major enhancements and additions. They support probing, subroutines, data logging and communications, and other key functions.

It's almost a truism, but top-level commitment is essential to success with BCL, CIM, or any new technology.

It was not blind faith, Mr. Harris believes, that inspired this confidence. Rather, it was the widely shared view of what CIM was all about. Not a Utopian dream of the factory of the future, but a clear set of principles with identifiable criteria was at work here—flexibility, standardization, open systems, streamlining, and unified data sources.

BCL contributed to or enhanced all of these goals. Steve Harris sums it up: "With BCL we got what we were looking for, a CIM glue that makes NC programming and machining an integral part of the future. For us, BCL is truly the NC interface language."

Source: From "Interfacing NC with CIM" by M. Albert, July 1990, *Modern Machine Shop*, pp. 66–79.

The most important benefit of BCL input is part program exchangeability. Figure 8.1 illustrates how BCL input approach does away with postprocessing. The CL file is converted into a binary format, resulting in an exchangeable part program.

BCL advocates' efforts resulted in the EIA's BCL standard RS-494, released in 1983. This standard provides for part program interchangeability between machines of different configurations or controls. According to Harris (1991), "Any plant's CIM strategy should include a neutral language as part of its architecture and BCL provides the best solution available for achieving neutral data exchange. It is easier to implement than any other NC language regardless of the size of the shop" (p. 152).

Figure 8.1
Traditional postprocessing (left) yields NC workpiece programs that are not exchangeable. To machine an identical workpiece on another machine, the program must be postprocessed again unless the machine and control are exactly alike. To create exchangeable NC programs (right), the cutter location data in the CL file is not postprocessed in the traditional sense. Instead, the data is converted to a binary format and remains independent of the machine tool.

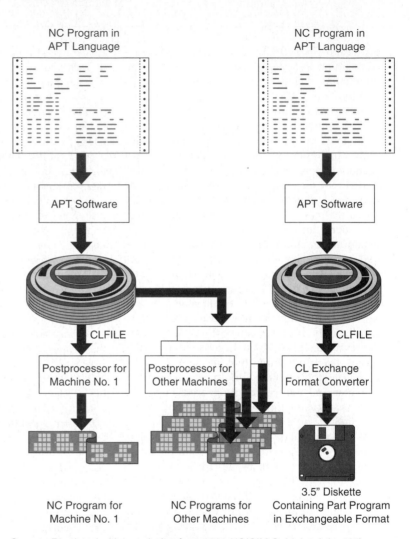

Source: Reprinted with permission from 1989 *NC/CIM Guidebook* (p. 165). Copyright 1989 Gardner Publications, Inc.

RS-494 has been updated in 1992 as EIA standard 494B. The latest version includes ASCII text representation of BCL, called ACL. ACL provides for the conversion of BCL into text form that can be edited and converted back to BCL.

BCL input is likely to get a boost from both the PDES and the U.S. Air Force initiative on the **next generation control** (**NGC**). PDES is expected to enable data exchange for parts and assemblies under the control of various manufacturing standards (see Box 8.11). The NGC initiative is aimed at developing a **neutral manufacturing language** (**NML**) to address communications among various NGC modules, machine management, real-time process controls, and so on. Though the three concepts—BCL, PDES, and NML—have yet to interact, their interaction would no doubt accelerate the pace of CIM implementations at the plant level.

| | **BOX 8.11**   *Manufacturing Standards* |

Many people are not aware that standards exist for things like postprocessor commands or output CLDATA. In fact, there are many standards dealing with manufacturing details such as these, and not all of them are in agreement. Manufacturing standards loosely related to postprocessing are available from ISO, ANSI and EIA.

International Standards Organization (ISO) Standards

ISO 4342-1985—Numerical Control of Machines—NC Processor Input—Basic Part Program Reference Language. This standard defines a very basic APT input language. The tool movement is restricted to three degrees of freedom (such as no rotary movement). Part geometry is also simple. The next revision of this standard will be based on the work being done by the American X3J7 committee, which is responsible for the ANSI X3.37 APT Standard.

ISO 3592-1978—Numerical Control of Machines—NC Processor Output—Logical Structure (and Major Words). This standard defines the output CLDATA format for CAM systems. The current standard defines the logical format of the output file, not the physical format. A new revision of 3592 is being voted on in early 1992. It will define for the first time the physical method of encoding CLDATA. Standard compliment CAM systems and postprocessors will not require additional software to physically communicate (for a time at least). The new standard also defines new record types to handle more sophisticated NC machines and machining techniques.

ISO 4343-1978—Numerical Control of Machines—NC Processor Output—Minor Elements of 2000-type Records (Postprocessor Commands). This standard defines vocabulary for postprocessors. A new revision of the postprocessor vocabulary is in the

works and should reach the preliminary voting stage in 1993. It defines language for many machining technologies not covered in 1978 revision.

ISO 6983-1988—Numerical Control of Machines—Program Format and Definition of Address Words. This three-part standard defines the NC machine input language format details, including standard (G) and (M) codes. Work continues on an extended program format that will provide access to advanced controller features.

American National Standards Institute (ANSI) Standards

ANSI X3.37-1987—American National Standard for Information Systems—Programming Language—APT. This standard defines an APT input language, output CLDATA format and postprocessor APT and CLDATA. The postprocessor vocabulary is extensive and quite different from ISO. The ANSI X3J7 committee continues to enhance this standard, with new revisions expected sometime in 1993.

ANSI/CAM-I 101-1990—Dimensional Measuring Interface Standard. This standard defines the logical and physical format of bidirectional communication inspection data between computer systems and inspection equipment. There is no ISO equivalent standard. However, efforts are now in progress to adopt the DMIS standard internationally.

Electronics Industries Association (EIA) Standards

EIA 494—32-bit Binary CL Exchange (BCL) Input Format for Numerically Controlled Machines. This standard defines a neutral language for NC machines as well as the physical representation of that language. Postprocessing, in the traditional sense, is performed at the NC machine. A preprocessor converts the CAM system output to the

BCL format. The NC machine vocabulary is extensive and different from the ISO and ANSI postprocessor vocabularies.

Source: From "NC Postprocessing: An Important Part of Factory Automation" by B. R. Francis, July 1992, *Modern Machine Shop*, pp. 98–99.

8.3   FMC AND FMS

When a machining or turning center is further automated for material handling, its sophistication approaches a flexible machining cell (FMC). Usually, a dedicated robot handles the needs of the cell for tools and workpieces along with loading and unloading, as required. A FMC is defined as a group of dissimilar machines and processes located in close proximity and dedicated to the manufacture of a family of parts that are similar in their processing requirements. The control of the FMC is maintained through a cell controller, which is a specialized software that integrates the controller platform (hardware and system software) with the cell-specific application programs. PCs or workstations are used as cell controller platforms. They require cell-specific application software to work as cell controllers.

The degree of automation for the various tasks that produce the part determines whether the facility is an FMC or just a machining center. To be designated FMC, the facility should be able to handle machining of any part within the family for which it has been developed, and changeover time between the parts should be brief and software-controlled. An FMC comprises a machining center, a robot or similar device for loading and unloading, and auxiliary devices or systems that enable it to machine the group of parts belonging to a family.

When material handling devices such as AGVs or automated conveyers combine several FMCs, the resulting production facility is called a flexible manufacturing system (FMS). Thus, an FMS consists of several machine tools along with part and tool handling devices such as robots, arranged so that it can process any part of the family for which it has been designed and developed. The operator's role is reduced to filling and emptying the part handling devices, and overall supervision.

An FMS can comprise nonmachining equipment, such as inspection or heat-treatment stations. A computer—usually a set of computers hierarchically linked—plans, executes, and controls the entire production. FMSs are sophisticated and expensive, costing millions of dollars. Their complexity and cost are two major reasons for their slow acceptance by industry, which seems to favor FMCs. When properly implemented, however, FMSs have proven very effective as modern production facilities.

The difference between FMC and FMS is subject to some controversy. According to *Production* magazine (February 1990, p. 54), the two systems differ in the number of material handling devices (one in FMC and many in FMS):

A manufacturing cell is defined as one materials handling device servicing multiple machines which are coordinated and paced by a computer or programmable controller.

A flexible manufacturing system (FMS) is defined as multiple materials handling devices servicing multiple machines with a computer responsible for coordinating, pacing and scheduling. (p. 54)

The two systems closely resemble each other in the volume and variety of production, as shown in Figure 8.2.

8.4   TOOL MANAGEMENT

Machining centers supported with computer-controlled workpiece and tool supply systems are the building blocks of FMSs. Process control and monitoring systems supported by an automated workpiece supply are readily available, but tool supply continues to be a weak link in machining center, cell, and FMS operations. In modern plants consisting of FMCs and FMSs, the inventory of tools may be large, especially if the tools are dedicated to individual machines and not shared. To contain tooling cost, an area of recent interest called **tool management** has been developing.

In today's plants, many cells run trouble-free with benefits such as faster throughput, increased machine usage, better quality, low work-in-process (WIP), and reduced costs. The key task for the next phase of cell technology is process refinement. One important refinement is tool management, especially in operations involving several tools. A successful FMS manages its tools as well as its workpieces. New tools need presetting before being loaded, current tools require constant moni-

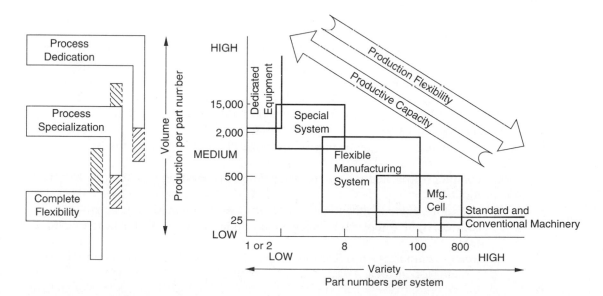

Figure 8.2
Capabilities of production systems on the basis of production volume and variety
Source: Reprinted with permission from *Computer-Integrated Manufacturing Handbook* (p. 2.136) by E. Teicholz and J. N. Orr (Eds.), 1987, New York: McGraw-Hill.

toring for wear and breakage, and dull and broken tools must be replaced so the proper tool for each job is at its place—and just in time.

Tool management can increase the productivity of cells and FMSs. It aims to optimize the use of all the tools—whether for cutting, forming, or any special purpose—in the plant. The objective of tool management efforts is to minimize tool change times and tooling inventory. In addition, tool management efforts can point out design deficiencies in production facilities.

Efficient tool management involves knowing what parts are to flow through the cell. It determines the inventory of the tools required. The next step is to prioritize the tools in terms of uses, locating the important ones near the workpiece. This is necessary for selecting the right tool changers.

Tool management software can provide:

random tool selection capability
tool delivery systems to the machine tool
tool condition sensing and **adaptive control**
retract and reentry routines
tool offset control
communication to the cell controller

Computer software can aid both in tool selection and timely replacement of tools too worn for the current job. PC-based tool control software programs track all tools, gauges, and fixtures throughout the shop. They locate critical tooling immediately, set up changes due to lack of tooling, and reduce machine and operator downtime. With such software, one can catalogue the tools needed, check their availability in the inventory, check which tools are at the station, create an exchange list, and output that data to the tool crib and presetter station for timely delivery of the tooling to the machine tool. Besides, it can provide other management functions as well. Future developments that will lead the progress toward untended production are automated tool delivery systems via AGVs and robotic gantry devices. Techniques that change the tool quickly have been reported to improve machine use by almost 10% (Hammer, 1989).

Tool management can be improved by one of the following methods:

1. computer-projected tool needs and printed preparation lists showing remaining tool life
2. operator-guided tool handling devices for fast, automated tool changing
3. computer-aided tool data management, updating and transfer to the CNC system
4. timely, computer-controlled tool preparation through an integrated tool presetter and storage area

A simple method to keep a watch on the tools in a cell is to monitor the horsepower and the time elapsed in the cut. Horsepower limits can be programmed into the APT NC programming macros for each tool. The elapsed time is compared with the tool life based on speed, feed, workpiece and tool materials, and so forth. The tool lives are entered and stored at the CNC of each machine tool in the cell. When the tool life is reached, the tool is changed and the flag is reset to track the life of the new tool. This approach creates difficulty when castings have extra stock or hard spots. The adaptive control approach takes care of the problem by automatically increasing or decreasing the feed rate based on the difference between the preset horsepower value and the actual reading during the cut.

Physical implementation of tool management involves a tool "highway" that carries tools to a battery of machines forming the FMS. Normally, tools are exchanged in sets when jobs are changed. Tools are preset manually according to the setup list. The manual method is time-consuming and error-prone. Since complete tool sets are changed along with the job, no one considers the useful life remaining in the various cutting edges or the fact that some of the replaced tools might have worked for the next job.

Some available systems—both hardware and software—offer integral control of all the tools and workpieces. They can also control, coordinate, and monitor the machine cycles and sequences. The resulting benefits are flexibility, increased system uptime, and higher usage with reduced labor and inventory—all leading to higher productivity.

Box 8.12 describes in detail the current practices in tooling management and the associated technology.

8.5   FLEXIBLE FIXTURING

The terms *jig* and *fixture* are generally used interchangeably, though there is a basic difference between the two. Both hold, support, and locate the workpiece. In the case of a cutting tool, however, the jig guides it, but the fixture references it. Thus,

---

■ BOX 8.12   *Tooling Management in an FMS*

Tooling management is one of the most essential aspects of an effective flexible manufacturing system (FMS). Implementation of an effective tooling management system will reduce tool cost and inventory, decrease machine downtime, and improve the quality of the parts being produced. In addition, proper tooling management is important because machine downtime often results from tooling-related problems. Moreover, the costs for perishable and durable tooling in an FMS may be greater than labor and raw material combined. Losses due to operator error, damaged tools, overstocking and understocking of tooling, lost or misplaced tools, underused inserts, and poor-quality parts can be avoided through the use of modern tooling management methods.

A proper tooling management strategy must contain elements that allow for monitoring and control of the tooling used in the FMS. These are:

A machine control unit (MCU) that provides for tool breakage detection, tool wear detection and compensation, redundant tooling, and adaptive control;

A tool identification system that is machine readable;

A tool preset area with tool assembly components, an automatic tool presetter, and a tool identification reader; and

Computer control software for tooling management under the FMS operating system.

While these items alone will not insure a complete tooling management plan, they are the essential components of a basic FMS tool management system.

Breakage Detection. Tool breakage detection can be performed by post-process or in-process methods. The post-process method normally utilizes a fixed sensor to determine whether the last tool selected is still good. This method is particularly useful when machining parts with small end drills, drills, and taps (less than 10 mm), where the breakage torque is very close to the maximum torque needed in the machining operation. Many tool breakage sensors are accurate enough (within ±0.05 mm) so that they can detect chips or notches in the tooling. These same sensors may be used as automatic tool-length measuring devices for setting the offset of the tool before it starts in the machining operation.

The in-process method for tool breakage detection normally consists of either a force transducer to monitor axis-cutting force, a torque monitor for sensing spindle torque, or an acoustic monitor for detecting large deviations in acoustic emissions generated by the cutting process. The force transducer and torque monitor are set to a predetermined limit based on test or history of the tool's cutting forces for the material being machined. When the tool exceeds this maximum, the spindle stops and a tool recovery sequence is initiated. Particular care must be exercised if the cutting tool material is ceramic, because a sudden decrease in cutting force can result in tool breakage.

An acoustic monitor provides an effective means for detecting tool breakage by sensing the acoustical emissions produced in the cutting of metal and chip formation. When the acoustic monitor detects signals that are outside of a preset value, the tool is classified as broken, and machining is immediately stopped. In practice, acoustic monitors function better than force or torque monitors for determining breakage of small tools because acoustical emissions are detectable even with tools as small as one millimeter in diameter. However, an acoustic monitor may not be effective when used in conjunction with advanced cutting materials or large multiple insert tools.

The three methods mentioned here for detecting tool breakage are usually implemented as software and hardware options that are internal to the machine tool. Nevertheless, separate units can be installed and integrated into many machines without these options.

There are two alternatives when breakage is detected. The first alternative is to replace the tool with an identical tool and repeat the cutting process. The second alternative is to classify the part as scrap or rework material and call for the next part to be machined. If the first alternative is selected, it is standard practice to skip the hole being drilled (or tapped) because the tool could be broken and become stuck in the workpiece. In the latter case, rework procedures would include using a mechanical method combined with the application of heat for extracting the tool and then manually repeating the operation.

Most FMSs apply one or more methods for detecting tool breakage. Spindle torque monitoring is the most popular method for detecting tool breakage. In situations where small tooling is used, tool length detection sensors are popular. Most sensors require additional processing time from the MCU; however, in cases where the parts are complex and costly the additional time is negligible.

Compensating for Wear. Tool wear detection and compensation is a technique for using probing devices and sensors to detect tool wear and adjust corresponding offsets to keep tolerances within specifications. The two most popular methods for tool wear detection are the machine base-mounted sensor and the touch-trigger probe. The machine base-mounted sensor is located away from turret and spindle movement. Prior to a selected machining operation, the tool will be moved to a programmed position and placed on the sensor. The deviation from the preset "new tool" value will then be transferred to the appropriate offset for compensation.

The touch-trigger probe differs from the base-mounted sensor by its attachment to the turret or spindle. The probe position is read after a machining operation is completed and the probe touches the part with a stylus at selected points that determine deviations from tolerance. As with the machine base-mounted sensor, the difference between the actual values and the standard values are relayed back to the appropriate offsets. Occasionally, the touch-trigger probe is also used as a broken tool detection device to check small drill and tapholes for broken tools.

Redundant Tooling.   The use of redundant tooling (duplicate tools) in the tool storage matrix is a basic strategy of tooling management if discernible tool wear or failure potential is present. The tool matrix will frequently be an automatic tool changer with the ability to hold sixty tools or more. The number of duplicates for any particular tool is determined by four variables: average cutting time for specific tools on individual parts scheduled for the FMS, cutting time allotted based upon wear history, tool breakage history, and tool positions available in the storage matrix.

Redundant tooling is commonly used in conjunction with a tool breakage detection device.

A popular strategy practiced by industry with advanced FMSs is to warn the operator when the next-to-last tool is worn out. The system will then tell the operator which tools are to be replaced or repaired.

The use of adaptive control techniques offers substantial cost savings for machining in an FMS. According to existing data, adaptive control has reduced machining times up to 35 percent. It has improved tool life, in many cases by as much as 50 percent. Through increased tool life and consistency in surface finish, it has improved quality by reducing scrap and rework by up to 25 percent.

Adaptive control methods attempt to maintain a constant spindle torque when the cutting tool encounters changes in hardness in materials or varied depths of cut. This is achieved through automatic regulation of the feed rate override. The range of override may vary from 35 to 150 percent of the preset value. Some adaptive control systems also have the ability to combine feed-rate override and spindle speed override, which will also help to improve surface finish and quality.

Identification Methods.   Tooling identification assists the overall tooling management program by providing pertinent information on the tool such as matrix position, length, components, and wear status. This is achieved through the association of a serial number with a particular toolholder. The identification device on the toolholder is normally either a bar code designation or an embedded computer chip.

The bar code method for tool identification can be achieved by using a printed label fastened to the tool with an adhesive, or by engraving the bar code on the tool. A human-readable character set is also usually printed on the bar code label for quick visual identification. Laser readers at the machine and wand readers at the tool preset area keep track of the location of the tool and prevent the operators from entering incorrect tool serial numbers. Laser readers at the machine also allow for tools to be placed randomly in the tool magazine.

The embedded-chip method for tool identification uses a microchip located in either the holder shank or the retention stud of the holder. The embedded-chip works in much the same way as the bar code identification. Microchips have been developed that can withstand harsh machining environments and some can read data as well as transmit it.

Most FMS installations use the bar code method for tool identification. A major reason for this choice is the low initial investment compared with the embedded-chip method. The greatest disadvantage in using bar code labels is their inability to stay attached to the tool and free of film in the cutting environment.

Tool presetting is the process of verifying or setting the length of a tool to be used in a machining operation. Most tool presetting will be performed in a designated tool preset area containing a terminal, printer, automatic tool presetter, a tool identification reader, and tooling components. The terminal will allow the operator of the tool preset area to review tooling status at the machines and will provide access to production information concerning the FMS. Sophisticated programs and control systems normally prevent the operator from manually entering any information to prevent operator error. Some displays will simply prompt the operator that the next tool should be preset and display the components used in assembling that tool.

In some cases, when a tool is called to be preset a CAD drawing of that tool will be displayed outlining all relevant information on the tool. The printer is used to automatically produce documentation of the important flags and warnings sent to the tool preset area from each MCU in the FMS.

The tool presetter is an optical comparator device that outputs the precision point of location of the tool nose radius with respect to the locating surface of the tool block. Accuracy within $\pm$ 0.001 mm ($\pm$ 0.00004 in.) is normally achieved with the tool presetter. After a tool is preset, its dimensions are automatically entered into the tool control system or the MCU.

The tool identification reader is normally a laser-reader wand or laser-reader gun that can be passed over the bar code label or embedded computer chip for tool identification.

The tool preset area should contain certain components in an organized tool cabinet that may be needed for repair, insert replacement, or assembly of tools.

Control Software. Most FMSs in use today are controlled by some type of hierarchical computer control system. In most cases, this control system is networked to a mainframe computer that supervises plant operation. The control system contains CAD/CAM and CIM software that includes provisions for scheduling, data distribution, process planning, and material handling. Software for tooling management and control is also an essential part of the FMS control system. Tooling management software usually accommodates operations such as the tracking of all tools loaded in the system, identifying specific tooling needs for setups and changeovers before they occur, and maintaining an active database which will contain all information on any particular tool.

The control system software for the tooling management strategy and the total system must be customized to the particular needs of the manufacturer. The base software may be a generic program or program package, but some modifications are always needed.

The use of total plant control software such as manufacturing resources planning (MRP II) has been a very effective method for achieving overall successful shop-floor management. There is a trend to integrate the computer control system for the FMS into MRP II. With such integration, a single tooling management package can control the FMS, machine cells, and individual machines, resulting in more effective control of tooling and reduced inventory due to unused duplicate tools.

A major reason that plants have not been able to complete the integration is the incompatibility between the hardware and communications protocols.

Source: From "Tooling Management in an FMS," March 1990, *Mechanical Engineering*, pp. 40–44.

---

the difference between the two is in their relationships to the tool. Figure 8.3 illustrates this difference, showing that in (a) the jig guides the tool with a bushing, while in (b) the fixture references the tool (i.e., locates it in relation to the workpiece). The term workholder refers to both jig and fixture.

The design of workholders is critical to the quality and cost of the parts machined. A poor design can undermine the capability of even a good machine tool.

Figure 8.3
(a) A jig guides the cutting tool, in this case with a bushing; (b) A fixture references the cutting tool, in this case with a set block.

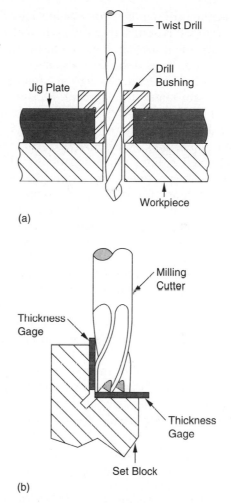

(a)

(b)

Moreover, a modification in part design may require a change in the jig or fixture. If the original fixture design did not contemplate likely modifications in product shape or size, then each major change in the product design would require a new fixture. Obviously, this is undesirable, especially since today's products have shorter design life cycles.

Jigs and fixtures play an important role in manufacturing. Their costs are easy to justify when production runs are long and repeat orders likely. With the recent trend of smaller batch sizes, the conventional one-of-a-kind or permanent jigs and fixtures seem expensive. The answer lies in developing "modular" designs that allow reassembly of standard components into different shapes and sizes. Modular fixtures are temporary setups (see Figure 8.4) of such components that offer the advantages of permanent tooling. Figure 8.5 shows some typical components that are used to build modular work-holding systems.

To modify fixture design at minimal cost, flexible or modular fixturing techniques are under development. **Flexible fixturing** implies that the workholder can adapt to cope with any part within a "family of parts." This is achieved through modular design, building workholders using Lego-like components that can be disassembled and reused. A **modular fixturing** system may have hundreds of components assembled in different combinations to produce an unlimited variety of workholders. "Modular workholders are particularly well suited for one-time jobs, infrequent production runs, prototypes parts, replacement parts, trial fixturing, and temporary tooling," according to *Jig and Fixture Handbook* (1992, p. 337).

Modular fixturing (Figure 8.6) is used when lot sizes and repeat job orders are small. One major drawback of modular fixturing is the large initial investment. But this cost is amortized over the period since the flexibility provided enables more use of the modular fixture. Readers interested in the technical aspects of modular fixturing should refer to chapter-end references, especially *Jig and Fixture Handbook* and Hoffman (1987).

Companies are always testing new ways for developing flexible fixturing. For example, in phase-change fixturing, a section of the part is submerged in a fluid-state material that solidifies to hold the part. In another method, called **potting**, a special adhesive holds the part in place during machining. The part is later removed with a solvent. Potting is based essentially on phase change, since the adhesive polymerizes. In fluidized-bed fixturing, a container filled with particles becomes the fixture bed for the part. This bed is porous, and controlled air is allowed to flow

Figure 8.4
Modular workholders can be assembled entirely from standard off-the-shelf components.

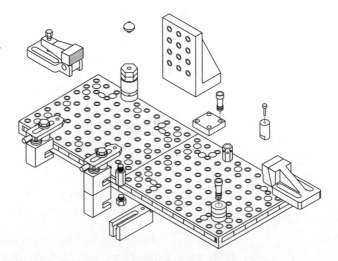

Source: Reprinted with permission from *Jig and Fixture Handbook*, 1992, St. Louis: Carr Lane Mfg. Co. Copyright 1992 by Carr Lane Mfg. Co.

Figure 8.5
Typical components of a
modular-workholding system

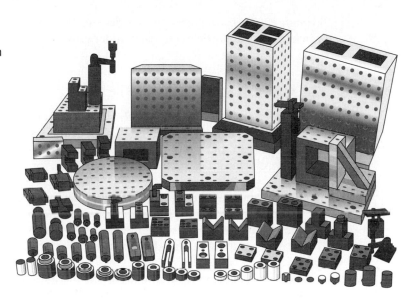

Source: Reprinted with permission from *Jig and Fixture Handbook*, 1992, St Louis: Carr Lane Mfg. Co. Copyright 1992 by Carr Lane Mfg. Co.

Figure 8.6
The choice between modular and permanent fixturing depends on the frequency of job runs and lot size.

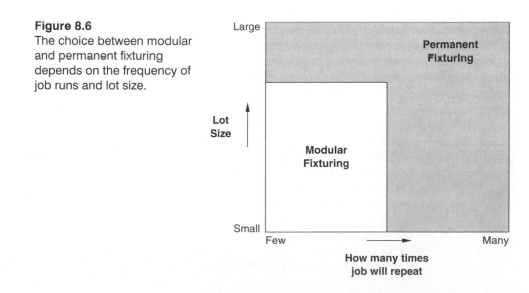

Source: Reprinted with permission from *Jig and Fixture Handbook*, 1992, St Louis: Carr Lane Mfg. Co. Copyright 1992 Carr Lane Mfg. Co.

through it. Air flow causes the particulate bed to act like a fluid. When the air is switched off, the particles compact under gravity, forming a solid mass that holds the part.

8.6   FLEXIBLE ASSEMBLY SYSTEMS

Flexible assembly systems automate assembly operations. They comprise a range of robots or programmable equipment and peripheral devices. The devices have their own controllers and accompanying programming languages. While installing an assembly cell by combining these devices, two major bottlenecks arise:

1. Programming languages of the various devices need to be mastered before a system can be developed, tested, operated, and maintained.
2. Functions needed on several devices may have to be rewritten in different languages, resulting in high-development and even higher maintenance costs.

Both bottlenecks essentially represent the antithesis of the CIM concept. Obviously, the need for standardization is urgent. Realizing that robot manufacturers have not yet agreed on such a standard, users often develop one of their own. Brussel and Valckenaers (1990), for example, used Modula-2 in developing an assembly cell; it allows application engineers to write off-line assembly tasks in a single programming language without the need for postprocessing. The approach also applies to other devices, such as robots or PLCs, that can be programmed to execute on-line operations requested over a communication link. The communication link is established between one of the computers of the control system and each special-purpose industrial device in the assembly system. The special device performs tasks imposed by the program in the corresponding computer. Conceptually, there is no difference between giving a task to a printer in a PC environment or to an industrial robot. The application program running on the cell computer activates the operational rules of the specific device through a communication link. The prevailing control situation is similar to a manufacturing manager giving instructions to a programmer, an engineer, or a secretary. Although using a single language, say English, the manager addresses their specific functions differently to get different jobs done.

Implementing a device driver comprises two application-independent programs. The first program is written in the device language, such as that of the robot. It runs in the device or robot controller and does nothing but look for and execute commands coming through the communication link from the cell computer. The second program is a library module that takes care of communication protocol, data representation and transmission, and reception of data to and from the device or the robot. Application programmers can access all device functions under procedures imported from the library module into application programs. The resulting environment is similar to that of a printer attached to a PC.

8.7   FLEXIBILITY

The term **flexibility** is often used to describe various manufacturing functions and facilities, for example, FMC or flexible fixturing. What exactly is flexibility? In the context of manufacturing, flexibility is a complex term. To most, it is the capability of producing different parts without major retoolings. To others, it may mean the ability to change a production schedule, to modify a part, to handle multiple parts, and so on. According to Lamb Technicon (*Production*, October 1988, p. 60), eight factors contribute to flexibility:

1. *Setup*—case and short-duration changeover of tools, fixtures, programs, and other issues affected by the changeover between parts
2. *Process*—the ability to produce parts in more than one way; also, the mix of parts the system can cope with
3. *Convertability*—the ability to change the system to handle new parts that may be similar or different
4. *Routing*—the ability to continue producing in the event of breakdowns
5. *Volume*—the ability to match market volume demands profitably
6. *Expandability*—the ability to expand the facilities easily and cost-effectively
7. *Operation*—the ability to shift the order of operations
8. *Production*—the range of parts that can be produced

A typical problem for manufacturing personnel is: How to compare the flexibility of candidate systems, especially when deciding to purchase one? The eight factors just listed can be considered, with their weightings reflecting the company's needs. A suggested weighting for these factors is 20%, 5%, 15%, 10%, 15%, 20%, 5%, and 10%, respectively. Next, the candidate systems are judged and rated on a scale of 1 to 10 for each factor. The one scoring the highest total is the most flexible, as the following example illustrates.

Example 8.1 _____

Consider the five FMCs (A, B, C, D, and E) listed in Table 8.1.

The flexibility points are obtained by multiplying the weighting for each criterion with the score and adding the results. For example, the value of 780 for C resulted from the following calculations: $20 \times 7 + 5 \times 5 + \ldots + 5 \times 6 + 10 \times 8$.

Next, the throughputs and capital costs for the candidate FMCs (or other flexible equipment) are compared with their flexibility points. Usually, the ratio between the capital cost and throughput is considered. The FMC (or flexible equipment) with the smallest ratio and an acceptable value of flexibility points is selected.

_____

Table 8.1
Five FMCs rated according to flexibility criteria

Criterion	Weighting Percentage	A	B	C	D	E
Setup	20%	8	3	7	6	9
Process	5	7	0	5	10	8
Convertability	15	8	10	9	5	10
Routing	10	3	8	4	8	7
Volume	15	5	7	10	6	9
Expandability	20	4	7	9	7	8
Operation	5	9	8	6	8	10
Production	10	4	6	8	7	9
Flexibility points:		585	635	780	665	875

TRENDS

❑ Modern machine tools are becoming more multifunctional. For example, machining centers replace boring, drilling, and milling machines.

❑ DNC's (distributed numerical control) broadened scope and use represent a significant step toward CIM.

❑ CNC systems are becoming more open and accessible to the host computer or cell controller.

❑ The implementation of "full-blown" FMSs in industry is slow. What has proven to be a practical technology is the machining or turning cell comprising one, two, or three machine tools with capabilities of palletizing, effective tool control, 32-bit CNCs, debugged software, fast servo and spindle motors, and SPC techniques. Additional cells are set up to suit other part families. Thus, the concept of unmanned machining is currently limited to manned untended machining cells.

❑ The number of FMCs in use is increasing much more rapidly than FMSs. Both are used primarily in metal cutting and assembly. About 40%–45% of FMCs and FMSs in use are for metal cutting, another 40%–45% for assembly, and the remaining 10%–20% for metal forming. About a quarter of U.S. manufacturing plants use FMCs, while about two-thirds use NC or CNC equipment.

❑ The number of expert systems for automating NC programming tasks continues to increase. Such systems accept part information in solid modeling formats, then generate machine instructions for both producing the part and verifying its integrity. One such system is CAM-I's Expert Manufacturing Programming System.

❑ CNC machines are available with positional accuracy of ±0.1 mm and a smallest programmable increment or decrement of 0.01 mm.

❑ A modern precision machining facility comprises CNC machines, CMM, computerized scheduling, and SPC.

❏ Machine control devices such as CNCs and PLCs have begun to use fiber-optic cabling. This allows connection to remote I/O devices, as far as eight miles.

❏ The inherent limits of machine tools are being compensated for by using probes and control software. Such techniques are called "an electronic pill for a mechanical ill."

❏ Machine tools are able to probe the workpiece and correct for locating errors, thus bypassing the concern about the last increment of drive or table positioning errors.

❏ Until the 1980s, the main goal in manufacturing management was to minimize direct labor in machining. Today, the trend is to shorten the time for setup, tool change, chuck jaw change, and the like. Tools can now be changed in seconds rather than in minutes. Chuck jaws can be changed in one minute rather than the usual 30 minutes. The axes can be traversed rapidly at speeds as high as 1,000 inches per minute, which is achieved in 60 microseconds. Spindle speeds of 5,000 RPM can be reached in 2.5 seconds. These reductions in setup and cycle times mean quicker response to orders, reduced inventory levels, and smaller economic order quantities.

❏ Newer CNCs feature more sophisticated diagnostics and self-repair capability, which are initiated each time the control is started up. The diagnostics searches out and reports on screen any defective modules. If the CPU board is faulty, the self-repair feature finds it, removes it from the system, and configures the backup board in its place. Expert systems-based troubleshooting and guidance are also becoming more available.

❏ Manufacturers of modern machine tools incorporate 32-bit CNC controllers, design them for multipurpose operations, and use more laser technology.

❏ Lathes and milling machines are converging into single units; CNCs are also following this trend. In the high-end market, one control is likely to suit both lathe- and mill-based operations.

❏ Newer CNC controls are interactive. Developing programs with them is, according to Hall (*Modern Machine Shop*, July 1992), "similar to answering questions and entering numbers on an automatic teller machine at the bank" (p. 58). Figure 8.7 shows an example. Hall predicts: "The days of mysterious programming languages and codes are virtually over" (p. 65).

❏ Shop-floor programming systems provide more user assistance, based on expert system philosophy, to simplify the fixturing and location tasks of the operator. Part programs get modified automatically to account for the difference in part location.

SUMMARY

A number of production-related issues concern CIM. Advances in CNC and DNC, for example, have enabled integration of plant facilities with the other functions of the business. This chapter described such developments, along with those in machining and turning centers. The technologies of FMC and FMS were also presented.

Figure 8.7
A DOS-based control screen as an example of highly prompted fill-in-the-blank formats that simplify CNC programming

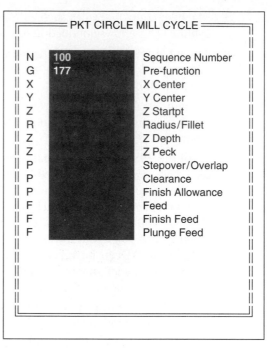

PKT CIRCLE MILL CYCLE		
N	100	Sequence Number
G	177	Pre-function
X		X Center
Y		Y Center
Z		Z Startpt
R		Radius/Fillet
Z		Z Depth
Z		Z Peck
P		Stepover/Overlap
P		Clearance
P		Finish Allowance
F		Feed
F		Finish Feed
F		Plunge Feed

Source: Reprinted with permission from "The ABCs of CNC" by A. K. Hall, July 1992, *Modern Machine Shop*, p. 59.

The chapter also addressed two other areas critical to CIM, namely tool management and flexible fixturing. The meaning and quantification of flexibility and selection of equipment based on this criterion were discussed. Several boxes highlighted current applications in industrial case studies. The coverage in this chapter has concentrated on the latest shop-floor practices.

KEY TERMS

Adaptive control

Binary cutter location (BCL)

Direct or distributed numerical control (DNC)

Flexibility

Flexible fixturing

Forming

Machine tool

Machining center

Modular fixturing

Mean time between failures (MTBF)

Next generation control (NGC)

Neutral language

Neutral manufacturing language (NML)

Override

Part program compatibility

Potting

Quick die change (QDC)

Tool (or tooling) management

Turning center

Universality

EXERCISES

Note: Exercises marked * are projects.

8.1   A pair of scissors is a simple manufactured product. Describe how it could have been produced, listing the basic processes that might have been used.

8.2   Comment on the statement, "The modern machining center is to the engine lathe what the F-16 is to the camel."

8.3   Explain the difference between FMC and FMS in approximately 200 words. Why is FMS's acceptance by industry slow?

8.4   DNC technology's concept has changed over the years. Discuss both the concepts, namely direct numerical control and distributed numerical control. Which one is more compatible with the scope of CIM and why?

8.5   Write a 300-word essay on parametric programming.

8.6   Explain in 400 words how tooling management contributes to CIM.

8.7   Explain the meaning of FMS. What role does it play in CIM? Limit your answer to approximately 300 words.

8.8   The following data pertains to three FMSs available in the market. Which of these should be selected?

FMS type	A	B	C
Cost (in millions)	$ 10	$ 8	$ 14
Throughput (in thousands)	250	215	386
Flexibility scores (scale 1–10):			
Setup	4	6	8
Process	5	7	3
Convertability	6	3	2
Routing	2	8	4
Volume	3	4	5
Expandability	5	4	3
Operation	9	8	7
Production	7	8	9

The weighting percentages for the flexibility factors are 15, 20, 5, 10, 20, 10, 15, and 5, respectively.

8.9*   Visit a local manufacturing facility and study the operation of an FMC or FMS. Ask someone familiar with the facility whether the investment is paying for itself.

8.10* Study the most modern machining center at your organization or at a nearby manufacturing facility and list its major features. Compare your list with those of other class members. (If a machining center is unavailable, then obtain relevant brochures from the vendors and use them in this project.)

8.11   Circle T for true or F for false or fill in the blanks.

a.   Cellular manufacturing is concerned with the manufacture of products made of cellular materials.   T/F

b.   FMS technology is limited to machining of metal components and their assembly.   T/F

c.   COMPACT II is a software for material handling.   T/F

d.   CNC software is another name for part programming.   T/F

e.   The first NC machining center comprising ATC and APC is reported to have been operated unmanned in:

        1950   1975   1986

f.   APT is an acronym for _____

g.   Name five important manufacturing tasks that fall under CAM:

        _____

        _____

        _____

        _____

        _____

h.   Name three companies that produce and market machining centers:

        _____

        _____

        _____

SUGGESTED READINGS

Books

Drozda, T., & Wick, C. (1983). *Tool and manufacturing engineers handbook* (Vol. 1). Dearborn, MI: SME. (There are six other volumes in this comprehensive handbook series. These are: Vol. 2, Forming; Vol. 3, Materials, Finishing, and Coating; Vol. 4, Quality Control and Assembly; Vol. 5, Manufacturing Management; Vol. 6, Design for Manufacturability; Vol. 7, Continuous Improvement.)

Gilbert, R., & Llewellyn, J. (1985). *Programmable controllers—Practices and concepts*. Manufacturing Technology Book Press.

Hartley, J. (1984). *FMS at work*. London: IFS.

Hoffman, E. G. (1987). *Modular fixturing*. Lake Geneva, WI: Manufacturing Technology Press.

Kalpakjian, S. (1992). *Manufacturing engineering and technology*. Reading, MA: Addison-Wesley.

Luggen, W. W. (1991). *Flexible manufacturing cells and systems*. Englewood Cliffs, NJ: Prentice-Hall.

Monden, Y. (1983). *Toyota production system*. Norcross, GA: Industrial Engineering and Management Press, Institute of Industrial Engineers.

Owen, A. E. (1984). *Flexible assembly systems*. New York: Plenum Press.

Ranky, P. G. (1990). *Flexible manufacturing cells and systems in CIM*. Guildford, Surrey (England): CIMware Limited.

Seames, W. (1990). *Computer numerical control: Concepts and programming*. Dearborn, MI: SME.

Shingo, S. (1985). *A revolution in manufacturing: The SMED system*. Cambridge, CT: Productivity Press (Original in Japanese, translated by Andrew P. Dillon).

Webb, J. (1988). *Programmable controllers*. Columbus, OH: Merrill.

Wick, C., Benedict, J., & Veilleux, R. (1984). *Tool and manufacturing engineers handbook: Vol. 2. Forming*. Dearborn, MI: SME.

Wick, R., & Veilleux, R. (1985). *Tool and manufacturing engineers handbook: Vol. 3. Materials, Finishing and Coating*. Dearborn, MI: SME.

Monographs and Reports

Jig and Fixture Handbook. (1992). St. Louis: Carr Lane Manufacturing Co.

Mason, F. (Ed.) (1992, August). *Computerized tool management systems*. SME Blue Book series. Dearborn, MI: SME.

Modern Machine Shop. (1992, July). (Emphasis is on NC and CAM.)

Machining Data Handbook. Cincinnati: Metcut Research Associates.

Modern Machine Shop. (1994, March), 1994 Guidebook. Cincinnati: Gardner.

Articles

Albert, M. (1989, January). NC verification: Taking a closer look. *Modern Machine Shop*, pp. 82–89.

Albert, M. (1990, February). Integrating inspection and machining. *Modern Machine Shop*, pp. 80–90.

Beard, T. (1990, February). In touch with quality and productivity. *Modern Machine Shop*, pp. 66–77.

Brussel, H. Van, & Valckenaers, P. (1990, Spring). Hierarchical control of a generic flexible assembly cell. *Journal of Applied Manufacturing Systems*, 3(1), 25–33.

Charles, D. S. (1988, Spring). Broad scope DNC: Automated factory data. *COMMLINE*. 18(2), 15–16, 29.

Coleman, J. R. (1990). PCs make it with MEs. *Manufacturing Engineering*. (The three-part series appearing in February, March, and April look at how manufacturing engineers are using personal computers to gain an edge in various areas.)

Coleman, J. R. (1990, March). Machining centers cut it in job shops. *Manufacturing Engineering*, pp. 41–45.

Eade, R. (1989, May). For flexibility, you can't beat them. *Manufacturing Engineering*, pp. 49–52.

Gruver, W. A., & Senninger, M. T. (1990, March). Tooling management in an FMS. *Mechanical Engineering*, pp. 40–44.

Hammer, H. (1989, January). A new game plan for tool control. *Modern Machine Shop*, pp. 52–63.

Herrin, G. E. (1990, February). Today's CNC diagnostics (part II). *Modern Machine Shop*, pp. 142–146.

Horn, D. (1988, September). Optical fibers, optical sensors. *Mechanical Engineering*, pp. 84–88.

Huber, R. F. (1988). Machine tools: A global view. *Production. 100*(9), 40–45.

Kegg, R. L. (1988, October). FMS technology today: Friendlier, maturing, successful. *Modern Machine Shop, 61*(5), 80–88.

Lynch, M. (1992, June). Exploring distributed numerical control. *Modern Machine Shop, 65*(1), 126–128.

Meister, A., & Summer, E. (1988). Numerical control program verification: A productivity challenge. *COMMLINE, 18*(3), 14–18.

Martin, J. M. (1989, April). Managing tools makes the cells click. *Manufacturing Engineering. 102*(4), 59–62.

Murphy, P. (1988, July). PCs and PLCs: Partners in control. *InTech*, p. 33.

Noaker, P. M. (1988, August). Workholding: Firm but flexible. *Production*, pp. 50–55.

Pelkey, B. (1989, Spring). The need for flexibility in shop floor control systems. *COMMLINE, 18*(2), 17–19.

Randhawa, S. U., & Bedworth, D. (1985, June). Factors identified for use in comparing conventional and flexible manufacturing systems. *Industrial Engineering*, p. 40.

Tabenkin, A. (1990, February). Use of surface finish to get optimum production. *Modern Machine Shop*, pp. 52–62.

Vasilash, G. S. (1988, August). The elements of modern machine tools. *Production*, pp. 38–42.

Vasilash, G. S. (1990, February). The PLC: Still useful after all these years. *Production*, pp. 34–38.

Vasilash, G. S. (1990, May). Why you should consider small CMMs. *Production*, pp. 66–68.

CHAPTER 9
Shop-floor Control

"Intelligent measurement devices, or smart sensors, are important because when you think of a total plant information system, you need to consider the quality of the information entering the system, regardless of where it comes from. Intelligent measuring devices help ensure quality data at the data collection point. The information no longer has to work its way up to some kind of inferential machine to figure out whether the measurement was good in the first place."

Jack Clarke, Foxboro Co., in *Managing Automation*, January 1990

"Standards that promote integration may restrict innovation. Innovation is viewed as a key competitive advantage in the control system industry."

—*CIM Integration Tools* (SME Blue Book series)

"The gap between automation of the process and automation of the office is what's left. The trend is to fill this CIM gap between the corporate automation, which is done, and the plant or process level automation, which is just getting done."

—Scott McLagan, Fisher Controls, in *Managing Automation*, January 1990

"HP-Unix version 8.0 will allow a CIM system to respond to certain asynchronous events like equipment breakdowns by rerouting the path of products through the assembly line."

—Michael Puttre in *Mechanical Engineering*, July 1991

Manufacturing production represents a complex task. Seldom do things take place at the shop floor as planned. The answer to the prevailing reactive or "fire fighting" condition lies in proactive control of functions to minimize the difference between what was planned and what actually exists. Some typical shop-floor functions are:

1. data collection, both manual and automatic
2. daily information management, such as decision support and operator instructions and procedure

3. day-to-day resource management of material, WIP inventory, tools, and more
4. plant-floor scheduling, loading, and routing
5. monitoring and control of quality
6. process control of assembly, robotic, machining, and/or test cells

By following the right combination of materials, processes, and routing information, and by keeping track of the containers, tools, and material handling devices, it is possible to route workpieces efficiently through production, storage, and distribution. In this chapter, we consider the short-term control of shop-floor resources. By short-term we mean minutes, hours, and days. **Shop-floor control** is also called **production control**.

9.1   DATA LOGGING AND ACQUISITION

Shop-floor control depends on analysis of regularly generated data, which forms the basis of control decisions. Factory data collection and analysis are important tasks within shop-floor control. Traditionally, data has been collected using verbal, written, or semiautomated methods. In the verbal method, the foreman who knows how to build the product advises the operator orally. With sophisticated products, the written method—in which a documentation packet follows the product through its production route—works better. Although better than the verbal method, the written method may not be efficient. In the semiautomated system, terminals are located throughout the plant for collecting shop-floor data and also for downloading product data relating to materials, process plans, and labor contents. The system is semiautomated, since people must interface with the terminals.

In a CIM environment, the semiautomated terminal system interfaces with the plant's host computer and the company's mainframe-based MRP II software. Semiautomated systems are being further automated to allow a higher degree of integration. Newer systems rely heavily on various input-output (I/O) devices to keep track of ongoing shop-floor conditions. Table 9.1 summarizes the major I/O devices, which are basically electromechanical sensors or transducers. Data input technologies are keyboard, bar code, magnetic stripe, and **optical character recognition** (**OCR**) systems. Computers are increasingly used for collecting manufacturing data from the shop floor and elsewhere.

Data collection can be manual or automated (sometimes referred to as off-line and on-line, respectively). In the off-line method, data is recorded on a stripchart

Table 9.1
Data collection and control devices

Switch Inputs	Foot switch, limit switch, door switch, frequency counter, timer, event counter, etc.
Nonswitch Inputs	Keyboard, bar code wand, magnetic wand, bar code or magnetic stripe reader, etc.
Relay Outputs	Lights, alarm, door opener, etc.
RS-232 Serial Port	Weigh scale, laser scanner, printer, etc.

recorder and manually digitized and entered into the computer for analysis. This approach is inefficient and prone to errors. Moreover, only limited data points can be generated. In the on-line approach, a computer logs the data automatically— thousands of data points being practical. The resulting data is free of experimenter-induced biases, though not necessarily of experiment-induced errors. The data can be transferred directly to a disk for storage and subsequent analysis later. The computer can be an active participant in data collection by controlling the functions of the process, such as switching the units on or off and activating the devices.

Automatic factory data collection (FDC) is a necessity for manufacturers supplying to automotive companies and the U.S. Department of Defense (DoD). Other discrete-part manufacturing companies also find FDC helpful in controlling the information flow. FDC is not a cost-saving tool; rather, it is a tool for making timely decisions to enhance resource utilization. What type of data should be included in automatic FDC? A rule of thumb is that any data with a life cycle of 48 hours or less is a prime candidate.

Most sensors used in data collection produce either a voltage or a current that can be converted to voltage, or a series of pulses the data logging system counts. Output signals are also one of these two types. Software used in logging data consists of subroutines that can "talk" to specific devices connected to the system. The user might issue the following commands: Read the voltage on input at connector A, turn the stepper motor 500 steps in the clockwise direction, and so on.

In a modern manufacturing facility, shop-floor monitoring and control systems comprise a set of software tools that support the day-to-day execution of the plan. Such systems fill a gap between real-time, high I/O demands of low-level processing systems and an upper-level distributed processing environment.

Instrument Interconnection Standards

There are two common standards for connecting experiments or equipment in a computerized data acquisition environment. These are the Institute of Electrical and Electronics Engineers' standard IEEE-488 bus and the **Computer-Assisted Measurement and Control (CAMAC)** crate.

The IEEE-488 bus is an electrical standard that permits up to 14 devices on a single cable. Each device has an address for communication that is arbitrated by a master device. Typically, the master device is a computer. Other devices are independent and can be under computer control via a bus or manual control from their own switches. Thus, the plant devices to be run off an IEEE-488 should be designed as "slaves" to the computer with an IEEE-488 outlet, or operating with their own power supply and support electronics. These devices allow for two-letter abbreviations as control commands.

The CAMAC crate system takes a different approach to instrument interconnections. In this, a large box called a **crate** supplies power, interfacing logic, and other necessities for each device. The limitation of this approach is that devices are not operable independently, so flexibility is low. The benefit is that up to 25 devices can be connected in a single box, thus eliminating the need for 25 interconnecting cables. Besides, data transfer rates from or to the outside world are higher than with IEEE-488.

9.2   AUTOMATED DATA COLLECTION

CIM-oriented plants depend more and more on automated data collection. Collected data is analyzed, and the resulting information is distributed throughout the organization. According to a 1985 report by NASA, six different technologies are in use or under development in automated data collection for inventory management and other control needs of CIM. These are:

bar codes
optical character recognition (OCR)
vision or image processing
radio frequency identification (RFID)
magnetic
voice

The principles of these technologies are briefly described here; their advantages and limitations are summarized in Table 9.2 (pp. 309–311).

Bar Codes

Automatic data collection is essential for streamlining the flow of information in CIM environments. Bar coding is one technology suitable for this task. It is faster than keyboarding: Compared to 60–80 words per minute throughput of keyboarding with error rates as high as 40%, bar coded scanning technology can handle 1,400–2,100 words per minute with one error in 10 million reads. Bar coding suffers from one major limitation: Sometimes scanning fails to read a bar code. It is not an error of reading, it simply is a breakdown. You may have noticed this at grocery checkouts. While a cashier can solve the problem promptly by reattempting to scan the item, in automated manufacturing nonreads can halt the entire operation. RFID and voice technology are being developed as alternatives.

Bar code technology was invented in 1949 by Woodland and Silver (Box 9.1). In the late 1960s, according to Bert Willoughby (Doyle, 1985), "The two things that gave bar coding the shot it needed were the laser and microprocessor. In a technological sense, they were the driving force. The microprocessor even more so because it was instrumental in both printing and reading." The first bar code scanner was installed in late 1960 in a Kroger supermarket in Cincinnati.

The symbols used in bar coding correspond to specific combinations of digits, letters, or punctuation marks. Scanners read the symbols (codes) by shining light at them and monitoring the reflected light. Bar code scanners often employ a laser, since its concentrated coherent light maintains focus over a wider depth of field. Thus, labels need not be positioned precisely under the scanner. Figure 9.1 illustrates this benefit of the laser over conventional light sources.

Bar codes are alternating dark bars and light spaces. The individual widths of these bars and the spaces contain the coded information; their heights carry no interpretive information. The purpose of bar heights is merely to ensure that the

---

| BOX 9.1 | *The Fathers of Bar Code* |

October 20, 1949 was the birthday of bar code. On that date Norman J. Woodland of Ventnor, New Jersey, and Bernard Silver of Philadelphia, Pennsylvania, filed a patent application titled "Classifying Apparatus and Method." The inventors described their invention as relating "to the art of article classification. . .through the medium of identifying patterns."

The invention disclosed automatic apparatus for classifying things according to photo-response to lines or colors coded with classifying instructions, imprinted or attached to the things to be classified.

Woodland and Silver not only disclosed the first optical bar coding symbology, but they also described an apparatus that could automatically scan the symbol and take action based on the scanned information.

Source: From "The Fathers of Bar Code" by R. Adams, March/April 1986, *Bar Code News*, p. 18.

---

scanning device does not miss the barred area. In general, the wider the bar coded message, the longer the bars. There are several coding systems, each with a fixed set of dimensions for the bars and spaces. Both bars and spaces have certain tolerances.

Several bar coding schemes have been proposed. Figure 9.2 shows common bar code symbols and their orientations. The most common bar coding system is **Code-39**. Bar coding in manufacturing has evolved around this code. As a general-purpose system, Code-39 requires no special technology to produce; a full uppercase alphanumeric character set can be printed using any printing technology. Moreover,

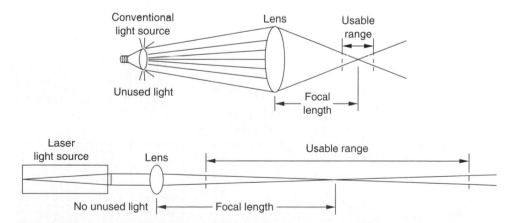

Figure 9.1
The usable range (depth of field) of laser light is significantly more than that of ordinary light, allowing even imprecise labels to be read.
Source: Reprinted with permission from *Robots and Manufacturing Automation* by C. R. Asfahl, 1992, New York: Wiley. Copyright 1992 by John Wiley & Sons.

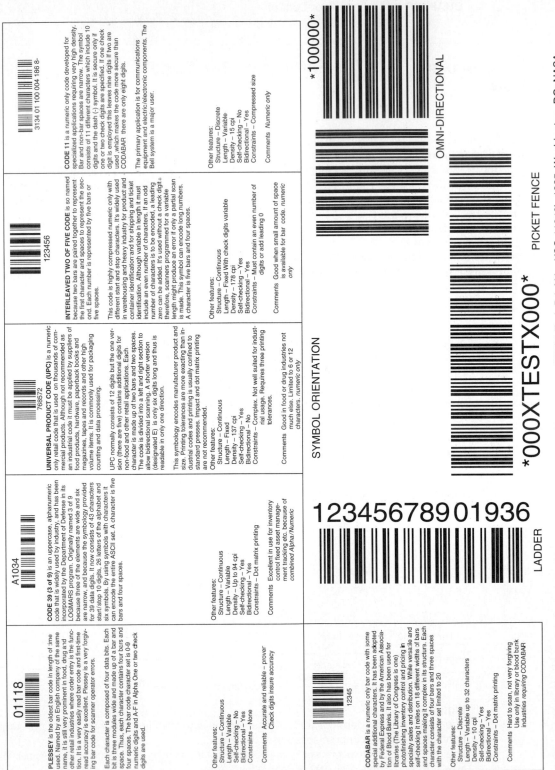

Figure 9.2
Common bar code symbols

01118

PLESSEY is the oldest bar code in length of time used. Named by an English company of the same name, it is still very prominent in food, drug and other retail industries where order entry is the function. It is a very easily read bar code and first-time read accuracy is excellent. Plessey is a very forgiving bar code for scanner operator errors.

Each character is composed of four data bits. Each bit is three modules wide and made up of a bar and space. Thus, each character contains four bars and four spaces. The bar code character set is 0-9 numeric digits and A-F in Alpha One or two check digits are used.

Other features:
Structure – Continuous
Length – Variable
Self-checking – No
Bidirectional – Yes
Constraints – None

Comments Accurate and reliable – prover Check digits insure accuracy

---

A1034

CODE 39 (3 of 9) is an uppercase, alphanumeric code that is widely used by industry, and has been incorporated by the Department of Defense in its LOGMARS program. Originally named 3 of 9 because three of the elements are wide and six are narrow, and because the symbology provided for 39 data digits. It now consists of 43 characters start/stop 10 digits, 26 letters of the alphabet and six symbols. By using symbols with characters it can encode the entire ASCII set. A character is five bars and four spaces.

Other features:
Structure – Continuous
Length – Variable
Density – Up to 94 cpi
Self-checking – Yes
Bidirectional – Yes
Constraints – Dot matrix printing

Comments Excellent in use for inventory control fixed asset management tracking etc. because of *combined Alpha/Numeric*

768572

UNIVERSAL PRODUCT CODE (UPC) is a numeric only retail code that is used on thousands of commercial products. Although not recommended as an industrial code it must be applied by suppliers of food products, hardware, paperback books and magazines, tapes and records and other high volume items. It is commonly used for packaging counting and data processing.

UPC normally consists of 12 digits but the one version (there are five) contains additional digits for non-food and other retail applications. Each character is made up of two bars and two spaces. The code is divided into a left and right section to allow bidirectional scanning. A shorter version (designated E) is only six digits long and thus is readable in only one direction.

This symbology encodes manufacturer product and size. Printing tolerances are more exacting than industrial codes and printing is usually confined to standard presses. Impact and dot matrix printing are not recommended.

Other features:
Structure – Continuous
Length – Fixed
Density – 137 cpi
Self-checking – Yes
Bidirectional – Yes
Constraints – Complex. Not well suited for industrial usage. Requires three printing tolerances.

Comments Good in food or drug industries not much else. Limited to 6 or 12 characters, *numeric only*

123456

INTERLEAVED TWO OF FIVE CODE is so named because two bars are paired together to represent the first character and spaces to represent the second. Each number is represented by five bars or five spaces.

This code is highly compressed numeric only with different start and stop characters. It's widely used in warehousing and heavy industry for product and container identification and for shipping and ticket identification. Although variable in length it must include an even number of characters. If an odd number of characters is to be encoded, a leading zero can be added. It's used without a check digit therefore, scanners programmed for a variable length might produce an error if only a partial scan is made. This symbol can encode long numbers. A character is five bars and four spaces.

Other features:
Structure – Continuous
Length – Fixed With check digits variable
Density – 178 cpi
Self-checking – Yes
Bidirectional – Yes
Constraints – Must contain an even number of digits or add leading 0

Comments Good when small amount of space is available for bar code, *numeric only*

3134 01 100 004 186 8-

CODE 11 is a numeric only code developed for specialized applications requiring very high density. Bar and non-bar spaces are narrow. The symbol consists of 11 different characters which include 10 digits and the dash (-) symbol. It is secure only if one or two check digits are specified. If one check digit is employed this leaves nine digits If two are used, which makes the code more secure than CODABAR there are only eight digits.

The primary application is for communications equipment and electric/electronic components. The Bell system is a major user.

Other features:
Structure – Discrete
Length – Variable
Density – 15 cpi
Self-checking – No
Bidirectional – Yes
Constraints – Compressed size

Comments *Numeric only*

12345

CODABAR is a numeric only bar code with some special additional characters. It has been adopted by Federal Express and by the American Association of Blood Banks. It also has been used for libraries (The Library of Congress is one) photofinishing inventory control and pricing in specialty sales and distribution. While versatile and self-checking it relies on 18 different widths of bars and spaces making it complex in its structure. Each character consists of four bars and three spaces with the character set limited to 20

Other features:
Structure – Discrete
Length – Variable up to 32 characters
Density – 10 cpi
Self-checking – Yes
Bidirectional – Yes
Constraints – Dot matrix printing

Comments Hard to scan, not very forgiving Use only in library or blood bank industries *requiring CODABAR*

1234567890 1936

SYMBOL ORIENTATION

000XTESTX000

LADDER

100000

OMNI-DIRECTIONAL

PICKET FENCE

Reprinted with permission from *Advanced Data Collection for Inventory Management* (NASA Tech Brief No. KSC-11349), 1985, Washington, DC: NASA.

Code-39 is easily read and reliable. It is specified by most manufacturing companies, including those in the automotive industry, as well as by the U.S. Department of Defense (DoD) and the Department of Energy (DoE). DoD's **LOGMARS (logistics applications of automated marking and reading symbols)** bar code affects 50,000 suppliers, who are required to use this standard's symbols on their products.

The major limitation of Code-39 is the large surface area it requires, resulting in low density (information per unit area). Alternative codes provide higher density, that is, more information in a limited space.

The universal product code (UPC) on grocery labels differs from Code-39. Retail sales have standardized around UPC, while meat packing and fiber box industries are based on Interleaved-2-of-5-Code (I-2/5-Code). These codes cannot be printed using movable type systems, have poor read-reliability, and do not offer a complete alphanumeric character set.

The initial attempt at using bar codes in manufacturing was made by the Automotive Industry Action Group (AIAG). It first published a symbol specification for shipping and parts identification based on Code-39. Since then, five additional standards have been published for primary metals, individual parts, containers, data identifiers, bar code evaluation guidelines, and vehicle identification numbers. Many suppliers to automotive companies are complying with these standards. AIAG specifies a four-tier format: The first tier is for the part number (type), the second for the unit of measure, the third for the supplier (source), and the fourth for the serial number.

Bar code systems are as good as the printed symbol. Therefore, the label material and printing technology used for labeling are important considerations. The material may be paper, mylar, plastic, metal, cloth, or foil. Printing technologies include dot matrix, ink jet, formed character impact, thermal transfer, photographic, metal etching, laser print, and ion deposit.

Modern bar code systems based on large-scale integration (LSI) technology ensure reliable reading. PC-based systems enable more than one code to be read, and programmability options allow more flexibility.

Scanners.   A basic element of any bar code system is the **scanner**. Scanners emit a light beam and detect its reflection to differentiate between dark bars that absorb light and light spaces that do not. Signal durations determine the widths of the bars and spaces. The information is sent to the reader, which uses decoder logic to validate the existence of the code based on a comprehensive decode algorithm.

Bar code scanners are either contact or noncontact type. Contact-type scanners physically touch the symbol or the protective covering over the symbol. Such scanners are typically **wands** (pens) or **slot readers**. Noncontact scanners, on the other hand, are activated even when held some distance away from the symbol. They are either a fixed-beam or moving-beam type.

In wand scanning, the wand is placed in the margin at one end of the symbol and whisked smoothly across the symbol to the margin at the other end. Any failure to scan across all the bars and the spaces results in a no-read. Slot readers are scanners designed specifically for reading bar codes on identification cards, credit cards, or personnel badges. The card is passed down the slot reader track to read the code. Figure 9.3 shows some contact-type wands and slot readers.

Portable fixed-beam noncontact scanners can read the codes without touching them—even through a thick glass covering. To ensure that codes will always be read, these scanners have a high depth of field. They can also be swept manually across the bar code, much like the wands, from one margin to the other. In the case of portable moving-beam noncontact scanning devices, a moving mirror system automatically seeps a light source or a laser back and forth across the symbol. The user only needs to point the scanner at the code and pull the trigger to read it. Depending on the ambient light, symbol resolution, and bar-to-space width, the depth of field varies from one inch to two feet. Figure 9.4 shows two common types of hand-held noncontact scanners.

Stationary models of these scanners can be used at fixed locations in a plant, for example, at workstations or for **work-in-process** (**WIP**) monitoring. When symbols are affixed to containers, totes, cartons, or pallets, such scanners can read them without human intervention. Stationary fixed-beam models project a focused light beam normal to the motion of the object being scanned. The moving-beam type focuses a moving spot in an area called the scanning curtain. The symbol entering this area is scanned several times depending on its size and how fast the object is moving. Figure 9.5 shows stationary bar code scanners in use.

Portable hand-held readers are battery powered. They store scanned data in a solid-state memory and later download it to a computer via direct link, phone line, or FM transceiver. Stationary models can be stand-alone or on-line under the control of a host computer engaged in real-time data collection. Serial ports can interface with an array of digital devices such as CRTs or printers, which may all be networked. If required, both types of readers can be equipped with a variety of options, including keypad and display.

Scanners and readers are not the same. Scanners are front-end data collection sensors for readers, which power them and to which they are attached. Scanners interface with and provide data to the reader which, in turn, interfaces with and provides data to the computer. Together they constitute a bar code data collection system.

Optical Character Recognition

Like bar codes, optical character recognition (OCR) recognizes and processes symbols. But unlike the bar code system, which interprets data coded in a series of bars and spaces, OCR devices interpret human readable characters for computers. As wand or slot readers, OCR scanners collect character information in the form of pixels. Data can be scanned either from the left or the right. Typical character fonts are OCR-B, found at the bottom of the UPC symbol on grocery items, and OCR-A, found on DoD bar code labels, paperback books, and retail clothing tags. The average OCR scanner can read 20-200 characters per second, high-speed systems as many as 1,200 characters per second.

Feature extraction or template comparison are the two techniques for decoding data scanned by OCR systems. The former method compares character features

Figure 9.3
Bar code contact wands and
slot readers

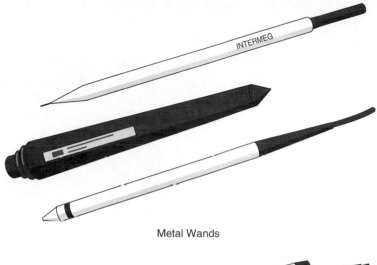

Metal Wands

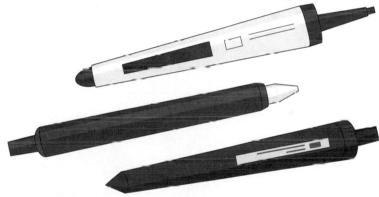

Plastic Wands

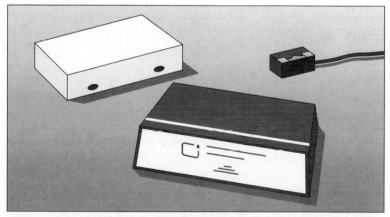

Slot Readers

Source: Reprinted with permission from *Advanced Data Collection for Inventory Management* (NASA Tech Brief No. KSC-11349), 1985, Washington, DC: NASA.

Figure 9.4
Hand-held bar code scanners

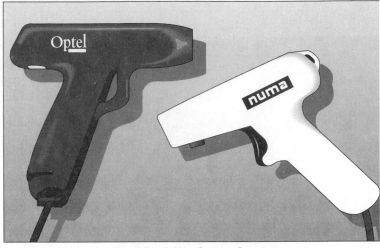

Fixed-Beam Non-Contact Scanner

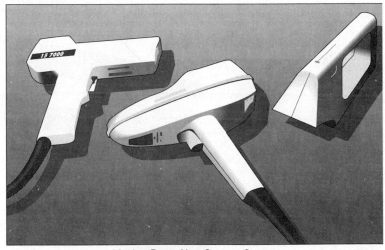

Moving-Beam Non-Contact Scanner

Source: Reprinted with permission from *Advanced Data Collection for Inventory Management* (NASA Tech Brief No. KSC-11349), 1985, Washington, DC: NASA.

such as vertical, horizontal, or diagonal lines and loops with those stored in the computer memory. The latter method compares the character, pixel by pixel, after it has been decoded in binary data form. For each character, the computer memory contains an array of pixels called a character template.

Vision or Image Processing

In vision or image processing systems, computers analyze and interpret images. Though there may be different approaches to analysis, most vision systems begin the

Figure 9.5
Stationary bar code scanners

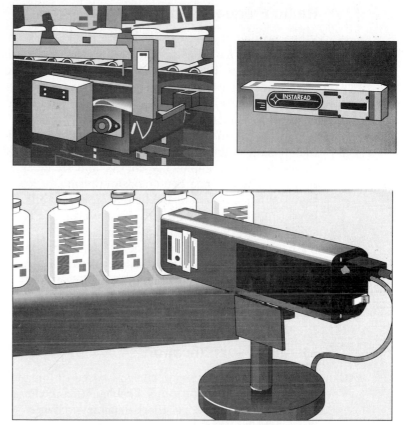

Source: Reprinted with permission from *Advanced Data Collection for Inventory Management* (NASA Tech Brief No. KSC-11349), 1985, Washington, DC: NASA.

task with a camera scene divided into pixels. The computer compares the pixels to identify prominent object features such as edges or holes. Comparing these features with those of the images stored in memory allows recognition.

Cameras used in vision systems are either vidicon, similar to a commercial TV camera, or **CCD** (**charge-coupled device**). A third type is the CCPD (charge-coupled photo-diode) image sensor. Both CCD and CCPD cameras are based on solid state electronics.

Vision systems can carry out a variety of tasks in seven general categories: (a) gaging, (b) verifying, (c) identifying, (d) recognizing, (e) locating, (f) detecting flaws, and (g) multimedia integrating. Identifying includes all the tasks in which symbols determine an object's identity, whereas recognizing uses observed features of the object. Multimedia integrating combines the image data with word processing, database, graphic and communication systems. This versatility of integration is attractive for CIM applications. Vision system applications in manufacturing include sorting, material handling, process control, machine monitoring, safety, guidance, and more.

Radio Frequency Identification

When no line of sight exists between the scanner and the identification tag, as in some material handling applications, or when read/write capability is required, **radio frequency identification (RFID)** is the answer. The object being tracked has a transponder that transmits a specific radio frequency, representing a unique signature or data stream that the transmitter or reader can interrogate. The antenna picks up the signal.

RFID systems are based on transmission of a radio signal and its obstruction by the object if in the capture window. The principle is illustrated in Figure 9.6, which shows the basic components: control unit (reader), antenna, and coded identification tag. The antenna continuously transmits a low-wattage (1–7 mW) microwave signal. When a tag enters the field of view, the reflected signal gets frequency-modulated. Coded tags may be active or passive. The active tag is battery-operated, can store several Kbytes of data, and can add, delete, or change the data in the tag on the basis of the key code received. RF tags carry predetermined process information that local scanning logic stations at each workstation can read. After completing the programmed work, the station adds a relevant message to the tag. At the end of an assembly line, a computer can off-load all the information, thus producing a complete record of the assembly and test functions.

Magnetic Identification

Magnetic identification systems are based on magnetic stripes, similar to those on the back of most credit cards, or magnetic ink characters, similar to those on bank checks and deposit slips. Electromagnetic charges encode the information on the stripe, which is then decoded for the computer. This technology is in wide use for ID badges and time and attendance records, since large amounts of data can be stored in a single stripe.

Voice Technology

Sometimes voice technology is used in conjunction with other systems to enhance identification. This technology falls into three categories:

1. speech synthesis—enables computers to "talk" with people (a C-P type of communication, Figure 4.1, Chapter 4)
2. voice input—converts spoken data into digital form for storage and processing by computers
3. voice recognition—enables computers to understand the human voice

The operator speaks into a microphone connected to the voice recognition system, which recognizes words by referring to a preprogrammed vocabulary. The system is initially taught to understand words spoken by the person in the prevailing environment, so background noise is not a problem once the system has been trained.

Figure 9.6
Principle of radio frequency
identification

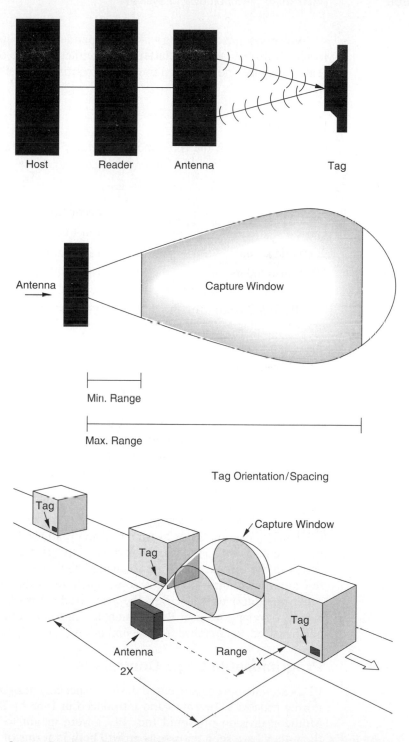

Source: Reprinted with permission from *Advanced Data Collection for Inventory Management* (NASA Tech Brief No. KSC-11349), 1985, Washington, DC: NASA.

Voice systems may be speaker-independent or speaker-dependent. While the former is discrete only, the latter can be either discrete or continuous. Discrete systems recognize one word at a time, while continuous systems understand even fluent speech.

Comparison

The error rate with the three most commonly used automatic identification devices are:

Device	Error Rate
OCR	1 in 1,000
Bar code readers	1 in 3 million
RF transponders	1 in a billion

Table 9.2 compares the advantages and disadvantages of the automatic data entry technologies discussed in this section.

9.3   CONTROL TYPES

CIM operations are based on various types of control such as on-off, sequential, or motion. The controllers vary from a simple sensor-based switch—for example, for turning a heat-treatment furnace on and off—to sophisticated devices involving complex logic. Sequential control is used for operations that must be carried out in a given sequence. Each operation or event lasts a fixed time or until a trigger signal from the sensor controlling the process changes its state. Motion control based on the feedback principle increases precision, an example being spindle positioning on a machining center.

On the basis of the energy source, control systems may be mechanical, electro-mechanical, pneumatic, or hydraulic. Usually, an electronic or computer control aids the system. In a CIM plant, computer control is predominant since its logic being coded in the software is easier to change. General-purpose computers offer only a few input and output ports, however, which limits the extent of interfacing with the other plant devices. Moreover, the associated software is comparatively more difficult to develop and use. Programmable logic controllers (PLCs) were developed, and are used, to circumvent these limitations.

Programmable Logic Controllers

PLCs are microprocessor-based devices especially designed for controlling and monitoring processes. Invented and introduced in 1968 by Bedford Associates, now the Modicon division of Gould Inc., PLCs were meant to replace relay panels. Since then, they have seen enormous growth both in terms of sophistication and popularity in industry. The inventing company also coined the term PLC. Today PLCs are almost a necessity for CIM, since they are used not only to control a process but also

Table 9.2
Advantages and disadvantages of automatic data entry technologies

Terminology	Advantages	Disadvantages
Bar Coding	• Ability to read from a distance (location flexibility—a few inches to 5 feet) • Inexpensive labels • Only computer printable and computer readable language • Speed—easy and fast • Low susceptibility to misreads • Compact size • Light weight • Fixed-beam scanners are unaffected by bar codes in different colors • Symbol placement is not critical for stationary moving-beam scanners • Many bar code labeling options are available	• No universal code for all users • Space (label size) required by bar code itself is sometimes a problem
OCR	• Same printed characters are read by people and machines	• OCR has 300 times greater probability of data entry errors than bar coding. • Scanners must make contact with label/characters—no flexibility • Wand scanning requires careful orientation—scanner must be aligned with a row of small characters • High-speed scanning of moving items is impractical • Hardware is expensive—has not come down in price as expected • Manufacturers/vendors have not been cooperative in source-marking
Vision	• Can handle tedious, repetitive tasks better than human operators • Can dimension the scene/device • Can read bar code labels • Can read OCR labels/characters	• Each system must be designed for the application—each application has its own set of unique characteristics • Amount of data to process is *very* large • Requires larger/faster computer

Table 9.2 *continued*

Terminology	Advantages	Disadvantages
Vision *continued*	• Can handle a wide variety of input provided the software is developed for the type of input • Input is real time	• High data storage requirements dictated by the image resolution (number of pixels) required • Extensive software development required—each specific application usually requires its own type of image enhancement and/or pattern recognition algorithms • Well-defined image presentation required for maximum speed and reliability • Multiple cameras are required when there is no way to predict orientation • Image distortion may cause template matching problem when an object being scanned is undergoing acceleration/deceleration in the imaging cycle • Color processing compounds the amount of data and processing required • When the distance from the camera varies by large distances, the apparent size of the image is changed, causing pattern recognition problems
RF	• Provides highly directional code reading • Has the ability to read through solid, non-conductive objects (does not have to be "seen" to be read) • Tags are not affected by harsh environments • Potentially, SAW Tags may be available for pennies when mass production problems are solved	• Errors will occur when two or more tags are in the same capture window at the same time • Tag costs are very high • Readers are stationary—not portable

Terminology	Advantages	Disadvantages
Magnetics	• Can be recorded at higher density than bar codes • Offers the opportunity to alter the data	• Cannot be read from a distance (must make contact) • Cannot be reproduced • Susceptible to electromagnetic interference (important where data security is concerned) • Expensive
Voice	• Simplification and ease with which data can be collected • Speed—real time as user speaks • Operator friendly • Can handle large amounts of *variable* input data • No label or tag required on item/part • Frees hands of operator • Enhances data entry when used along with other automatic identification technologies such as bar coding • Computer generation of speech from text now practical because of high speed/real time processing integrated chips	• Speaker-independent: — high cost — highly susceptible to background noise interference — *very* limited vocabulary — may not be able to detect word endings (must distinctly pause between words) • Speaker-dependent: — must be trained for individual speakers (this is an advantage for security applications) — cannot distinguish between acoustically similar words — limited vocabulary

Source: Reprinted with permission from *Advanced Data Collection for Inventory Management* (NASA Tech Brief No. KSC-11349), 1985, Washington, DC: NASA.

to monitor it. According to Asfahl (1992), the PLC is "one of the most genius devices ever devised to advance the field of manufacturing automation. So versatile are these devices that they are employed in the automation of almost every type of industry." He adds, "thousands of these devices go unrecognized in manufacturing plants—quietly monitoring security, managing energy consumption, and controlling machines and automatic production lines."

A formal definition of a PLC provided by the **National Electrical Manufacturers Association (NEMA)** is

> a digitally operating electronic apparatus which uses a programmable memory for the internal storage of instructions by implementing specific functions such as logic sequencing, timing, counting, and arithmetic to control, through digital or analog input/output modules, various types of machines or processes. The digital computer which is used to perform the functions of a programmable controller is considered to be within this scope. Excluded are drum and other similar mechanical sequencing controllers.

Thus, a PLC is an interfacing device that enables logic operations on input signals to generate output responses according to the set of programs in its memory. An array of devices and sensors such as pushbuttons, timers, counters, photocells, limit switches, thermocouples, and potentiometers can provide the input signals.

PLCs have replaced mechanical relays by logic gates. The principle of a PLC can be understood better by considering electromagnetic relays and their operation. Circuits based on relays switch devices on and off in a given sequence, with the timing controlled in a way similar to the wash/dry cycle in household dishwashers. The relay consists of a coil and contacts that open or close the circuit. The contacts are made or broken when the coil is energized electromagnetically (electric current flowing through the coil develops temporary magnetism in the core, which is attracted or repulsed, and the resulting movement closes or opens the circuit).

Operating and programming a PLC are easier to comprehend with a ladder logic diagram, also called a **ladder diagram**. The terminology is most appropriate, since the diagram, as shown in Figure 9.7, does look like a ladder having two rails (vertical lines) and several rungs (horizontal lines). Ladder diagrams have been used in industry for several years to document the connection circuits for switches, motors, heaters, timers, relays, and similar devices that control the operation of a process such as a parts-cleaning station. The rails represent electrical power and electrical return, while the rungs are the relay and other components. The left rail is the "hot" line of an electric circuit, usually energized at 120 volts AC to ground, while the right rail is the neutral line. For logic circuits, the left rail is energized at lower voltages—5 volts DC for logic "1" and 0.7 volts for logic "0." The relays are made up of logic instructions in the PLC.

The following example (Pond, 1993) illustrates the principle underlying PLCs. Consider the requirement of automatic on/off switching for an electric motor. (The motor is shown in the lowest rung of the ladder diagram in Figure 9.7.) The vertical rails are the power and return lines. Operating the circuit involves the following sequence:

Figure 9.7
A ladder diagram that explains its use in automatic on/off switching of a motor

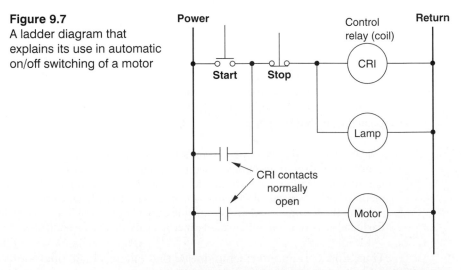

Source: Reprinted with permission from *Introduction to Engineering Technology* by R. J. Pond, 1993, New York: Merrill/Macmillan.

1. *Initial conditions.* The power line is continuously energized. The motor is off. All rungs are open and no current flows through them. The stop pushbutton is in N.C. (normally closed) state.
2. *Press START.* The control relay coil in the upper rung is activated. The current energizes the relay and the motor-run lamp switches on. An energized relay closes the normally open (N.O.) and opens the normally closed (N.C.) contacts. In this case, its effect is to close the N.O. switch, which is in parallel to the START button. This keeps the current flowing through the relay for energization even when the START is released back to its initial open condition. Further energization closes the other N.O. switch (in the motor rung), which allows current flow through the motor to turn it on. Thus, the motor gets switched on and stays so until the STOP button is pressed. In other words, the control relay is holding itself in.
3. *Press STOP.* The N.C. stop button opens. This cuts off current to the control relay, thus de-energizing it. The lamp switches off, the two N.O. contacts are opened, and the motor switches off. This returns the circuit to its initial conditions. Another cycle is set off when the START is pressed again.

Figure 9.8 shows the keys of a typical PLC keyboard. The keys depict the components that are used to construct ladder diagrams. Using such a keyboard, the operator programs the PLC to digitally control a specific manufacturing process.

Figure 9.9 shows the role of a PLC in a more sophisticated task. In this application, it automatically controls the process comprising robots, machine tool, assembly operation, and material movement to and from the workstation.

When PLCs were introduced to replace the hard automation of relays, a prime design requirement was that they be comprehensible to their users in the plant, especially maintenance technicians. Since the users already understood the ladder diagram associated with relays, PLCs were developed on ladder logic to achieve smooth transition to PLC technology. While this approach had its merits in the

Figure 9.8
Keyboard of a typical PLC
depicting the components
used in constructing ladder
diagrams

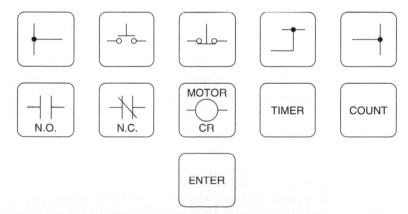

Source: Reprinted with permission from *Introduction to Engineering Technology* by R. J. Pond, 1993, New York: Merrill/Macmillan.

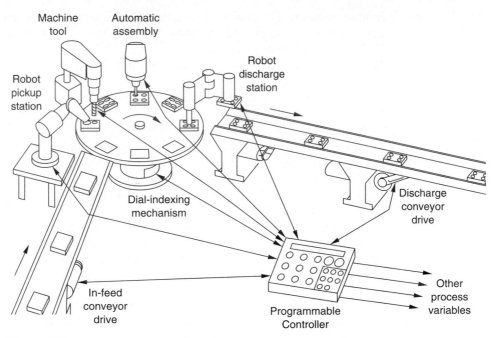

Machine
tool

Automatic
assembly

Robot
discharge
station

Robot
pickup
station

Dial-indexing
mechanism

Discharge
conveyor
drive

In-feed
conveyor
drive

Programmable
Controller

Other
process
variables

Figure 9.9
PLC as a controller of a sophisticated workstation
Source: Reprinted with permission from *Robots and Manufacturing Automation* by C. R. Asfahl, 1992, New York: Wiley. Copyright 1992 by John Wiley & Sons.

beginning, for CIM it created another barrier due to the special way of programming PLCs. This drawback has lately been realized and overcome; modern PLCs can be programmed using common programming languages as well.

Unlike PCs, PLC technology was developed exclusively for manufacturing. PLCs' design took into account the harsh environment in which they were to operate, the skill level of the users, and the need for more I/O ports. All these resulted in robust design and user-friendliness, giving PLCs an early edge over PCs. Box 9.2 brings out the differences between PLCs and PCs.

PLCs are also effective in real-time control, a common requirement with manufacturing processes that allows the PLC to collect data and control the process. Their response in milliseconds is fast enough for most processes, for example, activating an emergency stop. PLCs are relatively inexpensive, user friendly, and easily interfaced with manufacturing workstations and other devices at the shop floor.

Fully electronic and with no moving parts, PLCs offer the following advantages:

longer life
higher reliability
lower power consumption
lower cost relative to the mechanical relay-based systems
flexibility of modification and adaptability
less space requirement due to smaller size

BOX 9.2 *PLCs and PCs*

"Low-end controllers represent unique solutions for many industrial applications," says Andrew Chatha, president of Automation Research Corp. (ARC). "Using these units as simple electro-mechanical relay replacements is the largest application. Most are stand-alone configurations. An increasing number, however, can exchange data with other PLCs and personal computers on a network."

"Add powerful instruction sets, math capability, and analog I/O," he says, "and these systems allow economical phased automation in applications that used to require larger, more expensive PLCs.

Phased automation is redefining the domain of mid-range PLCs. One controller features 640 I/O points, a 0.49 microsecond scan rate, embedded diagnostics that allow 98 preprogrammed English-language error messages, and structured programming with Machine Stage Language (MSL).

"A cultural change on the factory floor is giving machine tool operators more control over their equipment, says Stewart Walton, a Texas Instruments engineer. "Ladder logic programming has traditionally been a time-consuming and difficult process. Structured programming, however, breaks complex tasks into several simple ones. For instance, using MSL, the operator can build the PLC program directly from the application's flow chart one step at a time, without worrying about the interlocking necessary with traditional ladder logic. Programming time is 50% faster and debugging is simpler. You test one small module at a time without worrying about changes affecting other program modules."

Integrated systems are likely to grow in number at the cost of high-end PLCs. Integrated systems are hybrid controllers that combine a personal computer and PLC. "Combine integrated capabilities with application software tools for effective communication and database management," says Chatha, "and you can bridge the gap between plant-floor PLCs and plant management systems, such as MRP II and quality management."

ARC believes that industrial personal computers, such as the on-board microprocessor technology, will continue to complement PLCs, not replace them. The following areas will be affected.

Supervisory Control. PLC data storage and manipulation capabilities are usually limited. PCs offer substantial memory for data storage and manipulate data. As a result, they can be powerful tools for analyzing plant data and performing statistical process control.

Operator Interface. PLCs are designed for high-speed machine control, so workers traditionally interact with the system through pushbutton panels and lights. As machine operators become increasingly involved in plant operational decisions, pushbutton panels will become unacceptable.

Programming Terminal. Most PLC companies are selling PCs to program their PLCs. PCs are easier to use than hand-held terminals. Users also can program in high-level languages and receive on-line documentation.

Source: From "PLCs and PCs," May 1990, *Manufacturing Engineering*, pp. 26–28.

◻ **BOX 9.3**	***The PLC: Still Useful After All These Years***

Quite simply they're (PLCs—programmable logic controllers) computers optimized for factory floor control. They have been around since 1969. After all, what kind of electronic control technology lasts 20 years?

According to John McElfresh of Modicon: "The PLC is a special-purpose, industrially hardened computer for the control of machines and processes. 'Special-purpose' means the programming and maintenance characteristics are specially suited to the engineering and maintenance personnel found in industrial environments. It's modular and rugged in design, built to live on the plant floor or industrial environment, which is harsh, electrically noisy, and caustic."

Pat Babington of Allen-Bradley, a manufacturer of PLCs, provides a broader definition: "The PLC is a networked, intelligent, highly reliable, highly productive control device designed to do information gathering, diagnostics, and analysis for factory automation. It is not a discrete relay replacer."

A fundamental difference between a PLC and a PC is that the PLC is specifically suited, not adapted, to and for factory use. It is not a disc-based system like a PC is. In effect, with a disc-based system, there is a constant checking of what to do next. With a PLC the answer of what to do next is inherent; there is no consulting. The program—and it's interesting to note that many PLC manufacturers are using PCs as the front-end for programming, because it's easier for someone to work with—just runs in sequence. That's all there is there. People in the PLC business talk about the PLC being "deterministic," which means that its next step is plotted: no deviation, no free will. Scanning is being performed at a rate of 10 to 20 msec.

The PLC program is stored in battery-packed RAM or EEPROM. It's not going anywhere. And the electrical fields generated in factories from walkie-talkies to induction heaters don't affect it. There is custom silicon in PLCs—counters and timers, for example—which isn't found in a PC.

Essentially, a PLC consists of a number of rectangular containers that are attached to a rack and fitted into an electrical connector known as a "backplane." One of the containers is the CPU, another a power supply, and then there is an assortment of modules for various types of analog and digital input/output (I/O). The size of a PLC is measured in terms of the I/O it can handle. The range can be considered to span from 24 to something like 64,000 I/O.

A couple of things were recognized a few years ago. One is that the PLC is a tremendous wellspring of real-time factory data. The other is that the factory is an important element within a manufacturing business, one that should be connected to other aspects of the business, such as purchasing, accounting, and so on. These two—and perhaps others (for example, the concept of CIM)—led to the PLC-computer connection.

Source: From "The PLC: Still Useful After All These Years," February 1990, *Production*, pp. 34–38.

Box 9.3 points out PLCs' continued usefulness even after a quarter of a century, a rare occurrence with electronic devices.

Like microcomputers, PLCs have proven to be a boon to manufacturing. Over the years, they have evolved beyond recognition. Today, they work in concert with various overlapping functions that support CIM.

9.4   SENSOR TECHNOLOGY

The developments in sensor technology play a major role in achieving CIM at the operational level in manufacturing plants. In this section, touch probes and fiber-optic sensors and their networking are briefly discussed.

Touch Probes

Touch probes are electromechanical sensors that can locate the position of tools, workpieces, or other objects by touching them. They are one of the basic building blocks for untended production in FMCs or FMSs. A probe system comprises three subsystems: (a) probe head and associated hardware, (b) probe interface into the CNC, and (c) software. The software handles the probing cycles, as invoked by programmers, to operate the probe system. Examples of probing cycles are generation of machine axis moves, storage of captured data in proper registers, mathematical analysis of data to extract information, and so on.

Fiber-optic Sensors

Fiber-optic sensors are used to measure a variety of physical phenomena such as pressure or vibration. They are useful in automated factories because of their small size and immunity to noise.

Fiber-optic sensors offer several advantages over other sensors. The major advantages are immunity to electromagnetic and radio frequency interference (RFI), electrical isolation, and operability at high temperatures over a wide bandwidth. While these characteristics are especially important in structures such as space vehicles, where fiber-optic sensors embedded in composite materials emit frequency, phase, and intensity data, they are also desirable for manufacturing.

A typical use of fiber optics in manufacturing is as displacement transducers. Fiber-optic displacement transducers are basically amplitude sensors in which transmitted and received light signals interact as light is reflected off a target. As the target moves away from the sensing fibers, the intensity of reflected light increases until a maximum is achieved. Beyond that point, the intensity diminishes. These transducers have been used in repeatability testing of precision lead screws of machine tools and in modal analyses of the read/write heads for hard disk drives. Fiber-optic sensors are the sensors of the 1990s and beyond. Lower prices have made them more attractive recently.

Box 9.4 describes applications of smart sensors in modern manufacturing.

Sensor Networking

One of the main advantages of fiber-optic sensors is that the fiber can both sense and transmit. This results in lower cost per element, which leads to cost-effective networked sensor systems. The typical fiber-optic network consists of a common light source, an optical fiber carrying light to each sensor, and another fiber carrying the returning modulated light to a photodetector and signal-processing circuitry.

BOX 9.4   *Smart Sensors Respond to Factory Stimuli*

Our quest for better quality, lower costs, and faster production is behind the use of high-performance sensors in manufacturing. Machine vision systems, smart sensors, electronic and electromechanical gages, and coordinate measuring machines now make all sorts of measurements in factories. And there are plenty of parameters to be measured—liquid and gas flow and the level, motion, pressure, temperature, and position of an object. These sensors are made possible by fiber-optic, microwave, capacitance, infrared, ultrasonic, laser, magneto-restrictive, and inductive techniques. Thanks to the microprocessor, they are getting smarter.

While everyone agrees that highly instrumented production systems improve product quality and reduce costs, measuring this improvement is not easy. The task has led to the development of the Sensor Center for Improved Quality (SCIQ) at the Industrial Technology Institute (Ann Arbor, Mich.). The Center's goal is to justify the cost of sensor-generated information by isolating and evaluating costs that result from not using sensors, such as when bad parts go through a production process undetected, and dealing with sensors in a CIM environment.

Pressure Is Most Common.   Measuring pressure is one of the most common forms of sensing in factories. The three most popular techniques are the traditional piezo-resistive (strain-gage), piezo-electric and linear variable differential transformer (LVDT) methods.

Pressure sensors are becoming more accurate and are integrating more electronics. The use of thick-film and thin-film deposition techniques, the availability of complex microprocessor chips, and modified strain-gage materials all contribute to greater accuracy at higher temperatures. Many sensors now have on-board amplifiers for more usable outputs and temperature-compensation circuitry for increased stability with changes in the ambient temperature. Digital signals are also being used more to improve reliability.

Smart sensing plays a key role in field calibration of pressure sensors. But with a pressure transmitter that contains a microprocessor, this can be done remotely, either from a control room or using a hand-held terminal in the field. The microprocessor can further reduce the need for recalibration by compensating for changing calibration parameters and it allows remote diagnostics.

There's a natural alliance between silicon pressure sensors and the microprocessor because they are made of the same material. Eventually, a single chip will have both sensing and control elements. In fact, present efforts to make micromachined silicon actuators (i.e., valves, nozzles, and switches) could lead to an entire closed-loop control system on a chip. The chip that sensed and processed the pressure or any other parameter also would act on it by activating the appropriate controls. This would be true distributed control. For now, the trend is to use a number of sensors, all feeding to a central processing unit.

Optical glass fibers can also be used as pressure sensors. In fact, they can sense about any parameter on the factory floor. These sensors will assume an increasingly larger role in manufacturing operations, especially in process control. Because they're essentially passive devices, they are useful in hazardous environments where a spark cannot be tolerated or electromagnetic interference might hamper the sensing operation.

Optical fibers need not be made of glass. One company has produced a rubber optical fiber, which has a number of potential uses, including safety switches for doors and emergency stops and grip regulators for industrial robots. The pressure-sensitive fiber consists of a highly refractive silicone rubber inner core, a low-refractivity silicone rubber middle layer, and a thin sheath of fluorinated rubber.

Temperature Measurements.   Temperature is another important parameter in factories. Technologies for measuring it include resistance temperature detectors (RTD), thermocouple (TC), thermistors, germanium resistance sensors (GRT), diode and capacitance sensing, and infrared (IR) and optical pyrometers.

As with pressure sensing, electronics has contributed to smaller and more accurate sensors with greater intelligence. And digital output has made signal conditioning and processing easier. Sophisticated programmable temperature transmitters use microprocessors at the sensing point and in the signal-conditioning, control, and monitoring systems.

Mass-Flow Sensors.   Their cost is going down, their reliability is improving, and the bells and whistles are being added.

Motion Sensing and Object Detection.   A variety of motion-sensing techniques exist, making use of inductive, photoelectric, capacitance, magnetic, microwave, ultrasonic, and laser methods. The simplest one uses a presence sensor—usually a magnetic, inductive proximity, or photoelectric device coupled with a timer.

A variation is to use a motion detector with sensor, timer, and comparator in one package. This is useful for conveyer lines, feeders, fans, blowers, pumps, grinders, crushers, blenders, and mixers.

The trend among photoelectric and proximity sensors is toward smaller, plug-in packages. By eliminating the need to rewire sensors to the controller, plug-in units allow rapid and easy installation and replacement; it should be noted, however, that there's no U.S. interchangeability standard.

An exciting development is the use of laser technology in photoelectric sensors. These sensors emit an intense monochromatic light, which is highly collimated. . . . to qualify small parts, like fasteners and screws, at up to 300 parts per minute.

Source: From "Smarter Sensors Respond to Factory Stimuli" by D. Horn, September 1989, *Mechanical Engineering*, pp. 64–67.

The simplest network involves a single fiber-optic sensor. In multiple sensors, the responses of individual sensors need to be distinguished. This is achieved either by time-division or wavelength-division multiplexing (see Chapter 4). In the former, each sensor operates over a specific time-delay slot, whereas in the latter, each operates over a specific wavelength band.

For CIM, the possibility of distributed processing achieved through a hybrid approach to networking is attractive. This approach involves feeding passive fiber-optic sensors into an electro-optic node through fiber cables. Within the node, the optical signals are converted to an electronic form, processed, reconverted back to the optical form, and then transmitted over a fiber pathway to the central control processor.

TRENDS

❑   A trend toward embedding real-time systems into CAM hardware has been found (Puttre, 1991). Such systems respond to factory problems such as machine breakdowns by rerouting the product flow.

❑   Shop-floor control of production facilities is enormously enhanced by microprocessor-based technologies. The developments in this area continue unabated.

BOX 9.5   *Electronic Tool Identification*

For the past decade, shops striving for unattended machining have courted automatic cutting tool identification. While machine tool builders and control builders have generally not shared the same enthusiasm with the users about the importance of having this capability installed on the machine, they have worked together to implement a number of viable methods. These methods range from interfacing an actual tool gage unit into the CNC to labeling the toolholder with a bar-coded tag. Most of these methods faced limitations when functioning in an adverse shop environment, and were unable to provide all of the needed information. These obstacles have been overcome with today's Electronic Tool Identification Systems, which store tool data on an integrated circuit (IC) chip embedded in the toolholder.

Electronic tool identification systems consist of three basic parts: the code carrier that stores the information (commonly referred to as the tag), a read/write head which transfers data to and from the tags, and the interface unit consisting of a processor that manages the transfer of data.

The tag is an IC chip packaged in a variety of forms to meet a wide range of applications where computer-based management systems are employed. When used for tool identification, the tag typically selected is a small button approximately 0.5 inches in diameter and about 0.25 inches thick, which is epoxied flush into the tool holder (Figure 9.10a). The tag in this form can easily store up to 256 bytes of data (up to 7 Kbytes for larger tags). Information is stored on the chip utilizing EEPROM (electrically erasable programmable read only memory) technology, which provides an effective means of nonvolatile storage. Nonvolatile storage is a primary requirement since it would be impossible to maintain continuous power to the tags when installed in tool holders. EEPROM devices can store data for up to ten years without being connected to a power source.

When information is written to or read from the tag, its internal logic must be activated by power from an external source. This power is received inductively from the read/write head when the two units are aligned to each other. Power is passed inductively between the tag and the head without physical contact. In a similar manner, data is transmitted inductively across the same air gap. These devices are generally immune to oil, water, chemicals, grease, and metal chips, making them desirable for identification of cutting tools.

A typical block diagram of an electronic tool identification system is illustrated in Figure 9.10 b. The tool's identification tag is read as the tool is loaded into the tool chain. The data is transferred by the interface unit over a serial data line to the computer numerical control (CNC) unit. With the aid of special applications software, which manages the transfer of tool data inside the CNC, the tool data is received and loaded into the proper fields in the CNC's tool data table. Data read from the tag includes tool length, tool diameter, accumulation of cutting time, number of teeth, tool ID number, tool pocket number, and torque limit. The flexibility of the system is further enhanced by utilizing both read and write capability, which permits a tool's information to be updated. Data such as accumulated cutting time and adjusted length may be updated in the event a tool is removed and taken to another machine.

Electronic identification of cutting tools presents a bright future in production control, inventory control, and shop-floor data collection systems. The technology is available now to permit descriptive information to be permanently attached to tools, fixtures and even workpieces. This information records where they have been, what has been done to them and what needs to be done. Electronic identification is a viable technology and it is available now!

Source: From "CIM Perspectives" by G. E. Herrin, July 1992, *Modern Machine Shop*, pp. 140–142.

Figure 9.10
(a) The identification tag consists of an IC chip encased in a variety of shapes and sizes. It is 0.5 inches in diameter and 0.25 inches thick; (b) Typical method of implementing electronic tool identification, from the tool to the table.

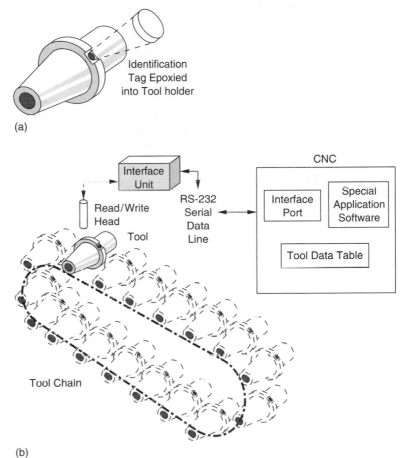

(a)

(b)

Source: Reprinted with permission from "CIM Perspectives" by G. E. Herrin, July 1992, *Modern Machine Shop*, p. 142. Copyright 1992 Gardner Publications.

- ❏ Newer sensors are smart—they are intelligent measuring devices that ensure quality data at the data collection points.
- ❏ PLCs' use in CIM plants continues to proliferate. Along with PCs, PLCs are the basic tools for flexible automation and thus are essential to the realization of full-blown CIM facilities.
- ❏ Tool identification techniques (Box 9.5) play a crucial role in enhanced control of plant operation under CIM.

SUMMARY

Shop-floor control is an important activity in manufacturing plants. Such controls attempt to bring discipline in an otherwise confused environment of the shop floor. Major considerations for achieving effective CIM shop-floor control were presented in this chapter. Without relevant data, shop-floor control is difficult. Data acquisition and logging techniques were discussed along with the role of automation in this task. The major technologies of bar codes, OCR, and others were described in detail. Programmable logic controllers and sensor technology were also discussed, since they are crucial to shop-floor control.

KEY TERMS

Charge-coupled device (CCD)

Code-39

Computer-assisted measurement and control (CAMAC)

Crate

Ladder diagram

Logistics applications of automated marking and reading symbols (LOGMARS)

National Electrical Manufacturers Association (NEMA)

Optical character recognition (OCR)

Production control

Radio frequency identification (RFID)

Scanner

Shop-floor control

Slot reader

Wand

Work-in-process (WIP)

EXERCISES

Note: Exercises marked * are projects.

9.1   Compare the CIM-related advantages and limitations of the PC with those of the PLC.

9.2   Compare the technologies of bar codes, OCR, RFID, and vision on the basis of five major criteria important in CIM.

9.3   Discuss the role of PLCs in CIM plants.

9.4*   Visit a local manufacturing plant to observe PLCs in operation. Report your observations to the class or the instructor.

9.5*   Visit a nearby manufacturing facility to study its shop-floor control needs. Discuss the strengths and weaknesses of the prevailing control methods from a CIM viewpoint, and suggest some better ones.

9.6   Circle T for true or F for false or fill in the blanks.

    a.   Sensor and transducer are merely two names for the same device.   T/F

    b.   Even a modern PLC is not a microcomputer.   T/F

    c.   The bar code commonly used in manufacturing is:

        Code-39   UPC   2-of-5-code   None of these

        (Circle your answer)

    d.   Name an appropriate standard for connecting data collection equipment in a computerized data acquisition environment of FMS.

        _____

    e.   Name five major data collection and control devices commonly used in modern factories.

        _____

        _____

        _____

        _____

        _____

SUGGESTED READINGS

Books

Asfahl, C. R. (1992). *Robots and manufacturing automation*. New York: Wiley.
Pond, R. J. (1993). *Introduction to engineering technology*. New York: Merrill/Macmillan.

Monographs and Reports

Advanced data collection for inventory management (NASA Tech Brief No. KSC-11349), 1985, Washington, DC: NASA.
Tavora Calos, J. (1988). *CIM integration tools for programmable controllers and flexible manufacturing*. SME Blue Book series. Dearborn, MI: SME.

Journals and Periodicals

Bar Code News. A magazine exclusively devoted to bar code technology and other identification techniques.

Articles

Doyle, P. C. (1985, November). Bar coding—past, present, and future. *Bar Code News*, p. 14.

Horn, D. (1988, September). Optical fibers, optical sensors. *Mechanical Engineering*, pp. 84–88.

Harding, K. G. (1991, April). Sensors for the '90s. *Manufacturing Engineering*, pp. 57–61.

Puttre, M. (1991, July). Real-time systems tackle tough tasks. *Mechanical Engineering*, pp. 55–58.

Puttre, M. (1988, March). Monitor grinding wheel dressing via sound levels. *Modern Machine Shop*, p. 252.

CHAPTER 10
Robotics and Material Handling

"Robot programming languages are almost as varied as the robots they are designed to manipulate."
— Wesley E. Synder in his book *Industrial Robots: Computer Interfacing and Control*

"Reconfigurable robots that could be assembled from a small set of general purpose components are one of the keys to lower cost and greater flexibility."
— Raj Reddy (director, Robotics Institute, Carnegie Mellon University) in *Mechanical Engineering*, June 1990

"The word robot epitomizes the public image of industrial automation. The image is partially correct because: (1) industrial robots are only part of the total automation picture, and (2) the public idea of an industrial robot is highly glorified."
— C. Ray Asfahl in his book *Robots and Industrial Automation*

"Only 25% of U.S. plants use a robot, the most common use being load/unload tasks."
— Anonymous

"For every 10,000 workers, Japan has 170 robots and the US only 20."
— *Production*, February 1991. (This statement may be misleading, since in Japan both programmable and nonprogrammable manipulators are considered robots, whereas in the U.S. only programmable manipulators are.)

In discrete-parts manufacturing, movement and handling of materials and tools represent a major activity in terms of time and cost involved. As Figure 10.1 shows, the average workpiece is actually on the machine only during 5% of the door-to-door time, and only during 30% of this time is it actually being machined. Thus, workpieces are either waiting or being transported most of the time. Reducing the moving, waiting, positioning, loading, gaging, and other idle times can increase production rate many-fold.

Figure 10.1
The actual machining time of an average workpiece in a conventional plant is a very small fraction of the door-to-door time.

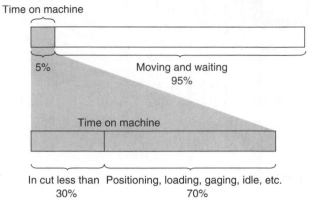

Source: Reprinted with permission from *Computer-Integrated Manufacturing Handbook* by E. Teicholz and J. N. Orr (Eds.), 1987, New York: McGraw-Hill.

Facilities planning and design, computerized plant layout, and, more recently, simulation as an analysis tool, are some of the attempts to reduce material handling costs on a long-term basis. Design improvements in the existing material handling facilities or implementation of new technologies are other alternatives. The application of robotics in loading and unloading at workstations is one such example.

The recent practice of combining operations so most can be completed in one setup, as in modern machining centers, eliminates or reduces setup and material handling costs. FMCs offer similar benefits, since most operations to process a component within the part family occur at one place. The technology of robotics as flexible automation is another tool to minimize material and tool handling costs. In a CIM environment, material handling is more streamlined due to better tracking of work-in-process inventory and tools.

This chapter discusses the role of robotics in CIM. Beginning with an overview, recent developments in robotics are presented. The fact that different languages for programming robots hinder the progress of CIM is highlighted. The value of robotic vision technology to CIM is discussed. This chapter also includes material handling considerations that pertain to CIM.

10.1   ROBOTICS

Over the years, the technology of robotics has matured. The understanding of the term robot, even with laypersons, provided it an unusually high visibility that has both helped and hampered the field of robotics. It has been helped because it has attracted the attention of both experts and novices. This resulted in enthusiasm by production managers as well as upper management toward the introduction of robots in their operations. It has been hampered by vendors' "tall" promises about robots' benefits to manufacturing. Moreover, most applications of robotics in manufacturing, especially in the beginning, created islands of automation with little or no attention to integration. Since a robot is basically NC equipment, the programming methods and languages for both machining centers and robots could have been more similar. In the late 1980s, robotic applications in manufacturing began to reflect the needs of CIM.

Overview

The history of robotics is as fascinating as the robot itself. The term comes from the Czech-Polish word *robota*, which means work. It was coined by Czechoslovakian dramatist Karel Capek for the stage theater play R.U.R. (Rossum's Universal Robots), which appeared in the U.S. for the first time in 1922. As an expression of disenchantment with technology following World War I, Capek's play gave robots a bad name; as man's servant it was shown to revolt. Later, science fiction writers, notably Issac Asimov with his 1950 novel *I Robot*, changed the robot's image to that of a friend rather than a foe of human beings. This work set the psychological and societal tone for robots' use in industry.

In the early 1970s, Unimation was founded as the first industrial company to concentrate solely on robotics. Since then, several companies have been incorporated; some survived but more failed due to the tough competition in this new, evolving technology. Now, the field has stabilized somewhat, primarily because the initial enthusiasm over robots has tempered.

A robot is basically an automated arm that can be programmed to carry out certain tasks of material handling within its geometric and power constraints. The Robotic Institute of America (RIA) defines it this way:

> A robot is a reprogrammable multi-functional manipulator designed to move material, parts, tools, or specialized devices through variable programmed motions for the performance of a variety of tasks.

From a CIM viewpoint, the key word in this definition is *reprogrammable*. This feature offers the robot flexibility, making it a CIM-conducive material handling device. The older technology for automating the tasks a robot can do today relied on fixed mechanical manipulators based on cams and gears. In fact, as mentioned earlier, in Japan, the term robot includes such manipulators within its definition.

Mechanical manipulators based on hard automation do not provide much flexibility; even minor changes in part design requires time-consuming modifications to these automatic devices. The advent of robots in the 1970s allowed dramatic improvements in this area.

Discussions of robots' historical development, types, performance specifications, application potentials, cost justification, and similar issues are outside the scope of this text. Interested readers should refer to the chapter-end titles on robotics such as Asfahl (1992). This chapter is limited only to issues of robotics that have direct bearing on CIM.

Programming

Programming is the process that enables robots to cope with the variations in tasks within the limits of **payload** and **work envelope**. Payload is the amount of weight (including the end effector's) the robot arm can handle safely, and the work envelope denotes the space boundary it can reach. Robots are programmed in exclusively developed special languages such as Victor's Assembly Language (VAL) or A Manufacturing Language (AML), or in a general-purpose language such as BASIC. Developed by Unimation for its PUMA series of robots, VAL is a popular language. Small robots such as Microbots are programmed in ARMBASIC.

Most robots include a hand-held controller called a **teach pendant**. Teach-pendant programming involves manually pressing the required buttons on the controller to enable the robot to follow a desired path. Teach pendants look similar to the controllers used for guiding cranes, except that they are sophisticated enough to remember the robot's path. While being programmed, the robot is moved slowly. Once programming is complete and error-free, it is set at the desired production speed at which it repeats its path. Programming a robot using its teach pendant suffers from the following major limitations:

1. The work cell is idle while the robot is taught.
2. Teaching in congested workspaces or changing surroundings (as in warehouses) may cause accidents.
3. The effectiveness of the moves taught reflects operator's visual acuity, complexity of tooling, and so on.
4. The robot is normally unadaptive and, hence, cannot cope with perturbations due to path, geometry, or part delivery. This results in frequent interruptions that can become costly; for example, in the automotive industry the line downtime cost is as high as $10,000 per minute. This lack of adaptivity holds true for other programming methods as well.
5. In close-tolerance assembly operations, as in microelectronics, where tolerances approach 0.05 mm, the teach pendant method is just not capable.

Obviously, teach pendant programming has limited use in CIM environments. However, teach pendants prove handy as an override tool to get the system back in operation in case of jams or interruptions.

The next less-sophisticated programming is one in which an operator holds the robot end arm and takes the robot through a dry run of the task. Again, the robot remembers the path, which it can repeat later. Commonly used in spray painting and welding, this method is also limited from a CIM viewpoint for the same reasons as teach pendant programming.

In the preceding two methods, the robot cannot do the actual work while being programmed, and hence these methods are less cost-effective in low-volume production. In one alternative, **off-line programming**, programs are developed on a simulated arm. The simulated arm may be physical or graphic. The simulated physical arm is lighter and easier to manipulate. The graphic arm is simply a computer model on the screen. With the aid of simulation software, programs are developed for the specific tasks the actual robot would perform at the shop floor. For CIM, off-line programming based on graphic models is ideal, since the task of programming can be carried out upstream at the product design stage itself, thus contributing to integration efforts.

Design and implementation of robotic systems (as well as of other flexible automation systems) often benefit from simulation. But in an integrated environment, the lack of a programming language common to both the robot and the simulation system is a drawback. When the simulation language is not a version of the robot language, translation is necessary.

Sensor-Controlled Robots

The effectiveness of robots in material handling tasks depends on the positioning and orientation of the items as they are fed to the robot. Due to its limited precision, the robot system may not be capable of positioning its **end effector** at the same point, especially after several cycles. Moreover, something may go wrong when the robot is repeating its cycles. The process is more effective if the robot can "feel" its payload and "see" its surroundings. Recent developments in robotic vision and tactile sensing provide these capabilities, which facilitate robots' integration with other equipment of the plant (see Box 10.1). One basic improvement is mounting a mechanical device to measure the gripper pressure, along with a limit switch that trips at a certain specified pressure. Photoelectric sensors are also used around the robot to stop it if an intrusion, which may be a person, occurs within the work envelope.

In certain situations, robotics may not be the appropriate technology for material handling automation. As an alternative, the possibility of using a **modular handling system** should be considered. Modular handling systems are formed by assem-

---

BOX 10.1   *Robots Tend to Business*

Industrial robots do one thing well—repeat programmed moves without tiring. Call them manufacturing's packhorses. The glamour is gone, but features and performance are better than ever.

"The new-generation systems are easier to maintain," says Mark Handlesman of GMF Robotics. "Optical encoders or resolvers provide fairly accurate positioning. Direct-drive or coupled motors link straight to robot joints, instead of through a gear or drive-chain. This reduces backlash."

Programming also is easier. Languages are powerful and user-friendly. Tools range from teach pendants to off-line simulation. You can juggle speed, accuracy, work envelope, cycle time, programming, and type of robot—from single loaders to six-axis pedestal robots, gantries, and integral arms — to fit almost any budget.

Source: From "Robots Tend to Business" by P. M. Noaker, March 1990, *Manufacturing Engineering*, pp. 74–77.

bling the required elements to achieve just the right amount of flexibility. An example is movorobot, developed by SKF Automation Systems, based on rails called movorail, linear actuators called movopart, and belt-driven carriages called tollobelt.

10.2   AUTOMATED GUIDED VEHICLES (AGVs)

Automated guided vehicles (AGVs) are modern material-handling and conveying systems. For the basic principles of AGV systems, their types, system components, and other technical considerations, readers should refer to books exclusively on AGVs (e.g., Hammond, 1986). This text covers only AGV issues that relate to CIM.

An AGV is like a forklift truck, but without the driver. It is a driverless vehicle that can follow guided paths, and an off-board controller that is basically a computer. The vehicle is capable of selecting its path to reach the required destination. It is this capability that provides AGVs flexibility and superiority over conventional systems such as monorails or conveyors. The brain behind AGVs' automated operation is the controller that receives information through a host computer. AGVs are sometimes used as mobile assembly platforms to provide flexibility and computer control.

Types and Technology

Once, AGVs were simply called wire-guided vehicles. Recently, alternative technologies have been developed. But the wire-guided systems still remain the most popular, operating 90% of all AGVs. A wire, embedded about an inch deep in the floor, emits low-level signals (0.5 ampere current), which the antenna of the carrier picks up and the on-board controller analyzes to determine the route. Wire-guided systems work best on floors with uncomplicated paths and limited distances. Smooth concrete flooring is desirable. It is better to avoid embedding the wire near metal reinforcements in the floor.

Some recent developments are taped or striped paths with painted lines or metal film defining the route. The carrier's ultraviolet (UV) light source illuminates the painted lines and reads the brightness of the reflected light to estimate its distance from the path. Another recent technology is a chemical strip that is laid over any surface and needs little maintenance. Painted, taped, or chemical-based paths have no distance limits. Route changes can be made easily without interrupting production. These paths work best, however, on clean floors, absent in most manufacturing plants.

As an important element of FMSs, automated guided vehicles represent a revolution in material handling. They have undergone two generations of development since their debut in the late 1970s. The first-generation AGVs based on wire-in-the-floor technology are outmoded now, primarily due to their inflexibility and stringent floor requirements for operation. The second generation AGVs, currently in use, are based on bar codes sited at strategic locations along the route and read by lasers. Newer AGVs based on vision technology are capable of picture processing any obstacle in the path, similar to what humans do. This "smartness" or "intelligence" results from the AGV's CCD (charge-coupled device) camera, which follows any light-colored path.

When considering AGVs for use in an FMS, the three primary issues are: (a) type (size and weight) of raw materials and parts between cells and to the assembly station, (b) fixturing, and (c) chips and swarf disposal systems. The AGV should, of course, be interfacable with the other equipment, such as CNC machine tools, robots, assembly workstations, and the associated material handling systems such as AS/RS or conveyors. In a CIM plant, AGVs are integrated with other plant resources and equipment through their controllers. The controller links the vehicle with the guidepath and is thus the "brain" of the entire AGV system.

Control

The control functions of AGVs fall into two categories: internal and external. Internal control is intended to ensure error-free operation of the vehicle, whereas external control communicates with the outside "world" via the host computer. It is the external control that renders AGVs appropriate for CIM environments. The control usually takes place at three levels as shown in Figure 10.2. Level 1 is the system host, which takes care of external communications. The floor controller represents level 2, while the vehicle processor operates at level 3. Box 10.2 describes the functions of the controller at these three levels.

10.3   AS/RS

Normally, the raw materials and tools are stored at a location central to the various shops of the plant. They are delivered to the workstation as and when required, which takes time and effort. The current practice, especially under CIM, is to store the materials and tools near the workstation in computer-controlled aisles under the control of an automated storage and retrieval system (AS/RS).

Figure 10.2
Three levels of vehicle control in AGVs

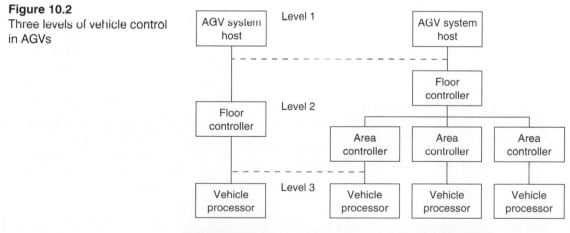

Small AGV systems   Large AGV systems

Source: Reprinted with permission from *AGVs at Work* by Gary Hammond, 1986, New York: IFS and Springer-Verlag. Copyright 1986 IFS (Publications) Limited and Springer-Verlag.

▢	**BOX 10.2** *Floor and System Controls*

The AGV system itself will usually contain three levels of control architecture: vehicle control system, floor control system, and vehicle on-board processor (Figure 10.2). The following description of the functions and tasks performed by each level is general in nature and may vary from vendor to vendor.

Vehicle Control System

The top level of the vehicle control system often communicates with, and is under the control of, the facility's host computer. This system control level is often referred to as the AGV controller or as the main-level computer. Most of the decision making takes place at this level as it oversees the system operation. As the knowledge center of the system, it provides many services, including:

Monitoring floor equipment status.
Reporting inventory levels and vehicle utilization.
Supervising overall vehicle traffic.
Tracking loads.
Assigning destinations.
Providing workstation definitions such as type, height, and lane selection.

The vehicle control system stores in memory exact vehicle locations at all times and provides network access. For integration into manufacturing environments, the vehicle controller needs to communicate information such as vehicle location and type of load to the facility's host computer. The final task for this control level is to act as the interface to allow users to generate reports, change product data, or make on-line system queries.

Floor Control Unit

This level is also referred to as the data concentrator and acts as the 'traffic manager,' communicating directly with the vehicles and providing them with formatted, detailed commands. It, in essence, processes the communication between the central controller and the vehicles. Its functions or tasks are as follows:

Provides lane selection based on such factors as style restrictions and work content.
Makes carrier selection based on the closest available vehicle that is of the appropriate type.
Sends information to the vehicle such as exact destination, build time, lift/lower heights.
Provides system block and path selection for certain conditions, such as high priority moves, sequencing requirements, or failed guidepaths.
Generates required guidepath frequencies.
Provides collision avoidance.
Controls factory-floor equipment (as designed into the system), collecting all floor inputs/outputs, such as AS/RS, conveyors, and push-button panels.

Vehicle Processor

Generally the vehicle processor knows the vehicle location, can interpret commands received from the floor control unit, and can monitor the on-board safety devices. There are two basic types of vehicle processor: the intelligent type (automatic) and the non-intelligent type (semi-automatic). The non-intelligent type is not able to make decisions and requires such information as start, acceleration/deceleration commands, and external control to direct the vehicle to its final destination. This is usually accomplished by turning the guidepath on and off at decision points.

An intelligent or automatic vehicle contains a microprocessor, which gives it many capabilities not exhibited by the non-intelligent type. Some of these extra on-board vehicle capabilities are as follows:

Blocking between vehicles.
Automatic routing permitting vehicles to find their own path to the final destination.

Controlling speeds, accelerations, and decelerations.

Interfacing with other types of equipment.

Controlling the vehicle platform, such as raising/lowering a work platform or activating a load/unload mechanism.

Releasing a vehicle from a workstation at a prescribed time.

Uploading time spent at a workstation if the vehicle is released manually (this information is then used to select lanes and monitor throughput).

Displaying job information to worker, eliminating lost manifests.

Providing built-in diagnostics to help reduce troubleshooting and repair times.

Finally, since the on-board vehicle processor has all these capabilities, some systems have the ability to run in a degraded mode should the floor control unit or the central control computer fail.

Source: From *AGVS at Work* (pp. 38–40) by G. Hammond, 1986, New York, NY: Springer-Verlag.

AS/RSs are based on a concept developed in the late 1970s. They are tall, vertical storage spaces, much like a huge bookshelf, near the process stations. At the beck and call of a terminal, an unloading device tracks through the aisles to reach the appropriate stack and retrieves the item stored below or above and hundreds of yards away. AS/RSs help reduce the amount of labor and floor space for maintaining inventory. The high-density AS/RS gives rise to an efficient inventory—everything is in its place, ready to be accessed when needed. With items replenished properly, AS/RS manages material like clockwork.

What role does AS/RS play in CIM? It facilitates computerization of material management at the shop floor, leading to further integration. For example, the status of AS/RS inventory can be made available on-line to the designer or quotation manager. NC programs can be provided with built-in modules for material and tool retrieval and machine set-up. Thus, the auxiliary function of setting up the machine is also automated and integrated.

10.4   PALLETIZATION

Palletization is the process by which workpieces and tools are set up on appropriate pallets. The workpiece can be mounted directly on a pallet, or on a fixture that is then mounted on the pallet. Direct mounting suits contract shops because it avoids the cost of fixtures, since the same job may not be ordered again. Palletization eliminates the use of a machine tool as a setup device or bench, thus freeing it to continue machining uninterrupted. Palletization may not involve high technology and does not require installation of FMC or FMS. It can benefit even stand-alone machining centers or workstations.

A pallet and its receiver are precision devices, with repeatability as precise as 4–8 thousandths of an inch (0.1–0.2 mm). A pallet system comprises three basic elements:

the pallet, on which the workpiece or fixture holding the workpiece is mounted

the loader, which transfers the pallet from the carrier such as AGV or load station to the receiver, or vice versa the receiver, which is mounted on the machine tool table to grip the pallet, when it arrives, and hold it

Palletization addresses the need for better utilization of expensive equipment such as FMC or FMS. Setting the workpiece on a pallet rather than on the machine itself has been reported to improve usage of machine tools by 50–100 percent. Other advantages of palletization are:

1. Some parts require processing on several machines. Palletization eliminates the need to set up several times.
2. Pelletized workpieces are easier to move and set up, which reduces the total setup time.
3. Setup errors are minimized, resulting in fewer rejected parts.
4. Palletization promotes automation.
5. Permanently bar-coded pallets facilitate data collection and integration. For example, on reading the code, the CNC can request downloading of the part program appropriate for the job on the pallet.

For CIM, palletization requires flexibility and standardization. The ISO is working to introduce standardized pallet designs. The proposal is to provide at the base one T-slot for pallet sizes up to 800 mm, and two for 1000 mm and above. Designs that allow use of the same pallet and fixture for a variety of parts, at least for those belonging to a family, are preferred. Box 10.3 presents a case study on a robotic palletizer.

TRENDS

❏ Industrial robots, AGVs, and programmable conveyors are establishing themselves as important modules of CIM. Two distinct trends are occurring: (a) design for automation—not only from the viewpoint of product design but also from the facilities viewpoint, such as plants, processes, and distribution systems; and (b) user-friendliness and intelligence of robots and newer automated material handling devices, some working as adaptive control systems.

❏ AGV guidance systems are becoming more capable due to on-board microprocessors. New developments are doing away with wires. One system based on odometry is known as "dead reckoning." In this technique, the controller keeps track of the turning of wheels and refers to the starting point to assess the current position. In another system, based on vision, a strobe light bounced off the reflectors mounted on factory ceilings allows the controller to locate the AGV's current position.

❏ There is a definite trend for expert systems in robotics. Equally strong is the trend that allows part data from CAD systems to be downloaded and then used to "figure out" what robots must do and how they must be arranged to process the parts.

◻ **BOX 10.3** *Getting a Robotic Palletizer Up and Running*

To keep pace with manufacturing, engineers at Compaq's distribution center in Houston, TX, decided to automate the facility's palletizing operation. Previously, recalls materials handling manager Dale Lockamy, vacuum lifters had been installed to supplement a manual operation. That decision was based on ergonomic considerations. But throughput and consistency still weren't adequate.

So engineers drew up a list of palletizing alternatives. Both sequential, or row stripper, palletizers and gantry robot units were rejected because they required too much floor space. A mobile robot, where the robot was mounted on a rail, was also considered. "But we felt that the loads were too unstable and needed to be secured by banding as soon as possible. The mobile robot configuration would not allow that," says Lockamy.

Instead, two stationary, Cartesian robots (Pacific Robotics) were selected. A third robot feeds empty pallets to the system. The resulting system leaves the open space that Compaq required and is flexible enough to handle the required variety of patterns and products. Each robot occupies about 25 sq. ft.

Bar codes scanned on each carton tell a host computer the product type in the carton. The host then tells the palletizer the stacking pattern to use and the quantity to expect. After scanning, each carton is rotated 90 degrees by a turning wheel on the case conveyor.

One of the main difficulties in keeping the system running properly was maintaining a steady-state air supply in the system. The system is fed by plant air at 90 psi and any variations in pressure or volume affect performance. To solve the problem, Compaq added a dedicated air supply system to support the palletizing cell.

Lockamy sums up, "The system uses all off the shelf parts and is easy to maintain. We were concerned that we might have more problems with the robots, but things have worked out well."

Source: "Getting a Robotic Palletizer up and Running," September 1991, *Modern Material Handling*, p. 50.

❑ An increasing number of production data collection systems are being integrated with material handling and delivery systems.

❑ The use of palletization and AS/RS continues to expand.

SUMMARY

The roles of robots and other flexible material handling systems in CIM have been highlighted in this chapter. Earlier developments of robots as stand-alone automation systems has inhibited their integration within CIM. Standardization in programming and robots' integration with other devices are essential. AGVs, palletization, and AS/RS are appropriate ways to integrate the material handling tasks with other functions within CIM. Relevant discussions on these technologies have been provided in this chapter.

KEY TERMS

End effector

Modular handling system

Off-line programming

Payload

Teach pendant

Work envelope

EXERCISES

Note: Exercise marked * are projects.

10.1   Discuss the major problems in integrating industrial robots with another flexible automation system of a modern plant such as AGV.

10.2   How does the meaning of robot in Japan differ from that in the U.S.?

10.3   Write a 300-word CIM-oriented essay on AGVs.

10.4   How do palletization and its automation help achieve CIM?

10.5   Discuss the role of AS/RS in streamlining material handling. How do they contribute to integration at the shop floor?

10.6* Visit a local manufacturing facility and observe the effectiveness of AGVs, palletization, or AS/RS. Discuss with the plant manager any plan the company may have to further integrate the observed equipment with the machines it serves.

10.7* Visit a nearby plant where at least one robot is in production use. How can this robot be integrated further with other functions of CIM? Discuss your ideas with the supervisor or manager.

SUGGESTED READINGS

Books

Asfahl, C. R. (1992). *Robots and manufacturing automation*. New York: Wiley. (This book concentrates on applications rather than the design of robots. Robotics is treated as a module within the wider scope of flexible automation, not just as an isolated technology. The text emphasizes integration, a desirable feature for CIM.)

Ballard, D. H., & Brown, C. M. (1982). *Computer vision*. Englewood Cliffs, NJ: Prentice-Hall. (The book delves into mathematical analyses for image processing. The derived equations are written later as computer programs. The treatment is oriented toward artificial intelligence, and the concepts are presented through mathematical analyses.)

Hammond, G. (1986). *AGVs at work*. New York: Springer-Verlag. (As a primer, the book is aimed at managers and others new to the technology of AGVs. It covers AGV design, operating characteristics, and installation. Several case-studies on AGV installation are the strength of the book.)

Heath, L. (1985). *Fundamentals of robotics*. Reston, VA: Reston Publishing.

Hunt, V. D. (1985). *Smart robots*. New York: Chapman and Hall. (This book provides an overview of sensor-based robots and automation, including artificial intelligence considerations for robots of the next generation. Directed at manufacturing managers, its last chapter on CIM discusses through case studies the issues relating to acquisitions.)

Synder, W. E. (1985). *Industrial robots: Computer interfacing and control*. Englewood Cliffs, NJ: Prentice-Hall. (The book discusses the principles behind the control of industrial robots. Chapters on computer vision, computational architecture, and robot programming languages are included. Readers with a background in automatic control should find the book useful.)

Vukobratovic, M. (Ed.). (1989). *Introduction to robotics*. New York: Springer-Verlag. (Designed as a graduate-level text for mechanical and electrical engineering, this book presents mathematically rigorous coverage of the fundamentals and the kinematics of manipulation robots.)

Journals and Periodicals

Modern Machine Shop. November 1990 (emphasis on workhandling, workholding, and robots).

Modern Material Handling. (This monthly magazine covers current topics in material handling.)

Articles

Fahringer, B. J. (1989, July). AGVs and automation: Getting there. *Production*, pp. 35–39.

Gettleman, K. (1990, February). What's behind palletization? *Modern Machine Shop*, pp. 94–101.

Goldstein, G. (1990, June). Shaping the next generation of robots. *Mechanical Engineering*, pp. 38–42.

Krepchin, I. P. (1991, September). How to choose, justify a palletizing system? *Modern Material Handling*, pp. 48–50.

Noaker, P. M. (1989). Automate data collection. *Production*, *101*(4), 42–46.

Noaker, P. M. (1990, March). Robots tend to business. *Manufacturing Engineering*, pp. 74–77.

Thangaraj, A. R., & Doelfs, M. (1991). Reducing downtime with off-line programming. *Robotics Today*, *4*(2), 1–3.

Vasilash, G. S. (1990, April). Does material storage matter? *Production*, pp. 58–59.

CHAPTER 11
Quality

"If the toast comes out burnt every time, you don't fix it by scraping off the toast. You repair the toaster."
—Donald P. Ephlin, vice president, United Auto Workers

"Almost all of American industry suffered from the concept of 'inspecting in quality,' and only within the last decade has the concept of designing in quality begun to take hold generally. Designing for automatic assembly was another major force."
—William T. Sackett (president, XOX Corporation)
in *Mechanical Engineering*, June 1990

"What you find may be surprising. Most US firms pay a high cost for poor quality. Philip Crosby reports in Quality is Free *that this cost shouldn't exceed 2.5% of the final product price. Studies show, however, that the cost of poor quality ranges from 20 to 40% of total US sales. It averages almost 25% of the price of a new car, compared with 2.5 to 4% for Japanese autos."*
—J. Hugh Doston

"Americans used to fall into the trap of thinking that high quality costs more, but high quality and low cost go hand in hand."
—George Fisher, CEO, Motorola

"The whole quality issue in the U.S. machine tool industry has been pretty much misunderstood. Quality in this industry has traditionally meant make it to spec."
—Mitch Kurzawa, Chrysler Corp.

"Quality is being redefined. No longer just the elimination of mechanical defects, quality now means setting new performance standards and meeting customer preferences of style and comfort."
—Dr. S. Vajpayee, University of Southern Mississippi

"The quality is not a department but instead is a systematic process that extends throughout the company."
—Dr. A. V. Feigenbaum, General Systems

"Variation is sometimes called the root cause of poor quality, but it is also a fact of life."
—Hilario L. Oh, General Motors Corp.

In today's global market, product quality is a critical element of success in manufacturing. Most recent efforts to exploit new technology, including CIM, are aimed at enhancing product and service quality. The goal is to broaden quality control into quality assurance, which signifies a company-wide effort rather than a department function. In this chapter, we discuss issues of quality management and technology that have bearing on CIM.

11.1 INTRODUCTION

According to Besterfield (1990), "Quality is not the responsibility of any one person or department; it is everyone's job. It includes the assembly-line worker, the typist, the purchasing agent, and the president of the company. The responsibility of quality begins when marketing determines the customer's quality requirements and continues until the product is received by a satisfied customer" (p. 4). As Figure 11.1 indicates, several departments contribute to a quality product.

Quality is a cost issue, with far-reaching impact on company sales and profitability. The quality of manufactured products is an important factor in their marketability. In the past, quality was determined by a detection system comprising inspectors and their tools sorting out the nonconforming parts. As Box 11.1 describes, this off-line inspection approach involves several costs that on-line inspection can reduce. In recent years, the emphasis has shifted toward a preventive approach in which in-

Figure 11.1
Departments responsible for product quality

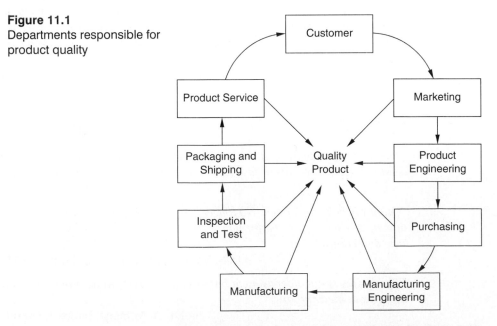

Source: Reprinted with permission from *Quality Control* by D. H. Besterfield, 1990, Englewood Cliffs, NJ: Prentice-Hall.

☐ **BOX 11.1**	*Eight Costs of Off-Line Inspection*

Mr. Bosch cites the following eight kinds of costs that can be sharply reduced or even eliminated by a change to on-line inspection.

1. Scrap losses—lack of timely feedback precludes process adjustment to avoid producing out-of-spec workpieces.
2. Rework costs include re-machining and reinspecting as well as lost income from machine time devoted to rework.
3. Machining center downtime while awaiting first part inspection to verify machine setups.
4. Staffing costs to run inspection routines manually instead of creating an inspection program on a one-time basis.

5. Material handling in sorting and delivering samples from the shop floor to the inspection area.
6. Greater in-process inventories from samples waiting to be inspected before the batches can be released.
7. Final assembly difficulties from nonconforming parts that slip through. Final inspection must be emphasized.
8. Warranty and liability claims from product failures and shortened wear life, and so on.

Source: From "Eight Costs of Off-line Inspection," February 1990, *Modern Machine Shop*, p. 272.

---

process inspection eliminates (if cost-justifiable) or minimizes the faults and errors in the product at the processing stage itself.

11.2   MODERN CONCEPTS OF QUALITY

Since the 1980s, product quality shifted emphasis away from checking for defects to designing. Crosby (1984) suggests four absolute concepts inherent in modern quality control management:

1. The definition of quality is conformance to requirements.
2. The system of quality is prevention.
3. The performance standard is zero defects.
4. The measure of quality is the price of nonconformance.

If, according to Crosby (1979), quality can be defined as conformance to requirements, then a measure of quality is the price of nonconformance. In other words, what does it cost to do things incorrectly? Crosby argues that the cost of quality can be divided into two—the price of nonconformance (PONC) and the price of conformance (POC). PONC is the cost of doing things wrong; it comprises all the unnecessary costs due to a product not being made correctly the first time. It has been estimated that in manufacturing PONC is as high as 20% of the total sales. In service industries, the figure is even higher—35%. The POC, on the other hand, is estimated to be 3–4% of the sales turnover. Thus, there is a lot of room for achieving quality through prevention.

Crosby advises following a zero-defect strategy. To achieve zero defects a company must work toward total quality control (TQC), which is feasible if the three Ss are implemented: *Seiri*, *Seiton*, and *Shitsuke*. These Japanese terms denote, respectively, three principles: elimination of the unnecessary, putting in order, and training and self-discipline. These principles should be followed not only in production but also in other areas of manufacturing.

Deming (1982) and Juran (1988)—the other two quality gurus—also consider the prevention approach superior. They treat poor product quality as an economic loss. These innovators argue that quality is rooted in the process and that management, not the workers, is responsible for poor quality. Quality begins at the top with the company president and chief executive officer and trickles down to all levels of the organization.

According to Feigenbaum (1983), the basic benchmarks of **total quality management (TQM)** are that:

1. quality is not a department but instead a systematic process that extends throughout the company.
2. this company-wide quality process must be correctly structured to support both the quality work of individuals as well as the quality teamwork among departments.
3. quality must be perceived in this process to be what the buyer says it is, not what an engineer or marketer or general manager says it is.
4. modern quality improvement requires new technology, starting with techniques for quality by design and DFM.
5. what makes this whole effort work is a clear customer-oriented management process and work process throughout the organization—one that people understand, believe in, and are a part of.

11.3   STATISTICAL QUALITY CONTROL (SQC)

Statistical quality control (SQC) means control of product quality using the principles of statistics. It developed as a powerful tool and contributed significantly to World War II efforts. Following the 1924 introduction of statistical charts for control of product variables (by W. A. Shewhart of Bell Telephone Laboratories), quality control became known popularly as SQC. H. F. Dodge and H. G. Romig (both also of Bell Telephone Laboratories) developed an additional application for statistics in quality control, known as acceptance sampling.

Of all the SQC techniques, control charts have been the most effective. **Control charts** are used to determine the inherent variation in a process. As long as the data pertaining to the process are within the upper and lower control limits and follow normal distribution, the process is considered stable and the resulting variation in the process acceptable. Any data outside the limits, or abnormal variations in those within, suggest assignable sources of error in the process, which should be set right to ensure quality production.

Let us consider the control chart technique through an example (Besterfield, 1990). The ABC company is interested in applying this technique to improve the power transmitting quality of a shaft. Let us assume that this quality is governed by the shaft's keyway depth. Table 11.1 shows the keyway depth data collected from machined shafts. Twenty-five subgroups, each with four samples, were measured to generate the data X_1, X_2, X_3, and X_4. The data have been coded to show the two places after the decimal, the base dimension being 6 mm. Thus, the value 42 for X_2 in subgroup number 15 means a keyway depth of 6.42 mm. From the measured data, the values of average, range, and control limits have been calculated as shown on page 345.

The values of the constants used in the equations depend on the sample size and are available from any statistics or quality control book. The resulting control charts are plotted in Figure 11.2 with their upper and lower limits. There are two points outside the $\bar{X}$-chart limits and one point outside the R-chart limits. When assignable causes are suspected, the data corresponding to the off-limit points is discarded and calculations are revised. The reasons for off-limit points are investigated and removed to improve the process. With time, when all the possible sources con-

Figure 11.2
Control charts for the data in
Table 11.1

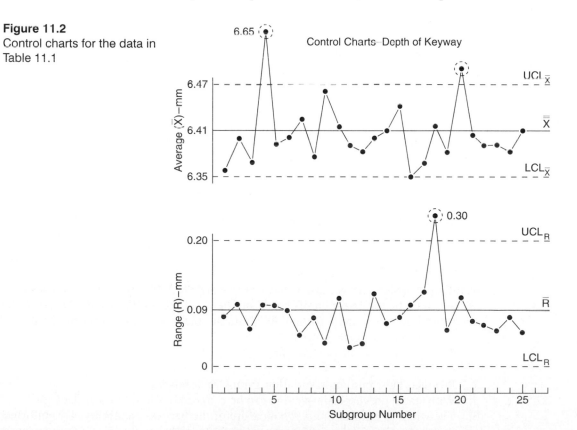

Source: Reprinted with permission from *Quality Control* by D. H. Besterfield, 1990, Englewood Cliffs, NJ: Prentice-Hall.

Table 11.1
Data and calculations for control charting

Subgroup Number	Date	Time	Measurements X_1 X_2 X_3 X_4	Average $\bar{X}$	Range R	Comment
1	12/23	8:50	35 40 32 37	6.36	0.08	
2		11:30	46 37 36 41	6.40	0.10	
3		1:45	34 40 34 36	6.36	0.06	
4		3:45	69 64 68 59	6.65	0.10	New, temporary operator
5		4:20	38 34 44 40	6.39	0.10	
6	12/27	8:35	42 41 43 34	6.40	0.09	
7		9:00	44 41 41 46	6.43	0.05	
8		9:40	33 41 38 36	6.37	0.08	
9		1:30	48 44 47 45	6.46	0.04	
10		2:50	47 43 36 42	6.42	0.11	
11	12/28	8:30	38 41 39 38	6.39	0.03	
12		1:35	37 37 41 37	6.38	0.04	
13		2:25	40 38 47 35	6.40	0.12	
14		2:35	38 39 45 42	6.41	0.07	
15		3:55	50 42 43 45	6.45	0.08	
16	12/29	8:25	33 35 29 39	6.34	0.10	
17		9:25	41 40 29 34	6.36	0.12	
18		11:00	38 44 28 58	6.42	0.30	Damaged oil line
19		2:35	35 41 37 38	6.38	0.06	
20		3:15	56 55 45 48	6.51	0.11	Bad material
21	12/30	9:35	38 40 45 37	6.40	0.08	
22		10:20	39 42 35 40	6.39	0.07	
23		11:35	42 39 39 36	6.39	0.06	
24		2:00	43 36 35 38	6.38	0.08	
25		4:25	39 38 43 44	6.41	0.06	
Sum				160.25	2.19	

Note: For simplicity in recording, the individual measurements are coded from 6.00 mm.

tributing to assignable variations have been corrected, the process becomes stable, as confirmed by the latest charts based on the most recent data. The data for a stable process approximate a normal distribution—the points are within the limits and their scatter is statistically normal.

The benefits of control charts are:

1. Products have minimum variation among themselves.
2. With stable processes, inspection can be reduced or eliminated altogether.
3. The width of control limits is a measure of the **process capability**. This information is used to enhance product quality through design by matching the process capability with design tolerances.

In order to illustrate the calculations necessary to obtain the trial control limits and the central line, the data concerning the depth of the shaft keyway will be used. From the data, $\Sigma \bar{X} = 160.25$, $\Sigma R = 2.19$, and $g = 25$; thus, the central lines are

$$\bar{\bar{X}} = \frac{\sum\limits_{i=1}^{g} \bar{X}_i}{g} \qquad \bar{R} = \frac{\sum\limits_{i=1}^{g} R_i}{g}$$

$$= \frac{160.25}{25} \qquad = \frac{2.19}{25}$$

$$= 6.41 \text{ mm} \qquad = 0.0876 \text{ mm}$$

From control chart tables, the values for the factors for a subgroup size (n) of four are $A_2 = 0.729$, $D_3 = 0$, and $D_4 = 2.282$. Trial control limits for the X chart are:

$$\text{UCL}_{\bar{x}} = \bar{\bar{X}} + A_2\bar{R} \qquad\qquad \text{LCL}_{\bar{x}} = \bar{\bar{X}} - A_2\bar{R}$$

$$= 6.41 + (0.729)(0.0876) \qquad\qquad = 6.41 - (0.729)(0.0876)$$

$$= 6.47 \text{ mm} \qquad\qquad = 6.35 \text{ mm}$$

Trial control limits for the R chart are:

$$\text{UCL}_R = D_4\bar{R} \qquad\qquad \text{LCL}_R = D_3\bar{R}$$

$$= (2.282)(0.0876) \qquad\qquad = (0)(0.0876)$$

$$= 0.20 \text{ mm} \qquad\qquad = 0 \text{ mm}$$

Source: Reprinted with permission from *Quality Control* by D. H. Besterfield, 1990, Englewood Cliffs, NJ: Prentice-Hall.

11.4   STATISTICAL PROCESS CONTROL (SPC)

Statistical process control (SPC) is an extension of the established techniques of SQC, with the difference being that SPC concentrates on the process rather than on the product. A process represents the cumulative effect of all the factors—machine, raw material, operator, work environment, and the like—that result in product variations. Recently, SPC has become popular, primarily due to the impetus it provided to manufacturing in Japan, resulting in higher-quality products. Product quality can be improved cost-effectively by implementing SPC. SPC is more than a production tool, it is in fact a management tool. The key word in SPC is process; effecting appropriate changes in the process at the right time improves both quality and productivity and reduces cost.

Although SPC is a new term, it is not much different from SQC, which has been in use since World War II. Both are based on the principles of statistics. The differ-

---

BOX 11.2   *SPC and SQC: NOT One and the Same*

There is a tendency to think of SPC (statistical process control) and SQC (statistical quality control) as being one and the same. Actually they are quite different. SPC is a measure of process capability while SQC insures the quality of the product being produced. Consider it this way: A powerful golfer may have the capability of hitting a golf ball, with a driver, 270 yards, plus or minus 10 yards, every time he steps up to the tee. This is one measure of his process capability. Unless he is able to place those golf shots directly on target, he will not be a championship golfer. Placing a defined SPC right on target and maintaining it is the essence of SQC. Thus, controlling and directing SPC is an important part of a successful SQC implementation, but it is not the same.

Source: From "SPC Will Work If:" by S. Birman, May 1989, *Modern Machine Shop*, p. 85.

---

ence is not one of essence, but of emphasis. SPC assumes that poor quality is due to the *process*, so it is the process that must be monitored to ensure product quality. The product is merely an outcome, an effect, and not the cause (see Box 11.2). Figure 11.3 illustrates how the concept of prevention is implemented by advancing inspection efforts up-front of the process. Note that inspection is a postprocess activity, whereas prevention is an in-process activity.

SPC is the sum total of all technical and managerial efforts to control the manufacturing processes for improving and maintaining quality. SPC techniques have demonstrated that it is possible to improve both quality and productivity simultaneously.

SPC should not be limited to control charts only. It is also a diagnostic tool; it tells us where the problems exist and provides hints on probable causes. Implementing SPC begins with the task of knowing the process in terms of what is called **process capability**. Process capability refers to the extent of variation in the product data when the process is stable. It is established by tracking the process spread for specific combinations of tooling, workpiece material, and sometimes the operator.

Figure 11.3
Prevention approach as the basis of SPC is better than the detection approach of SQC.

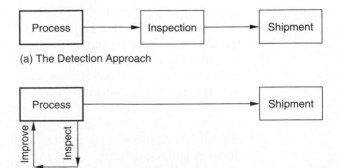

(a) The Detection Approach

(b) The Prevention Approach

To relieve the burden of analyzing process data, hand-held SPC data analyzers are used to generate control charts, which are statistical snapshots showing how the process is doing quality-wise.

The benefits of SPC are:

1. The process planner knows what the shop can do.
2. The QC manager is in a better position to manage.
3. The operator knows immediately when the process begins to behave abnormally.
4. Rejected products decrease.

11.5   PROCESS CAPABILITY

Process capability studies relate the needs of the design with the capabilities of the processes. The analysis begins with a test of the process (Smith, 1991) to determine its standard deviation (σ). Process capability is expressed as a ratio and denoted by PCR or C_r, yielding a number that quantifies it:

$$C_r = \frac{6\sigma}{U - L}$$

where $U - L$ is the difference, also called the tolerance, between the upper and lower specification limits set by the designer.

Some companies use the process capability index (PCI or C_p), which is defined as:

$$C_p = \frac{U - L}{6\sigma}$$

As can be noted, C_p is the reciprocal of C_r.

The relationship between the design specification and the process capability can be understood better with the help of Figure 11.4. This figure shows three different possibilities that can exist between the requirements of the design, in terms of upper and lower specification limits U and L, and the process capability 6σ. Note that case I ($C_p > 1.00$) is the best of the three, and that the process in case III is incapable of holding the required tolerance. In case II with $C_p = 1$, the process is just capable enough to hold the tolerance, provided the perfect centering between the specification width and the process variation width is maintained. If the setup disturbs this centering, defective items may be produced in case II. For this reason, a C_p value greater than 1 is desirable.

To ensure that all the parts produced by the process are within specification limits, a capability index (C_p) of 1.33 is usually used. In other words, the process is selected so that its capability matches the product tolerance to yield 1.33. Larger val-

Figure 11.4
Three possible scenarios
between design requirement
and process capability

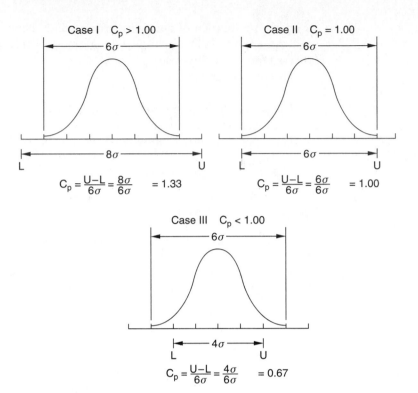

Case I $C_p > 1.00$

$$C_p = \frac{U-L}{6\sigma} = \frac{8\sigma}{6\sigma} = 1.33$$

Case II $C_p = 1.00$

$$C_p = \frac{U-L}{6\sigma} = \frac{6\sigma}{6\sigma} = 1.00$$

Case III $C_p < 1.00$

$$C_p = \frac{U-L}{6\sigma} = \frac{4\sigma}{6\sigma} = 0.67$$

Source: Reprinted with permission from *Quality Control* by D. H. Besterfield,
1990, Englewood Cliffs, NJ: Prentice-Hall.

ues of C_p are desirable, but they indicate the process selected is more precise than the given tolerances demand; in other words, the process is capable of holding closer tolerances than it is being used for. A less expensive process—for example, cheaper raw materials, less skilled operator, and the like—may achieve the desired tolerance.

Example 11.1 _____

The specifications for the machined diameter of a locating pin have been set at 2.87 ± 0.02 inch. A capability test has shown that the process standard deviation is 0.004 inch. Find the values of process capability ratio C_r and index C_p.

Solution

Upper specification limit, U = 2.87 + 0.02 = 2.89 inch
Lower specification limit, L = 2.87 − 0.02 = 2.85 inch
Thus, tolerance = 2.89 − 2.85 = 0.04 inch

Therefore

$C_r = (6\,\sigma) / \text{tolerance}$
$\quad = (6 \times 0.004) / 0.04$
$\quad = 0.60$

and

$$C_p = 1 / C_r$$
$$= 1.67$$

According to Besterfield (1990), "Using the capability index concept, we can measure quality provided the process is centered. The larger the capability index C_p, the better the quality. We should strive to make the capability index as large as possible. This is accomplished by having realistic specifications and continual striving to improve the process capability" (p. 108).

The discussion on process capability thus far has not considered one important factor: What happens to the match between the specification limits and the process if the initial process setting gets disturbed by, say, a change in operator or by tool wear? Figure 11.5 describes the result. On the left, the process and tolerance width $(U–L)$ are both centered. An upward shift in the process, on the right, has moved it closer to the upper specification limit, U. The same process may now produce nonconforming (e.g., oversized diameters or undersized holes) items, even though it is capable of holding the required tolerance $(U–L)$.

Thus, the capability index by itself is unable to measure process performance in terms of target or nominal value fixed by the designer. This performance is measured by another parameter, C_{pk}, defined as

$$C_{pk} = \frac{Z}{3\sigma}$$

where Z is the smaller of U-X or X-L and X is the target value.

Figure 11.5
Process shift affecting the central match between design specification and process capability

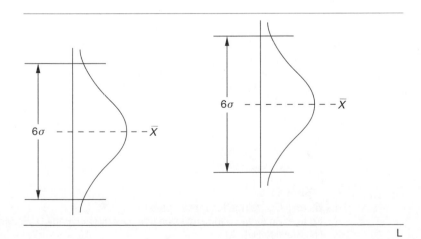

For a centered process, the C_p value is the same as C_{pk}. This is shown in the left column charts of Figure 11.6 for all three cases. When the process is not centered, values differ as illustrated in the right column charts; the difference between C_p and C_{pk} depends on the extent of off-centering.

Besterfield (1990) has compared C_p and C_{pk} to conclude that:

1. The C_p value does not change as the process center changes.
2. $C_p = C_{pk}$ when the process is centered.

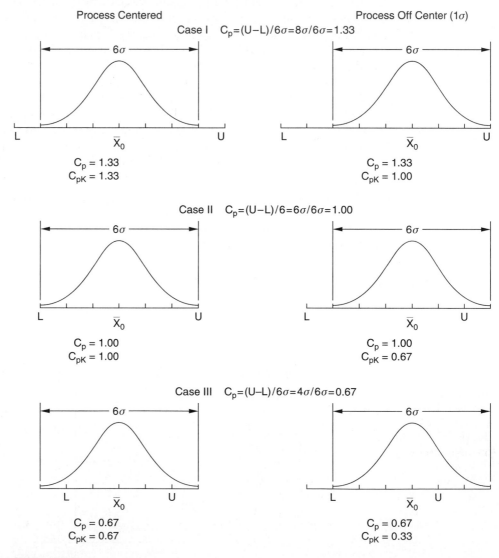

Figure 11.6
C_p and C_{pk} values for three cases
Source: Reprinted with permission from *Quality Control* by D. H. Besterfield, 1990, Englewood Cliffs, NJ: Prentice-Hall.

Table 11.2
Process groups

Process Group	PCR	C_{pk}	Control Chart
A	0.00–0.50	2.00–. . .	Recommended
B	0.51–0.70	1.42–1.99	Recommended
C	0.71–0.90	1.11–1.41	Required
D	0.91–. . .	0.00–1.10	Required

3. C_{pk} is always equal to or less than C_p.
4. A C_{pk} value of 1.00 is a de facto standard. It indicates that the process is producing a product that conforms to specifications.
5. A C_{pk} value of less than 1.00 indicates that the process is producing a product that does not conform to specifications.
6. A C_p value of less than 1.00 indicates that the process is not capable.
7. A C_{pk} value of zero indicates that the average is equal to one of the specification limits.
8. A negative C_{pk} value indicates that the average is outside the specifications. (p. 109)

Companies use one of the several process capability measures just discussed. For example, General Motors prefers PCR, whereas Ford uses C_{pk}.

On the basis of PCR, processes are classified into four groups, as shown in Table 11.2. Note that the use of control charts is mandatory for groups C and D.

Process capabilities are stored in a database for use by engineering and design personnel. This data indicates the product tolerance for various machines. Routing the parts through appropriate processes can thus ensure 100% conforming parts for the run. As an example, consider machines A and B with process capabilities of ± 0.01 and ± 0.02 inch, respectively. If workpieces must have a tolerance within ± 0.015 inch, then only machine A should be used to ensure 100% conforming parts.

Knowledge of process capabilities is also useful in areas other than process selection, for example, in analyzing the effects of work holding and fixturing on part production. If the part must be processed on three different machines, each requiring a separate fixturing, then the worst-case scenario may result in part spreads totaling the process capabilities of all three machines.

Example 11.2

The purpose of this example is to illustrate how the control chart technique is applied as an SPC tool.[1]

The coded data in Figure 11.7 is lengths of cam roller bushings with an expected tolerance of 1.328–1.334 inches. Measurements were made to the nearest ten-thousandth of an inch and the results were coded. The base measurement was 1.3300

---

[1] Adapted from *Statistical Process Control and Quality Improvement* (pp. 145–151) by G. Smith, 1991, New York: Macmillan.

Figure 11.7
SPC data and calculations for Example 11.2

$\bar{s} = \dfrac{\Sigma s}{k} = \dfrac{159.83}{25} = 6.39$

$\bar{\bar{X}} = \dfrac{\Sigma x}{k} = \dfrac{611.4}{25} = 24.5$

$UCL_s = B_4 \times \bar{s} = 2.089 \times 6.39 = 13.34$

$A_3 \times \bar{s} = 1.427 \times 6.39 = 9.12$

$UCL_{\bar{x}} = \bar{\bar{x}} + A_3\,\bar{s} = 24.5 + 9.12 = 33.6$

$LCL_{\bar{x}} = \bar{\bar{x}} - A_3\,\bar{s} = 24.5 - 9.12 = 15.4$

Specification: 1.328 to 1.334

Constants

n	A_3	B_3	B_4	C_4
3	1.954	0	2.568	.8862
4	1.628	0	2.266	.9213
5	1.427	0	2.089	.9400

Sample Numbers	1	2	3	4	5	6	7	8	9	10	11	12	13	14	15	16	17	18	19	20	21	22	23	24	25
Coded 0 = 1.3300	29	29	28	28	22	33	33	33	27	28	38	36	25	26	20	16	29	17	21	21	24	27	20	28	08
	30	19	27	27	26	36	27	20	21	21	29	27	25	35	31	19	29	04	04	16	10	25	21	20	30
	28	25	14	14	08	32	20	21	31	17	25	24	11	20	19	23	05	27	27	19	30	21	18	29	16
	37	21	18	18	27	27	33	27	30	34	26	26	27	25	30	16	19	26	26	33	16	27	26	27	09
	29	35	34	34	24	28	21	33	27	25	27	21	36	24	35	25	21	16	21	15	21	30	31	22	00
$\bar{X}$	30.6	25.8	24.2	24.2	21.4	31.2	28.8	28.8	27.2	25.0	29.0	26.8	24.8	26.0	27.0	19.8	20.6	18.0	20.6	22.4	20.2	26.0	25.2	25.2	12.6
s	3.65	6.42	8.07	8.07	7.73	3.70	5.50	5.50	3.90	6.52	5.24	5.63	8.96	5.52	7.11	4.09	9.84	9.30	4.72	8.11	7.63	3.32	6.06	3.96	11.26

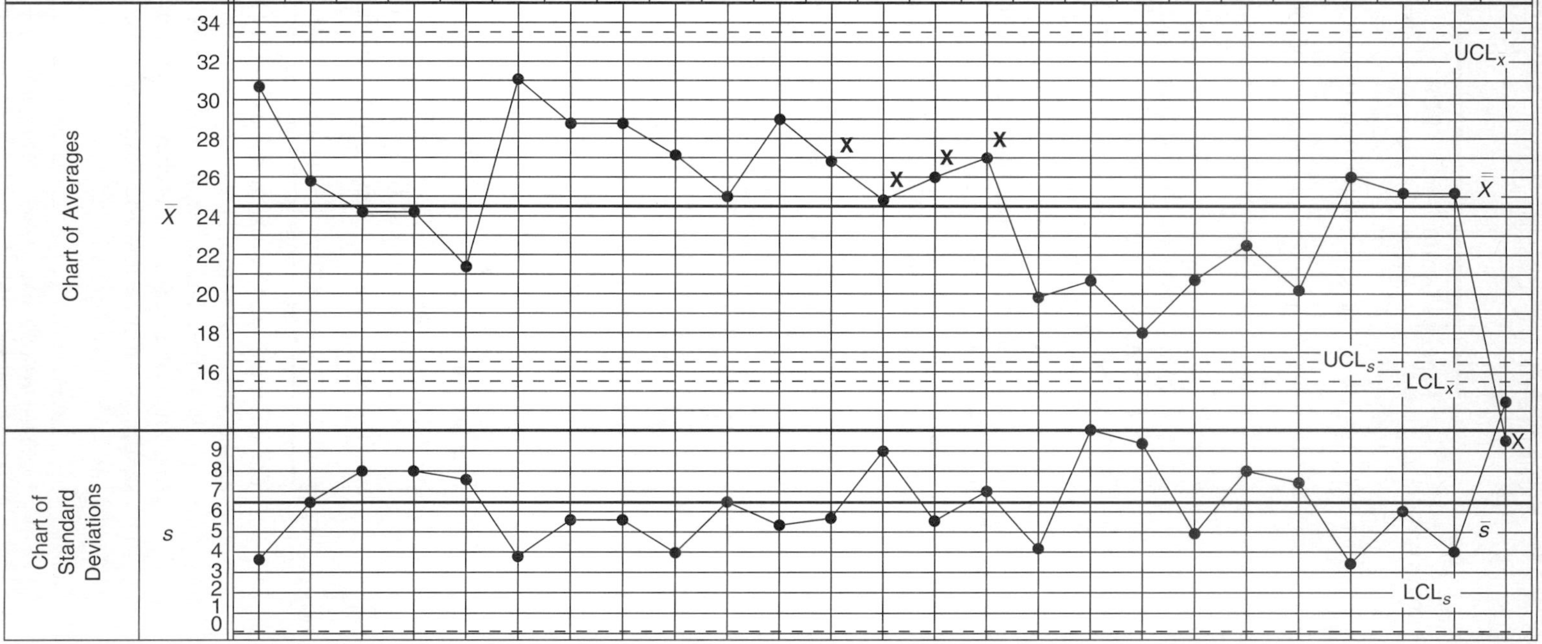

Chart of Averages — $\bar{X}$ (34, 32, 30, 28, 26, 24, 22, 20, 18, 16): UCL_x, $\bar{\bar{X}}$, UCL_s, $LCL_{\bar{x}}$

Chart of Standard Deviations — s (9, 8, 7, 6, 5, 4, 3, 2, 1, 0): $\bar{s}$, LCL_s

inches. The coded data for the 25 samples represents the last two digits; thus, a measurement of 1.3324 inches signifies a value of 24, and 1.3286 inches, 14. At the end, the results are decoded into physical dimensions. Thus, $X = 13$ means that the average value of a bushing length is 1.3313 inches, and s = 4 means that the sample standard deviation is 0.0004 inches.

Solution

The calculations for the first sample are illustrated.

> Coded values: 29, 30, 28, 37, 29
> Sample average, $\overline{X} = (29 + 30 + 28 + 37 + 29)/5 = 30.6$
> Sample mean (s):

$$s = \sqrt{[(29-30.6)^2 + (30-30.6)^2 + \ldots + (29-30.6)^2]/(5-1)}$$
$$= 3.65$$

Note that the calculation for sample mean involves division by $n - 1$, where n is the sample size.

Calculations for the other samples follow the same steps. The use of a calculator or computer is desirable.

The calculated values of $\overline{X}$ and s for all the samples are tabulated in Figure 11.7, where they have also been posted on $\overline{X}$ and s charts. Note that $\overline{\overline{X}}$ and $\overline{s}$ are obtained by averaging the 25 values of and s, respectively. These values are set as the charts' center lines. The upper and lower control limits have been set after calculating their values using the appropriate formulas and constants; these calculations are shown at the top of the figure.

Next, the charts are studied as a diagnostic tool. In this case, no apparent out-of-control points are evident in the s-chart. Moreover, the sample values of s are well distributed about its mean value; in other words, the process is behaving normally. In the $\overline{X}$-chart, however, sample number 25 is out of the limit, and the pattern of variation in individual sample values seems to violate normal distribution. As a rule of thumb, seven consecutive points on one side of the center line indicates a process shift. Thus, sample numbers 12 to 15 and 25 are suspect. The process condition corresponding to these samples should be investigated and the associated causes removed.

Another useful bit of information from this analysis is the average of the sample averages. Decoding its value of 24.5 into physical dimensions, we find the process average to be 1.33245 inches. Compare this with the average bushing length recommended by the designer, which is 1.331 [(1.328 + 1.334)/2] inches. The comparison shows that the process is producing longer bushings. This should be corrected by resetting the machine or other practical means.

As stated earlier, SPC analyses help only in diagnosing the problem. The user must devise and implement the solution. SPC should not be considered ineffective if the personnel using it are not skilled in reading the results.

Only after the assignable causes that give rise to out-of-control points and other abnormalities have been sorted out should process capability studies be made. Let us discard the suspect data in this example and proceed to determine the process capability, assuming that the remaining data represent a normal process (in fact, after correcting the process corresponding to suspect data, further bushings should be machined and used to generate new data). After discarding samples 12, 13, 14, 15, and 25, new values of $\overline{\overline{X}}$ and $\overline{s}$ are calculated as 24.71 and 6.07, respectively. New control limits are calculated, and the 20 remaining sample data are found to be within the control limits. It can thus be concluded that the process is operating normally and is stable (this should really be done with the new data).

The process standard deviation (σ) is found from the sample deviation (s) by

$$\sigma = \overline{s}/C_4$$

where the value of the constant (C_4) depends on the sample size (n) and is obtained from the table at the top right corner in Figure 11.7.

This works out to be 6.457 (since $\overline{s}$ = 6.07 and C_4 = 0.94). When decoded, σ = 0.0006457 inches.

If the given tolerance = 1.334 – 1.328 = 0.006 inches, then the PCR value = (6 × 0.0006457)/0.006 = 0.6457. A PCR value of 0.6457 means that 64.57% of the tolerance is used by the distribution of measurements.

To find the C_{pk} value, determine the difference between the nearer specification limit and $\overline{\overline{X}}$. This works out to be 0.001529 (= 1.334 – 1.332471) inches.

Thus, C_{pk} = 0.001529/(3 × 0.0006457) = 0.789. Since it is less than 1, it indicates that the present setting of the process is producing some bushings that are out of specification on the high side—longer than 1.334 inches. The user should reset the process, since it is not well centered. In a properly set process, the average of $\overline{\overline{X}}$ should equal the nominal dimension, in this case a bushing length of 1.331 inches.

As a technique, SPC demands careful application. Sometimes SPC is called ineffective, when in fact the mistake is in its use. Table 11.3 summarizes typical examples of mistakes in applying SPC.

11.6   MACHINING-INSPECTION CONTINUUM

Modern machine tools are expensive. It makes sense to ensure that they are in use most of the time. An idle machine tool can cost a substantial amount of money. To maximize usage, auxiliary tasks that hold up the machine should be carried out rapidly or in combination with other tasks. Rapid traverse features, scheduling efforts, pallet shuttles, quick die and fixture change, and similar techniques have been developed to reduce nonmachining time.

Inspection is one of the tasks that hold up the machine, if carried out conventionally with a workpiece on the machine. Several techniques can reduce inspec-

Table 11.3
Typical mistakes/miscalculations in applying SPC

Category	Typical Data Collection Procedure	Acceptance Parameter	Common Sources of Error
Machine Capability Study	30–50 consecutive pieces, one or two dimensions under study	C_p (C_{pk})	1. One or two dimensions do not represent variation of others. 2. The value of C_p or C_{pk} is not valid if frequency distribution is not normal. 3. Consecutive pieces do not represent potential variability. 4. Process variables will be less stable, more stable, or unknown. 5. Measurement error is unknown. 6. Measurement technique does not follow "rule of ten."
Process Capability Study	20 subgroups, five pieces in each subgroup	C_p, C_{pk} (Estimate of variation is "s-bar," average sample standard deviation)	1. Process is unstable on runs of over five pieces. 2. Target is not specified nominal. 3. Measurements are not taken in "dramatic" situations. 4. No time left to find the root causes of variation. 5. Subgroup is not rational.
	One or more subgroups, more than 50 pieces in each subgroup	C_p, C_{pk} (estimate of variation is sigma)	1. Product variation, not machine or process capability, is observed. 2. Control charting is meaningless after this "study" because sample variation cannot be controlled.
Statistical Process Control	Subgroup of five consecutive pieces, inspection frequency is one per hour	Control limits on X-bar, R chart	1. Control limits are not valid if no capability studies are performed for a given production run. 2. Control limits are calculated only to signal the effect of special cause of variation. 3. Sample size and sampling frequency do not represent true product variation. 4. Target for the mean is not the specified nominal. 5. Sampling frequency is not designed to catch accelerated (nonlinear) tool wear.

Source: Reprinted with permission from "Quality Is More Than SPC" by S. Birman, April 1991, *Modern Machine Shop*, p. 98.

tion time. Inspection may be NC-programmed as another operation of the machine tool based on touch probe technology. In such cases, the geometric database generates both the part program and the probing program. Thus, design, part programming, machining, and inspection become one function. Integrated software that accomplishes this comprises various modules to carry out individual tasks such as:

Verify that the dimensioning and tolerancing conform to the required standard, such as ANSI Y14.5-1982: Dimensioning and Tolerancing, and are consistent with workpiece geometry.

Create a **softgage** that provides on the screen a comparison between the geometric model (what is desired) and the inspection results (what has been produced).

Generate inspection paths for the CMM (coordinate measuring machine) or the touch probe mounted on the machining center.

Interface CMM paths and measurement data between CMMs or between machine tools and the host computer.

Execute the inspection path and collect data to create an as-machined model of the workpiece.

Compare measurement results with the softgage to make a go/no-go decision about the workpiece.

A softgage may be contrasted with the conventionally used **hard gage** for inspection. A workpiece that fails to fit the hard gage has out-of-tolerance features and is rejected. Similarly, a softgage rejects the part if it fails to fit into the softgage envelope. The only difference is that instead of the physical hard gage, the softgage overlays two CAD models on each other over the graphics screen to create the softgage. A softgage is created directly from the original design data, resulting in two obvious benefits: The softgage can be modified easily, and the cost of building and validating the hard gage is eliminated. A softgage can reduce inspection time by as much as 80%. The softgage approach allows machine tools to be used as CMMs, thus bringing CIM a step closer.

A trend evidently beneficial to production is to carry out as much inspection as possible on one machine. Small CMMs, which are manual machines, not motor-driven as are conventional CMMs, are appropriate tools (Vasilash, 1990) even for large production operations. In Vasilash's words, small CMMs "are peppered about on the floor, almost cheek-in-jowl with machine tools, and used to enhance in-process capabilities. Although many of these units were conceived as stand-alone units, networking capabilities are available today" (p. 67). Some of the small CMMs are termed personal CMMs, an example being Brown & Sharpe's MicroVal introduced in 1987. This machine is claimed to offer a payback period[2] as short as 1.4 months!

---

[2]The basic MicroVal (Vasilash, 1990) is equivalent to a conventional system comprising a 24-in. by 36-in. surface plate, 12-in. height gage, 22 calipers, two micrometers, a go/no-go pin gage set, a dial bore gage, an angle plate, and a dial indicator. The payback period calcula-

DMIS as a CAD-CMM Interface Standard

To facilitate transfer of inspection data between CAD systems and measuring equipment such as CMM, some sort of interface specification had to be standardized. One such interface is the **Dimensional Measuring Interface Specification (DMIS)** developed by Applied Automation Technologies, Inc. Inspection programs created at CAD systems and formatted according to the DMIS standard can be downloaded to CMMs having suitable processors for accepting such programs. The reverse process—uploading the inspection results into the design database—is also possible. Interfaces such as DMIS make the inspection task CIM-compatible.

DMIS is bidirectional. When the first produced part is taken to the CMM for measurement, the inspection procedure can be downloaded from the CAD system (see the latter part of Box 11.5). The procedure can be stored locally for inspection of the parts to be produced. In turn, the measurement results can be uploaded to the CAD system if it can accept DMIS. DMIS facilitates software compatibility as well. In a sense, DMIS is to CMMs what DOS is to PCs.

Approved by ANSI, DMIS facilitates communication of inspection data among computer systems and inspection equipment. DMIS version 2.0 became an ANSI standard in February 1990. It permits development of quality assurance program modules within an inspection system. With modules simulating execution, it is possible to watch a probe on the screen go around measuring the simulated part features before the part is actually produced. Thus, product design quality can be perfected in the CAD system itself. Such an approach is clearly compatible with CIM.

11.7   COORDINATE MEASURING MACHINE (CMM)

In discrete-parts manufacturing, the inspection task can be approached in one of two ways:

1. the hard-gage technique using surface plates, dial indicators, height gages, and so forth
2. CMM

Normally, CMMs take far less time for a given inspection task than the conventional hard-gage technique does. With their decreasing cost, CMMs are now

---

tions are based on an assumed scenario of the need to measure 12 characteristics on parts produced in a machining cell. Rate: 250 parts per hour; 5,000 per month are needed; the sampling rate is 30%, or 1,500 parts. Operator rate including fringes is $21 per hour. With the surface plate method, the time per piece is 19 minutes, or $6.72 per piece, resulting in a monthly cost of $10,080. With the MicroVal, it's 8.4 minutes, or $2.94 per piece, or $4,410 per month. The time it takes to transform the manual calculations into a report costs $252 per month (12 hours in terms of time), which is saved with MicroVal since it is done during inspection. Brown & Sharpe claim that the reduced inspection time/increased productivity, reduced downtime, and reduced scrap and rework all translate into a monthly savings of $10,061, which gives a payback period of 1.4 months in favor of MicroVal over the conventional surface plate inspection method.

more cost-effective than the hard-gage technique. Flexibility is the primary attraction of CMMs, along with high repeatability in the range of 0.0001 inches. Moreover, inspection results are accessible even to those in the shop who are not skilled in inspecting hard gages. The impact of CMM on inspection is the same as that of the computer on information processing. Both do their jobs rapidly and accurately.

A CMM consists of a structure having a base to support the workpiece (to be inspected), physical probes, and the software. The probes can move along the surface of the workpiece. The CMM is an engineered product (Box 11.3) designed so the weight of the part to be inspected does not deflect the machine excessively, and

---

BOX 11.3   *CMM Makes Its Point*

Like so many other pieces of industrial equipment containing computers, CMMs have evolved considerably in function as microcomputers have gained in power, yet their form has remained essentially unchanged. Like the first machines that appeared about 25 years ago, today's CMMs consist of a measuring machine structure, physical probes, microprocessors, and software. The measuring machine structure is essentially a crossbar to which an omnidirectional probe(s) is attached by low-friction air bearings. The crossbar is so constructed that it allows the probe to move along the surface of a workpiece within the measuring area, a cube defined by x, y, and z axes. (Some of the newer CMMs have motorized heads that can take measurements along five axes.) Typically, the workpiece is fixed to a granite surface plate. The crossbar and all guideways are also made of granite to ensure the stability needed for consistent measurements.

When the probe touches a point on the workpiece, the point is defined by its relationship to the axes. A succession of point location data is taken and directed to the microprocessor, which determines distances by using straight-line segment techniques and then compares the distances to nominals to calculate precise geometrical and dimensional part information.

Advances in microprocessor technology and software have increased the number of applications for which CMMs are suitable. Some of the first CMMs were developed to measure flat plates with holes. But as microprocessors that could handle larger arrays of data and perform more complex calculations became available, CMMs became especially useful for measuring objects with contours and irregular surfaces, like cams. Measuring these objects by hand can take hours, but with a CMM, measurements can usually be made in minutes. Typically, measurements are taken at several angular increments and compared with a mathematical curve describing the contour or a curve obtained from a master cylinder.

One of the primary factors limiting the performance of CMMs has been and will probably continue to be the physical probe itself. Since the machines were first developed, most CMMs have gathered data using contact sensing methods, which almost always rely on some form of touch trigger probe. These probes will still be used for the foreseeable future, since they can enter bores and other internal features of parts that optical instruments can't. But they are susceptible to vibration, humidity, heat, and other environmental conditions. As a result, inaccurate measurements may be made when a CMM is used on the factory floor.

Accordingly, CMMs until now have been used most often for off-line applications, especially those associated with tooling and quality control.

As a result, CMMs have been able to identify bad parts only after they have been produced. But as more companies adopt production processes that have little or no tolerance for bad parts, manufacturers perceive an opportunity to use CMMs for on-line manufacturing applications as well. This would involve the integration of CMMs with automated production equipment and would theoretically contribute to the prevention of production errors.

Manufacturers of CMMs have attempted to overcome the limits imposed by the use of touch trigger probes by using noncontact methods of inspection and measurement. For instance, some CMMs provide both a touch trigger probe and a machine vision system using high-speed cameras and image-processing software. While CMMs of this type can measure or inspect parts faster than can be done with touch trigger probes, they are limited by the cameras' 2-D capability.

This limitation may be overcome through the use of nondestructive imaging methods like magnetic resonance imaging (MRI) and electron resonance imaging (ERI). The methods operate in much the same way, although their applications are complementary: MRI can be used for inspection or measurement of nonmetallic materials, while ERI is appropriate for metallic parts. Due to their 3-D capabilities, CMMs using MRI and/or ERI would not only provide dimensional data, they could also detect material flaw and bonding voids.

The strategic purpose of this kind of equipment would be not only to prevent the production of bad parts, but also to aid in data collection and evaluation by engineering and corporate managers. Consumer and capital goods manufacturers believe that they could use the advanced data-gathering capabilities of these CMMs to better understand and control automated manufacturing operations. Indeed, some researchers and consultants have recently coined a new term, total control manufacturing, which provides for the collection and integration of sales, financial, engineering, and manufacturing data as an aid in managers' decision making. Even if such visions of manufacturing's future aren't fulfilled in the near term, CMMs with advanced imaging capabilities could virtually guarantee 100 percent on-line inspection of all critical parts for a wide variety of products, helping manufacturers reach more down-to-earth goals like the elimination of bad parts altogether.

Source: From "CMM Makes Its Point" by D. Deltz, December 1989, *Mechanical Engineering*, p. 54.

the natural frequencies of machine structures are not resonated (a dynamic consideration) by the disturbances (Box 11.4) likely to reach the machine. CMM control software includes a self-tutorial, a menu-driven procedure that offers basic training.

CMMs comprise three subsystems:

1. a mechanical subsystem and its base for controlled manipulation of the probe or probes
2. a microcomputer for number crunching
3. software that (a) handles the recording of data points, (b) assembles the data into verifiable features, (c) compares the inspected and desired features, and (d) prepares inspection reports

The control software allows the coordinate system to be adjusted relative to the workpiece position. A qualification screen helps to locate the probe within the CMM's soft coordinate system. With specific probe hits on the CMM's table-mounting qualification sphere, each probe tip is automatically located in the coordinate

BOX 11.4   *Measuring on Shaky Ground*

How do you measure to three-tenths accuracy surrounded by rumbling presses? The key, according to Fanamation, lies not in the elimination of vibration, but in the management of its frequency.

Total isolation from vibration is impossible. In fact, CMMs generate vibration internally largely due to all the moving components and their drive mechanisms. The secret is in separating the natural vibration range of the CMM as far as possible from that of the shop floor. Otherwise, externally induced shocks will resonate within the CMM, robbing it of its inherent accuracy.

Typical ambient factory vibrations are in the low range, one to twenty hertz (Hz). But due to an emphasis on high stiffness and low mass of moving components—amid a host of other design considerations—the lowest natural frequency of the Comero is 70 Hz. At Drawform, the relative frequencies are further separated by pneumatic spring mounts which dampen floor vibration down to the 2-Hz range. Thus, the "tightly wound" CMM floats as a unit, much like an ocean liner on a relatively calm sea, over the shaky floor below.

Source: "An Integrated Approach to Quality" by T. Beard, March 1990, *Modern Machine Shop*, p. 58.

system. Next, an alignment screen leads the user to establish a working plane with three hits on the workpiece surface. This establishes the absolute coordinates relative to the workpiece as laid on the CMM table. CMM thus does away with the need for locating fixtures that the conventional method requires. From this point on, inspection proceeds as long as all the features to be measured are within the measuring envelope. Measurement menus allow the user to select from a group of standard routines. To help the user select the routine, its function is graphically depicted as an **icon**. Measurement results are tabulated and displayed on the results screen that shows the deviations from the data previously entered into the system. These results can be printed, if desired.

Generally, a CMM can do as much inspection as seven or eight inspectors who are checking manually. Plants use CMMs in two primary modes. The manual mode, is slow and suits lab-type work, while the motorized or direct-computer mode is more suited to faster, production-oriented jobs with on-line inspection of similar parts.

Direct computer-controlled CMMs perform the repetitive tasks of inspection, including the moving of a touch probe from point to point, at high speeds. This provides the operator with information for a quick evaluation of the workpiece quality. In a typical session, the operator chooses the inspection program for the particular workpiece from a system menu. Once the program has been retrieved, a few keystroke commands start the inspection task by moving microprocessor-controlled probes in three directions. The coordinate data is converted into inspection data, which is compared with the design dimensions and tolerances. The results of the comparison are printed as reports, if desired. The inspection is fast, since probes can make a touch almost every second; thus, in 5 minutes some 300 points get measured.

Further sophistication is possible with triaxial CMMs, which are accurate sensors at the end of an arm capable of moving in, out, up, down, left, and right. The movements are computer-controlled and precisely measured. By mounting the component to be inspected, such as an engine block, on the platform and programming the CMM properly, any dimension on the part can be measured accurately. Once programmed, CMMs can measure an engine block in minutes. The traditional single-purpose hard gages and layout methods for the same measurements take hours.

In spite of the advantages of CMM, hard gages are likely to continue to be used in inspection for the simple reason that each method is good for certain types of measurements, and they complement each other. The hard gage is suitable for determining feature size, whereas form measurements such as roundness, flatness, or geometric relationships between part features are better handled by a CMM. But CMMs offer flexibility, speed, accuracy, and repeatability unmatched by the conventional hard-gage method.

CMMs also complement CNC capabilities. The CNC machine is sophisticated enough to machine extremely complex jobs. In the absence of an equally capable inspection device such as the CMM, it is difficult to express the CNC capability's quality traits in numbers.

Quality of CMMs

CMM quality depends on four characteristics: resolution, repeatability, accuracy, and linearity. Resolution specifies the CMM's finest incremental reading. The repeatability, also called **precision**, is the CMM's ability to duplicate a measurement. Repeatability is a function of machine stiffness. Accuracy is the lack of error in measurement, the difference between what the CMM measures and what a perfect CMM would have measured. A CMM can have high precision without being accurate, but it cannot be very accurate without being highly precise.

A good analogy is shooting a rifle at a bull's-eye target. If the shot misses the bull's-eye, then the distance between it and the point hit is a measure of accuracy (or inaccuracy). Precision cannot be measured with one shot. Suppose that five shots were fired. If the points hit were all tightly clustered at the top as in Figure 11.8b, then the shots had high precision or repeatability. Note that high precision does not mean high accuracy. The difference between the two terms is well illustrated in Figure 11.8.

Linearity is the least effective measure of CMM quality. It checks the machine for its movement along the three linear directions X, Y, and Z. It does not take into account straightness, squareness, and other characteristics that affect machine quality.

A better indicator of CMM quality is **volumetric accuracy**. It is determined by a **ball bar test**, which uses a metal rod with precision balls at each end. The rod is measured with the CMM under test, and the difference between the minimum

Figure 11.8
The difference between accuracy and precision

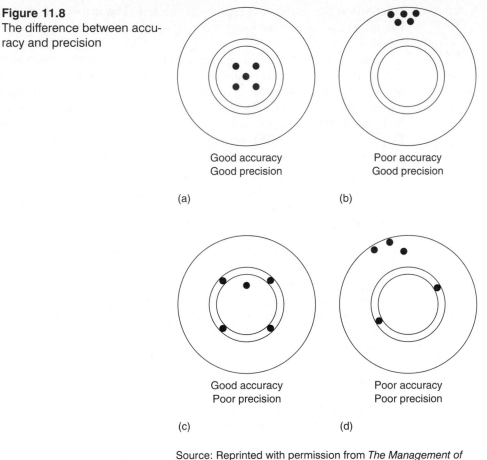

Source: Reprinted with permission from *The Management of Quality Assurance* by M. N. Sinha and W. O. Willborn, 1985, New York: Wiley. Copyright 1985 John Wiley & Sons.

and maximum observed measurements is the error. ANSI specification B89 for CMMs recommends determining volumetric accuracy by measuring at 40 specific points. The length of the bar is also important; short bars can yield an inflated value.

On the basis of volumetric accuracy, CMMs fall into three grades:

Volumetric Accuracy	Grade
0.0001 in. (0.003 mm)	very high
0.0004–0.0006 in. (0.010–0.015 mm)	high
0.0010–0.0015 in. (0.025–0.038 mm)	medium to low

Box 11.5 discusses some recent trends in CMMs.

BOX 11.5   *CMMs: Traits, Trends, Triumphs*

At Harris Graphics, it reduced the time it takes to inspect printing press parts by 94%. At Lucas Western, it cut the cost of quality to under 5% of sales.

"It" is a coordinate measuring machine (CMM), and it's making a name for itself in manufacturing plants large and small—in the R&D lab, the QC room, and on the shop floor.

The benefits are many. CMMs are flexible, measure almost any dimensional characteristic without special fixtures or gages, and require far less setup time than conventional surface-plate inspection devices. Increased accuracy is another plus.

Work is being done on many fronts to increase the effectiveness of these precision instruments.

Shop Hard.   Until recently, users assigned CMMs to protected areas because they were sensitive to foreign materials such as dust and oil mist. Several design changes, however, make it possible for CMMs to leave the lab. They include environmental enclosures that protect the mechanism from contaminants; increased speed and accuracy, which allow the CMM to keep up with production; and developments in the computer industry that have let CMMs integrate with faster, more powerful computers and be programmed in high-level languages like Fortran.

Some new features on recent CMMs to shop hard and render them faster are:

Scales and guideways have temperature sensors, and an automatic correction program (software) compensates for thermal influences.
To reduce the influence of floor vibration, CMMs are provided with antivibration mounts.
Air-conditioned enclosures are provided.
Special enclosures protect them from thermal changes and particulates.
Automatic probe change minimizes interruption of the measurement cycle. While one probe is being used the other is prepared.
The base is made of cast iron to provide damping.

Stainless steel bearing surfaces are precision ground for accuracy.

Moving the CMM from the lab to the shop floor means that industry can change its quality focus from defect detection to defect prevention. Instead of having the negative image of an error detector that renders three shifts of work unacceptable, the CMM can be seen as a production aid that verifies product quality in process and gives the final OK before the product leaves the door.

The CMM's center-driven ring bridge allows speeds up to 10 ips (254 mm/s) and acceleration of 80 in/s^2 (2 m/s^2), with volumetric accuracy of 0.0005" (0.013 mm) and repeatability of 0.00006" (0.0015 mm). Worktable capacity is 3000 lb (1364 kg). The CMM also has an eight-station probe changer and a two-axis probe articulator capable of more than 700 positions. The CMM can change probes automatically without establishing datums.

Inspection programs are loaded into the CMM's MicroVAX computer and stored in a directory accessible only by the cell's host computer. As a pallet loads into the CMM, proximity switches read a code on the side that identifies the parts. It then pulls the proper inspection program into the operating mode. The system will eventually report to the host computer (a trend in the CIM direction).

Standardization.   One action now under way to simplify the user's role in operating a CMM is CAM-I's Standard Operator Interface (SOI) project. Its goal is to let a skilled CMM operator activate any computer-controlled CMM and carry out part inspection using a previously developed part program. The SOI will specify standard screen icons and indicators that are independent of individual machine capabilities and language and terminology constraints.

After the CAM-I Quality Assurance Program approves the SOI, formal procedures will begin to issue it under ANSI procedures as a draft standard for a trial period of a year, says CAM-I's director of CIM technology, Bailey H. Squier. CAM-I may process it upward to render it an international standard as well.

"DMIS is a must in multivendor situations and for advanced CMM inspection solutions," says Fanamation's Maggiano. "Systems implementing DMIS on-line as the native language eliminate translators and provide a standardized link to other manufacturing systems."

"DMIS is a neutral interface," says Squier, "a pipeline big enough to carry all the required information from a CAD system to a CMM or other inspection device and back again. Without DMIS, every CAD system builder would have to create an interface to every type of measuring equipment with which it hopes to communicate. Conversely, every computer-controlled CMM would need an interface to every CAD system."

Another major consideration is communications—how to move the data from the CMM to other parts of the plant (and this is very important for CIM). Satellite links to other facilities around the world are already established for communicating CAD information.

"We're also looking at reverse engineering— the ability to accept a part from a customer and quickly and efficiently reproduce it," says Dan Morrissey of Bridgeport Machines, Inc. "The CMM has historically been the rear end of the process. It inspected the parts after they were made. But automakers often develop clay models before they know the actual dimensions. So the CMM can be both the front end and the rear end. With reverse engineering, it can digitize the unknown information to the CAD system, which generates the part programs, and then to the machine tool, which builds the component. Finally, the information goes to the CMM to check how well it turned out."

Source: From "CMMs: Traits, Trends, Triumphs" by R. R. Schreiber, April 1990, *Manufacturing Engineering*, pp. 31–37.

11.8   TOUCH PROBES

Touch probes on machine tools first appeared in the 1970s. Less sophisticated in the beginning, their potential was soon realized. In the 1980s, they became an essential element in untended machining on their own or as part of a CMM. Probes can aid the operation of cells, FMSs, or automatic work changing systems. It is estimated that three-quarters of the machining centers being sold come with spindle-mounted probes. Fixed probes are being used for measuring and correcting tool diameter and length, as well as for checking for broken tools. Touch probes substitute for many of the operator's functions.

Though the touch probe is quite effective as an inspection tool, its principle is straightforward. A touch probe is simply an extremely accurate limit switch. Whenever the probe stylus makes contact with a surface, it generates an electrical signal. This signal can be fed to the CNC that records the position with reference to the axes' coordinate system. Thus, the location of the surface contacted by the probe is determined. The surface in question is that of the target, which may be a workpiece or tool. Machine slides move the target.

The ability to inspect workpieces right on the machine tool is attractive for various reasons. For one thing, it allows correction of out-of-specification parts on the machine itself, saving the time of reloading that off-line inspection demands. In a recent test, Cincinnati Milacron, Teledyne, and the U.S. Air Force have demonstrated the feasibility of utilizing machine tool probes to:

machine a part
inspect it on-line while the workpiece is still in position
remachine to correct for any deviation
carry out a final inspection
generate an inspection report

The probe method is popular due to the following reasons:

1. Automatic workpiece setup. Probes allow location of the workpiece on a machine table and repositioning of the workpiece and/or program data to workable values. In the absence of probing, the workpiece position is established by time-consuming fixturing as was normally done in the past.
2. Tool setting. With the probe, this task is faster and more accurate than the manual method.
3. Verification. Inspection of the first part if carried out manually is time-consuming and generally involves additional offset calculations. The probe reduces this effort.
4. Closed-loop machining. By feeding probe data back to the CNC, offsets are automatically adjusted, particularly in turning operations where tool wear is a significant factor. In combination with verification capabilities, closed-loop machining encourages automated operations.

Depending on the purpose, touch probes are either the on-machine type, for workpiece setup and inspection, or the in-spindle type, for tool setting. The former is usually loaded into the spindle of the machining center or mounted on the turret of a lathe. The tool setting probes are stationary and mounted on the machining center table or nearby. The advantage of probe setting is time savings. For example, it takes only 15–20 seconds to probe and enter the offset for a turning tool. On the machining center it takes a few minutes, depending on the number of tools in the carousel, (e.g., 6 minutes for a 24-tool carousel).

Using a probe in inspection is more involved than in tool setting. The probe's accuracy is first determined by comparing results with a coordinate measuring machine. With programming ingenuity that can account for built-in errors, on-machine inspection routines can be made to rival the reliability of a CMM-based procedure. For turning operations, the probe-based inspection can compensate for tool wear. In some cases, the method has been proven to compensate for errors due to thermal distortion of workpieces. It must be remembered, however, that probe technology is still evolving and has not matured. But it is definitely out of the research lab and onto the shop floor.

The most common probe system is the inductive transmission type. Such probes need little maintenance and are extremely reliable. The other type is based on optical transmission principles. In this system the probe unit contains an infrared transmitter that sends coded information to an optical receiver mounted outside the tool/workpiece travel envelope. The receiver in turn transmits a signal through an interface to the CNC.

Besides the electromechanical components, touch probe technology also requires control software to be useful. CNC programming tools allow for standard probing cycles within control-resident conversational programming systems as well as using G codes. Various canned cycles are used along with this technology to find the location of geometry elements such as a point or surface. If the surface is not smooth or flat, as in the side of a rough casting, several points are measured and the workpiece orientation determined with reference to a plane defined through these points. The benefit can be extended to automatically adjust for any misalignment without having to reposition the workpiece. Some controls have a feature, called coordinate or axis rotation, that allows the CNC to rotate the program in the X and Y axes. This enables changes in the part program to adjust for workpiece orientation. The probe can also allow for program changes with reference to an already machined surface. Associated with the probe technology are digital readouts (DROs), an electronic display that replaces scales and verniers. As discussed in Box 11.6, DROs are low-cost devices that enhance productivity of machine tools.

□ BOX 11.6   *DROs' 16 Productivity Enhancements*

Today's digital readouts (DROs) can do just about anything handled by a computer numerical control (CNC) except move the table, saddle, or spindle.

Digital readouts, those wonderful electronic displays that replace the scales and verniers on machine handwheels, have come a long way. Today's DROs virtually duplicate modern CNCs except for directing the servos that move the table, saddle, and spindle. The operator still must perform this task when using readouts. For many shops, this situation is appropriate. DROs are relatively inexpensive, they can be retrofitted to nearly any existing machine with relative ease, they are a logical step to full CNC, and they can easily introduce electronic usage to the shop floor without upsetting the existing routine.

Many engineers and managers are unaware of the recent technological advances available in DROs. One of the most significant improvements is the incorporation of machine-mounted glass or electronic scales that are totally independent of the lead screw. This design eliminates the backlash problem found in early-model DROs that relied on resolvers mounted on the end screw. Readings from these DROs were totally lead-

screw dependent, and screws with any play at all gave bad readings. With machine-mounted scales, a saddle or table position is reported as it exists—not in relation to lead-screw turns.

Today's DROs also feature communication terminals for off-line programming and data transmission, memory for data storage, calculation capabilities, canned cycles, tool length offsets, cutter diameter compensation, probing capabilities, inch/metric positioning, and operation in either the absolute or incremental modes or a combination of each. These can be fitted to any machine axis of motion, whether it is linear or radial.

As a result of these features, DROs can be used in many creative ways to increase productivity and improve product quality. Sixteen different ways have been cited in the source.

A DRO with a resolution of 0.0005 inch is probably best for a working machine in the shop. A precision jig borer could be equipped with one that resolves to 0.0001 inch.

Source: "DROs' 16 Productivity Enhancements," May 1990, *Modern Machine Shop*, pp. 82–87.

Some people believe that a part cannot be inspected on the machine it has been machined on. This is not true. Most modern NC machines are accurate to within one-thousandth of an inch, which is comparable to that with CMM. The important consideration is the tolerance required on the machined surface. As a rule of thumb, if the NC machine is more than three times as accurate as the tolerance, which usually is the case, the machine can be used in conjunction with the touch probe (Owen, 1991) to inspect the part.

TRENDS

❑   Quality assurance is vigorously practiced in modern plants, though under different names such as TQM (total quality management), SPC, or quality circles.

❑   Rather than sort out defective products, today the trend is to find the sources of errors that give rise to poor quality, and to rectify problems in the process. The emphasis has shifted to SPC because quality through process control and product design makes more sense.

❑   Inspection is becoming a task within CAM, which brings inspection within the overall CIM concept through integration of CAD and CAM.

❑   Used originally for off-line measurements and machine tool positioning applications, CMMs now play an important role in on-line inspection of product and repair operations.

❑   The emphasis on product quality highlights the crucial role of metrology in today's plants.

❑   Some other recent developments include: (a) digital hand measurement tools used in conjunction with data collectors connected to computers; (b) sophisticated, yet relatively inexpensive, table-top CMMs for machine shops and tool rooms; (c) noncontact gaging based on optics and laser scanning; (d) two-way communications between CMMs and CAD systems; and (e) real-time monitoring and analysis of SPC data.

SUMMARY

The importance of quality and the role of metrology in CIM were presented in this chapter. The discussion focused on technologies that facilitate integration to reflect the proactive trend in quality assurance. Modern concepts of product quality and the associated tools and techniques were explored. SQC and its evolution into SPC were described, along with the role of process capability and its assessment. Finally, the technologies of touch probes and CMMs and related developments were described.

KEY TERMS

Ball bar test
Control charts

Dimensional Measuring Interface Specification (DMIS)

Hard gage

Icon

Precision

Process capability

Softgage

Statistical quality control (SQC)

Total quality management (TQM)

Touch probes

Volumetric accuracy

EXERCISES

Note: Exercises marked * are projects.

11.1   James R. Houghton, chairman and CEO of Corning Incorporated has said, "Our definition of quality is meeting customer requirements." Discuss the significance of this statement in approximately 200 words.

11.2   Name three individuals who have made extraordinary contributions to the advancement of quality during the last 20 years. What were their contributions?

11.3   Write a 300-word comprehensive essay on CMM, using sketches where appropriate.

11.4   Explain the role of CMM in CIM (limit your answer to approximately 200 words).

11.5   What is SPC and how does it help achieve the objectives of CIM?

11.6 $\overline{X}$ and s control charts are maintained on the thickness of a delicate strip for a transducer. The results of inspection are on page 369. Draw the control charts and discuss what they reveal.

11.7   If the specification limits for the strip thickness in the above exercise are 8.96 and 9.44 mm, find the values of PCR and C_{pk} and comment on them. Discard the out-of-control points before assessing the process capability.

11.8*   Obtain brochures or trade literature from the manufacturer or distributor of a CMM and study them. Identify the CMM's specifications and design features that would benefit CIM, and present them to the class or the company group you work with.

11.9*   Visit a local industrial firm (manufacturing or otherwise) known for its quality image, and study its efforts in TQM. On your return, write a 300-word report on your findings and discuss with the class (or company group you work with). Invite someone from the firm to participate in the discussions.

Sample Number	Thickness (in mm)			
1	9.97	9.87	9.88	9.79
2	9.93	9.78	9.90	9.89
3	9.87	9.69	9.89	9.99
4	9.83	9.98	9.80	9.87
5	9.89	9.78	9.90	9.89
6	9.89	9.88	9.93	9.98
7	9.78	9.68	9.93	9.99
8	9.78	9.59	9.78	9.46
9	9.93	9.98	9.90	9.59
10	9.87	9.88	9.89	9.85
11	9.67	9.78	9.78	9.86
12	9.98	9.68	9.88	9.87
13	9.94	9.98	9.99	9.99
14	9.95	9.78	9.88	9.89
15	9.83	9.88	9.68	9.79
16	9.83	9.73	9.89	9.69
17	9.93	9.77	9.88	9.87
18	9.97	9.78	9.89	9.79
19	9.96	9.88	9.90	9.85
20	9.89	9.68	9.99	9.69
21	9.78	9.79	9.95	9.87
22	9.78	9.88	9.80	9.79
23	9.93	9.87	9.88	9.85
24	9.89	9.78	9.95	9.89
25	9.90	9.79	9.92	9.87

SUGGESTED READINGS

Books

Besterfield, D. H. (1994). *Quality control* (4th ed.). Englewood Cliffs, NJ: Prentice-Hall.
Besterfield, D. H. (1990). *Quality control* (3rd ed.). Englewood Cliffs, NJ: Prentice-Hall.
Crosby, P. B. (1979). *Quality is free*. New York: McGraw-Hill.
Crosby, P. B. (1984). *Quality without tears*. New York: McGraw-Hill.
Deming, W. E. (1982). *Quality, productivity, and competitive position*. Cambridge, MA: MIT.
Feigenbaum, A.V. (1983) *Total quality control*. New York: McGraw-Hill.
Juran, J. M. (Ed.). (1988). *Quality control handbook*. New York: McGraw-Hill.
Ranky, P. G. (1990). *Total quality control and JIT management in CIM*. Guildford, Surrey (England): CIMware Limited.
Smith, G. (1991). *Statistical process control and quality improvement*. New York: Macmillan.
Wick, C., & Veilleux, R. (1987). *Tool and manufacturing engineers handbook: Vol. 1. Quality control and assembly*. Dearborn, MI: SME.

Monographs and Reports

ASQC Quality Cost Committee. (1986). *Principles of quality costs*. Milwaukee: American Society for Quality Control, Inc.

Guidelines for quality systems in the machine tool and related industries. November 1989. NMTBA (National Machine Tool Builders Association).

Quality assurance/quality control software directory. Published annually in one of the issues of *Quality Progress*.

Journals and Periodicals

Modern Machine Shop. April 1991 (emphasis on inspection, testing, and quality control).

Quality Progress. Monthly. American Society for Quality Control, Inc., 310 West Wisconsin Avenue, Milwaukee, WI 53203.

Quality Engineering. Quarterly. P.O. Box 697, Hauppauge, NY 11788.

The International Journal of Quality & Reliability Management. MCB University Press Limited, 62 Toller Lane, Bradford, England BD8 9BY.

The Quality Circles Journal. Quarterly. IAQC (International Association of Quality Circles), 801-B West Eighth Street, Suite 301, Cincinnati, OH 45203.

Articles

CIME staff report. (1989, December). A new dimension for automated inspection. *Mechanical Engineering*, pp. 52–55.

Beard, T. (1989). Managing the shop with SPC. *Modern Machine Shop*. *61*(9), 50–58.

Beard, T. (1990, February). In touch with quality and productivity. *Modern Machine Shop*, pp. 68–77.

Birman, S. (1991, April). Quality is more than SPC. *Modern Machine Shop*, pp. 96–103.

Horn, D. (1989, September). Smarter sensors respond to factory stimuli. *Mechanical Engineering*, pp. 64–67.

King, D. J., & Mariani, M. A. (1991, April). Smart height gage takes on the CMM. *Modern Machine Shop*, pp. 66–71.

Lavole, R. (1989, April). Shopping intelligently for CMMs. *Manufacturing Engineering*. *102*(4), 67–70.

Owen, J. V. (1991, August). CMMs for process control. *Manufacturing Engineering*, pp. 39–41.

Vasilash, G. S. (1990, May). Why you should consider small CMMs. *Production*, pp. 66–68.

PART IV

CIM Management

In part III we discussed the various production technologies conducive to CIM. These technologies and the associated personnel need to be managed. Most CIM failures have been attributed to poor management. These failures suggest that conventional practices of managing manufacturing operations and business must change to achieve CIM.

In this part of the text, we discuss management techniques and practices that are suitable for CIM. Also included are the educational and training needs of personnel, so essential to the successful implementation of CIM.

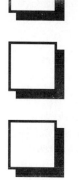

CHAPTER 12
Management of CIM

"CIM is working successfully for many manufacturers, but many more would benefit were it not for management's hesitancy or inadequate knowledge of CIM."
—Kenneth Bonnie, General Dynamics Corp.

"Carnegie Mellon surveyed 400 American corporate managers and found that a third of them distrusted their own bosses. More than half did not believe top management. Companies cannot expect the best commitment from employees who don't believe or trust their managers."
—Bert Casper, Remmele Engineering, Inc.

"Because CIM is a concept (and a radical one at that), not a highly defined and well-proven technology, the prospect of implementing a CIM strategy naturally arouses doubts and anxieties."
—Mark Albert in *Modern Machine Shop*, July 1990

"Today's successful company is a customer satisfaction system and the issue is 'who runs his customer satisfaction system the best.'"
—Hal Sperlich at a seminar focusing on the challenges manufacturers face in the 1990s

"Manufacturing competitiveness has its roots in management decision making, philosophies, and biases far more than in blue-collar work rules, technology challenges, and unfair competition. Additional overhead, slow decision making, and separation of product and process design are all symptoms of a bureaucracy that is eating itself alive."
—"Manufacturing Priorities for the 1990s," a Harbor Research Inc. report

"Flexible responsive manufacturing will be to no avail if it is not accompanied by flexible, creative, imaginative management."
—Robert Costella in *Mechanical Engineering*, June 1990

"Part of the U.S. difficulty in remaining competitive is that we have a generation of managers who don't understand the technology they are managing."
—Dr. Lester Thurow, dean, MIT Sloan School of Management, in *1988 ASEE Conference Proceedings*

"It makes little sense to install third-, fourth- and fifth-generation computer technology in second-generation organizations."

—Charles M. Savage, Digital Equipment Corp.

"W. Edwards Deming says if you're having a problem, 85% of the time it's management. I buy that."

—Don Petersen, chairman, Ford Motor Co.

"As nearly everyone knows, a manager has practically nothing to do except to decide what is to be done; to tell someone to do it; to listen to the reasons why it should not be done, why it should be done by someone else, or why it should be done in a different way; to follow up to see if the thing has been done; to discover that it has not; to inquire why; to listen to the excuses from the person who should have done it; to follow up again to see if the thing has been done, only to discover that it has been done incorrectly; to point out how it should have been done; to conclude that as long as it has been done, it may as well be left where it is; to wonder if it is not time to get rid of a person who cannot do a thing right; to reflect that he/she probably has a family, and that certainly any successor would be just as bad, and maybe worse; to consider how much simpler and better the thing would have been if one had done it oneself in the first place; to reflect sadly that one could have done it right in twenty minutes, and as things turned out, one had to spend two days to find out why it has taken three weeks for somebody else to do it, and that also wrong!"

—Anonymous, *Modern Machine Shop*, April 1989

"Speed often pays in product development even if it means going over budget. An economic model developed by McKinsey and Co. management consulting firm shows that high-tech products that come to market six months late but on budget will earn 33% less profit over five years. In contrast, coming out on time and 50% over budget cuts profits only 40%."

—a *Fortune* article quoted in the *Journal of Applied Manufacturing Systems*, Spring 1990

M anagement is the process of making decisions and directing the activities of personnel to achieve stated objectives. The objectives are successfully met when efforts are organized by communicating appropriate information for control and readjustment.

Large systems, such as a manufacturing enterprise, involve a hierarchy of management tasks, as shown in Figure 12.1. Management is generally grouped into three levels: top, middle, and lower. These levels consistently interact to optimize the use of available resources, which fall into three categories: capital, material, and human (Box 12.1). Even when the operations are run without CIM in mind, management involves interaction, and thus integration. What CIM does is reinforce this integration to create a real-time environment. In this chapter, we discuss the issues of management that directly affect or are influenced by CIM.

Figure 12.1
Planning and organizing hierarchy

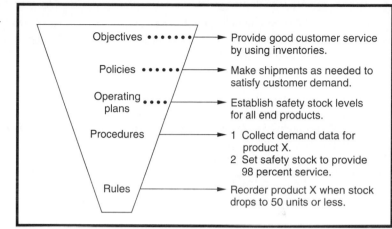

Source: Reprinted with permission from *Operations Management* by J. G. Monks, 1982, New York: McGraw-Hill.

■ **BOX 12.1**   *Resources*

An organization's resources are the material and nonmaterial assets available to achieve its objectives. Resource inputs to the production system may be conveniently classified into capital, material, and human resources.

Capital.   Funds are the immaterial store of value which permits the production and exchange of goods and services on a large scale. They are used to establish and regulate the amount of material and human inputs. In an aggregate sense they help to determine the level of technology and the tradeoff between the use of labor and the use of equipment. As more capital is allocated to a given phase of a production process, the level of technology typically rises and, via automation, equipment replaces human labor.

In free enterprise organizations, capital becomes available in the form of equity (stock) or debt (bonds) funds and is replenished via profits. In nonprofit organizations, taxes or contributions are a continuing source of funds to finance opera-

tions. Both profit and nonprofit organizations need to make efficient and effective use of capital.

Material.   Material resources are the physical facilities and materials such as plant equipment, inventories, and supplies. Included are control equipment like computers, electrical energy, and the raw materials used in operations. In an accounting sense, the material and equipment usually constitute the major assets of an organization.

Human Resources.   The human input is both physical and intellectual. Early production efforts relied heavily upon human physical labor, but as production technology and methodology advanced, a higher proportion of the human input became devoted to planning, organizing, and controlling efforts. Through the use of the intellectual capabilities of people, labor inputs are magnified many times, resulting in increased productivity as well as much closer worker-machine interface.

This closer integration of human and physical systems presents one of the major challenges of job design to achieve worker satisfaction.

The human resource, although it is often a key asset, is typically not accounted for in the balance sheet of the organization. Some attempts are under way to implement human resource accounting, but the problems are many, and the rate of progress has been slow. Nevertheless, it is the human resource, in the form of managerial talent, worker skill, and employee cooperation, that has stimulated the growth and development of the large-scale organizations that flourish today.

People also bring human values into an organization, and some of these values are institutionalized into the corporate organizational society. They become traditions, standards, and ethical guidelines, both for internal operations and for dealing with the public. These values often play a strong (but difficult to define) role in the organizational decision-making process.

Source: Reprinted with permission from *Operations management—Theory and problems* by J. G. Monks, 1982, New York: McGraw-Hill, pp. 11–12.

12.1   ROLE OF MANAGEMENT IN CIM

As mentioned in Chapter 1, CIM is not just a technology, it is a philosophy, a concept. Its reverberations spread throughout the entire organization. It may require dismantling some of the usual procedures and practices. CIM may bring departmental or group politics out into the open, since it may require demolishing the turfs that have developed over the years (logically, not physically). Responses such as "we never did it this way before" must be questioned.

The effects of this potential upheaval require full involvement by senior management; approving funds for CIM projects is not enough. CIM implementation, especially in the beginning, cannot be left to the middle and lower management.

The most important contribution senior managers can make to CIM is their wholehearted commitment. The chief executive officer must be involved directly or through immediate subordinates. A strong commitment ultimately creates a ripple effect that permeates throughout the entire organization. Such an atmosphere promotes rapid transition toward CIM by simplifying the tasks of middle and lower management.

In an article appropriately entitled "Integrating Islands of Automation Is Management, Not Technical Problem," Mehta (1987) identifies the following six tasks for the managers of CIM:

1. Develop a business model to understand the problem environment.
2. Develop a functional model for the processes, functions, and activities to describe both "as is" and "to be."
3. Develop an information model that identifies system interfaces, information exchange patterns, database requirements, and applicable technologies.
4. Develop a network model to identify communication and networking requirements.
5. Develop an organizational model to investigate the implications of integrating the various islands of automation (Figure 12.2) on the existing organization structure and culture, and how to safeguard against detrimental effects.

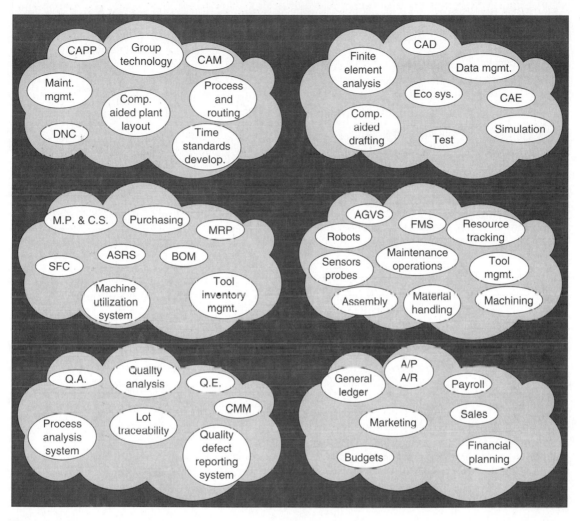

Figure 12.2
Clusters of tasks that evolve into islands of automation
Source: Reprinted with permission from "Integrating Islands of Automation Is Management, Not Technical Problem" by A. J. Mehta, November 1987, *Industrial Engineering*, p. 43.

6. Finally, develop the implementation plan which should take into account special features of the business and operations. (p. 42)

These tasks and their interrelationships are illustrated in Figure 12.3. Mehta's approach follows IDEF methodology developed by the U.S. Air Force Integrated Computer-Aided Manufacturing Project (Harrington, 1984), and Information Resource Management Aid (IRMA) developed by the consulting firm Arthur D. Little, Inc. Its impact on the classical management pyramid is illustrated in Figure 12.4.

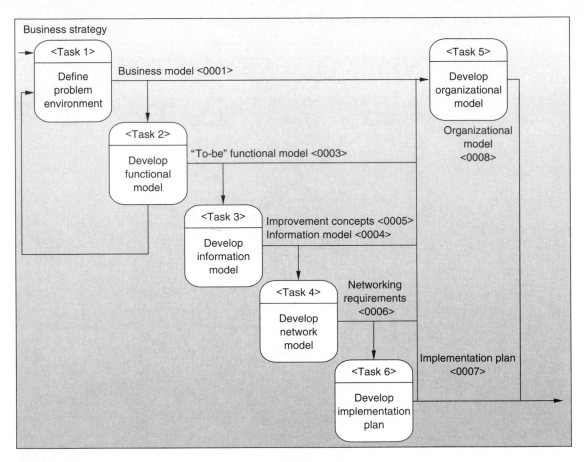

Figure 12.3
Task modeling for CIM at various levels: IDEF methodology
Source: Reprinted with permission from "Integrating Islands of Automation Is Management, Not Technical Problem" by A. J. Mehta, November 1987, *Industrial Engineering*, p. 49.

Figure 12.4
CIM managers' tasks and their interrelationships

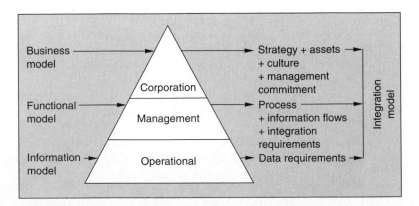

Source: Reprinted with permission from "Integrating Islands of Automation Is Management, Not Technical Problem" by A. J. Mehta, November 1987, *Industrial Engineering*, p. 44.

12.2   CONVENTIONAL WISDOM CHALLENGED

The growth of CIM since the 1980s has called into question the conventional wisdom on which we have based managing manufacturing since Taylor's scientific management principles of 1915. For example, conventional practices of excessive division of labor and labor-productivity-based management decisions are increasingly being reviewed.

According to Dr. Howard Aaron (Petrock, 1990), management should be working with modern wisdom rather than conventional wisdom. The following illustrates the major differences between the two approaches.

Conventional Wisdom	Modern Wisdom
Dictate the voice of the engineer.	Deploy the voice of the customer into every nook and cranny of the organization.
Automate for speed and reduced direct labor.	Automate to reduce variation and produce consistent quality at reduced cost.
Inventory is necessary to protect production.	Inventory hides problems, costs money, and should be targeted for eventual elimination.
Queues are required to maintain consistently high machine utilization.	Once in motion, always in motion. Production should be just-in-time. Eliminate queues. Reduce work-in-process.
Work means getting hands dirty.	Cleanliness is next to godliness.
Work harder.	Work smarter.
Solve problems.	Prevent problems.
Push material through factory.	Pull material through factory.
Lot sizes should be economical.	Lot sizes should be small, preferably ONE.
Overhead functions are essential.	Any labor that does not add value to the product or service should be eliminated.
Run machines at maximum speeds. Breakdowns are to be expected.	Endurance running is the game. Slower machine speeds and superb preventative maintenance are the keys.
Design and build to specified tolerances.	Design and build to target with minimum variation.
Purchase from multiple vendors.	Purchase from one supplier of proven performance.
Buy for minimum piece price.	Buy for the minimum product life cycle cost.
Reduce cost by reducing direct labor and increasing machine utilization.	Reduce cost by speeding product through the factory. Eliminate non-value-added operations such as handling, and inspection.

Expediting is a necessity.	Expediting wastes resources.
Quality is free.	Quality is not free, but it's the best investment we know.
Quality is the responsibility of the QC department.	Quality is everyone's responsibility.

In recent years, new ideas such as just-in-time (JIT) production, design for manufacturability (DFM), and quality circles have been experimented with and found more effective in managing manufacturing resources. Box 12.2 describes the idea of **lean production**, and Box 12.3 presents a case study on DFM, simultaneous engineering, and employee involvement at John Deere.

 BOX 12.2   *From Mass to Lean Production*

According to the authors of *The Machines That Changed the World*, the success of Japan's automobile industry can be attributed in large measure to its development and adoption of a new way of organizing to make things that they call lean production.

Lean production refers to a constellation of new organizational relationships both inside and outside the firm, to a different way of viewing workers, customers, and the environment, and to a different understanding of how technologies change and improve.

Lean production is a way to produce automobiles—in comparison with mass production—at lower cost, in less time, using less labor, achieving higher quality, with greater variety, and with fewer adverse impacts on worker health and the natural environment, all the while accommodating more rapid model change and including the latest in features and performance. Furthermore, the optimum scale of production is typically one-quarter of that in a mass production operation.

The managers of lean production-based organizations build manufacturing systems that seek perfection the first time, thus avoiding wasted time and materials, and that are populated by well-trained employees who are encouraged to measure their own performance and to suggest ways to regularly and continuously improve both product and process.

Womack and colleagues found that the key to lean production is new organizational relationships. Inside the firm, the division of labor is modulated by building cross-functional and cross-task teams at all levels, thus encouraging the sharing of responsibility and the close integration of disparate parts of the process. Information is widely shared and, in the hand of well-trained employees, becomes the foundation for both quality improvement and system flexibility. With less emphasis on hierarchy and more shared responsibility, greater labor-management cooperation follows.

In its relationship with external organizations, the lean production firm is focused on understanding and meeting customer needs, not on shaping or avoiding them. Suppliers and subcontractors are part of the production team, kept informed of the corporate plans, encouraged to build and maintain technical competence, and given long-term contracts that facilitate investment in people and technology for the long pull.

Located near the assembly plants and assured of their market, suppliers can be brought into the new product development cycle early, thus reducing the time to bring all the components of a new model on-line and increasing the chances that a part or component can fulfill its intended function at low cost and high quality.

Thus, the key aspects of lean production are organizational rather than technological. It is not that technology does not matter, but what is on the critical path is the structure and management of the organization. Technology that facilitates the management of information seems to be key to the functioning of a lean production organization, and a lean organization is probably also better situated to take advantage of advances in production technology than a mass-production-based enterprise. It is clear that introducing advanced automation technology into an enterprise managed on mass-production principles will not be nearly so successful as introducing the same technology into a lean production enterprise.

Finally, because it requires a change in a host of relationships and practices, Womack, Jones, and Roos suggest that the transition from mass production to lean production may be as difficult and take as long as the earlier transition from craft production to mass production, a transition that took decades. Developed to their highest levels in the Japanese automobile industry, the techniques of lean production are now finding their way into some major American manufacturing enterprises, but there is a very long way to go.

Source: "New Manufacturing Paradigms—New Manufacturing Policies" by C. T. Hill, 1991, *The Bridge*. *21*(2), 17–18.

BOX 12.3   *DFM, Simultaneous Engineering, Employee Involvement Aid Deere*

In the late 1970s, John Deere's manufacturing operation was characterized by the high-rise storage facilities it built to support its large-lot, high-inventory, long-lead-time system.

"Now we'd say they are monuments to waste," says C. Sean Battles, an executive consultant in Deere's strategic manufacturing planning operation. But when Deere's sales dropped an astounding 50 to 80 percent in the early 1980s as a result of agricultural and industrial equipment market decline, the company was "forced to change—and change rapidly—the way we were doing business."

Through a dramatic companywide effort, Deere did change. Today, the company has not only gained market share, it has also transformed itself into America's largest maker of farm equipment, and a major builder of construction and lawn care equipment. Last year, its 38,000 employees in twenty-nine plants in the United States, Canada, Mexico, and Europe helped produce $7 billion in total sales.

Perhaps the best and most satisfying proof that Deere has changed can be found in the Davenport, Iowa, factory where employees build four-wheeled, rubber-tired loaders, paint them orange (instead of Deere's trademark yellow) and ship them for sale to—you guessed it—Japan, under the Hitachi label.

How did they do it? Battles says that the turnaround is the result of a total manufacturing approach that encompasses a number of comprehensive programs, such as just-in-time and total quality commitment. Three programs, however, made and continue to make significant contributions to cost reduction and quality improvement: design for manufacturability, simultaneous engineering, and employee involvement.

In 1980, Deere's managers believed that product design represented a small percentage of a product's total cost. So they were surprised when an internal study revealed that a product's design influences over 50 percent of its manufacturing costs. Part costs, processing costs, quality costs, overhead costs—all were affected by product design. Battles says that the results of the study forced management to rethink its attitude about product design.

"We realize now that if we don't get a competitive product from design, we may never have a competitive product—no matter what we do later on the factory floor."

Battles credits design for manufacturability (DFM) in helping to make Deere's various product designs competitive. DFM is a design approach that emphasizes reducing the number of parts and fasteners needed to assemble a product.

"It's mind-boggling what you can accomplish with this program," Battles says. "When used early enough in the design process, it can help eliminate 50 percent of the parts, 50 percent of the assembly time, and 20 to 70 percent of the cost. Best of all, these savings are realized without compromising function, reliability, durability, serviceability, or appearance."

The Bigger Picture

DFM is one aspect of a larger design philosophy called simultaneous engineering (SE). SE, known at some companies as concurrent engineering, is a method of addressing early in the design process all issues that impact production of a product and customer satisfaction with the product.

To address such issues, teams of designers work with people from engineering, manufacturing, and even suppliers and customers to design the product. At Deere, the SE teams are formed within the new product and production development section. They are made up of product and manufacturing engineering representatives, as well as marketing, accounting, wage employees, outside suppliers, and customers.

"It's critical to bring everyone in early, right from the start of the project," Battles says. "That way, we get the voice of the customer and the voice of manufacturing in product development."

In the plants where SE teamwork was applied, Deere was able to reduce part numbers at least 50 percent and the number of operations by 50 percent. When a part's design requires half as many

operations, no one has to process, route, tool, schedule, or move material to or solve quality problems on the eliminated operations. This has a "ripple effect" in savings in both direct and overhead costs.

Battles admits there were a lot of challenges that had to be overcome to make DFM and SE work at Deere. These challenges included:

The "We're Different" Syndrome. Using the DFM-teamwork approach, Deere's lawn care division found it could save $1 million a year by reducing the number of nuts and bolts, the types of fasteners, and some of its operations. "We were able to design out parts that cause service problems and parts that don't enhance the value or function of the product," Battles says.

But when Deere's management tried to apply the principles that worked so well in the lawn care division to their heavy equipment operations, they encountered resistance. "We're different," the managers in the other divisions said, "We don't make lawn mowers." Battles says that, far from being different, the heavy equipment operations had 95% of the people and resources needed to implement DFM. All they needed was an attitude adjustment, which they made.

The Emphasis on Function Rather Than on Product. For DFM and SE to work successfully, people must be less concerned about how the process affects their job functions and more concerned about the impact it has on product and customer satisfaction. Too much focus on functions can throw up barriers to finding the most efficient way to design and manufacture a product. In the old days at Deere, Battles says, functions were primary. "The product was a second-class citizen. We had to get people to change their thinking from 'What is best for my function?' to 'What is best for business and how does it affect my function?'"

How-to Checklist for DFM

If your company is about to implement design for manufacturability and simultaneous engineering concepts, keep in mind the following tips from C. Sean Battles:

Keep people informed of the big picture. Employees are more committed when they are kept up to date on manufacturing, customers, competition, and technology that involves them.

Do a pilot project. People more readily accept new ideas when they see them working on their own product in their own plant. Make managers coaches, not just cheerleaders. Managers need to know and understand these new concepts and how they work. And they need to be committed to them.

Don't fall for the "We're Different" syndrome. Look at the principles and practices of design for manufacturing, simultaneous engineering, and employee involvement and see how they can work for you.

Get the machine operators involved on your DFM teams. Very often, they'll find a $25 solution to a problem that your engineers say can only be solved for $100,000 or more.

Design out new parts first before you modify existing parts. The quickest way to save money is to design out parts that cause service problems and don't add any customer value or function.

Get everyone involved up-front. The most successful teams are those that include the accountants, marketers, engineers and machinists right from the start of product development.

Hire an outside consultant. They can be very helpful to get training and a basic understanding started on new techniques.

Change your own attitude from "What is best for my function?" to "What is best for the business and how does it affect my function?"

Team-Building. With so many different functions working together, getting them to work as a team can be difficult. "The extent of teamwork used to be that our engineers would say, 'As soon as the product designer designs it, I'll tell him what's wrong with it.'" Now they say, 'How can I help with the design?'" Deere holds team-building sessions as needed to help employees keep working together efficiently and productively.

New Reporting Structures. Another element that was critical to teamwork and the overall success of DFM and SE was to have team members report to their teams, not to the departments they came from, such as accounting, marketing, or engineering.

While realigning reporting structures can be a daunting challenge, Battles explains that doing so is essential since it can take two to three years to get a new product from concept to the customer. It also makes it easier to establish priorities, especially in a large organization. Employees are less likely to be called upon to "put out fires" back in their old departments. Battles further asserts, "It creates ownership of the product, leading to greater commitment and pride from team members."

Battles notes that with any given team a different individual will rise to the top. For example, the team leader in a product development group for a new industrial transmission was an accountant.

Worker Skill Level. For the team approach to succeed, workers have to broaden their knowledge and abilities to be multiskilled. For example, managers of product engineering have to know and be responsible for industrial engineering, manufacturing, production control and quality. Assembly workers may have to know how to operate a press as well as weld.

Role of Employee Involvement

"Achievements like cutting overhead by 50% or more can only happen with the support of the people on the shop floor," Battles says. "Employee involvement is the key element for the continuous improvement of DFM, SE—or any other program, for that matter." Battles adds that employees will only participate up to the fences set by management. Therefore, it's the job of managers to widen the fence rows as much as possible. Here are some of the things Deere does to encourage maximum employee participation:

Information-Sharing. "We share more information today with our factory people than we did eight years ago with our salaried people," Battles states. Back then, the only two people who knew the strategic objectives of the factory were the general manager and the controller. Now Deere shares strategic plans and goals with everyone from the manager to the shop floor. "People in the know have more go," says Battles. "If you want people to make business decisions, you've got to tell them what the business is. They need to understand the concerns of the operations, how they impact them and the competition. They need feedback on quality and profits. That way, they can make the best business decisions."

Training. Battles estimates that a 30-year employee represents a $1-million-plus investment in wages and benefits. Therefore, just as you would maintain a machine tool, you should invest in improving the skills of your employees. Training should be regarded as an investment, not an expense. At Deere, employee training averages ten to twenty hours per employee per year.

Responsibility-Sharing. To encourage employee participation, Deere gets more people involved in all phases of product development. Employees now get involved in specifying, going to suppliers, even making "go/no-go" decisions when it's time to ship a product.

"Operators know how to do the job. They can simplify manufacturing, save time, suggest tooling, and improve quality all at the same time,"

Battles says. "Everybody wants to make a quality product and they know what's going on may affect their jobs for the next three to five years. So when given the chance to have their voices heard, 95 percent of your employees will get involved."

Battles says that these strategies are still being used and modified at most Deere factories. Simultaneous engineering is pretty much a way of life. DFM teams are being added. The factories that are moving the fastest are those in the industrial and lawn care equipment divisions, which compete directly with Japanese companies.

But already the bottom-line savings that have resulted from using these innovative practices are impressive. Battles states that Deere has reduced the number of parts by 50–60%, assembly time has been slashed 50–60%, and costs have been cut 20–70%, depending on the product or component.

In short, Deere had gone from building monuments to waste to building a foundation for a successful future.

Source: From "DFM, Simultaneous Engineering, Employee Involvement Aid Deere," January 14–28, 1990, *CIMWEEK* (The Newsletter for Computer-Integrated Manufacturing), pp. 1–5.

12.3   COST JUSTIFICATION

An important task of management is to keep facilities modern and up to date. To accomplish this, management budgets annually and spends the appropriated funds to acquire suitable resources. Irrespective of the resource type—labor, machine, computer, software, tooling, robot, building or any other—funding approval requires justification.

Traditionally, justifying investments in technology has been based on savings in direct labor. But in modern manufacturing plants, direct labor accounts for only 5%–10% of the production cost. As Ken Gettelman, late editor-in-chief of *Modern Machine Shop* observed, "As a percentage of total manufacturing costs, direct labor in many American and Japanese plants is well under ten percent. At the 1987 ultra-modern Mazak facility opened in 1987 in Worcester, England, the figure is less than five percent." Thus, management should deemphasize the traditional method of cost justification (and cost accounting) based on direct labor, and instead consider indirect savings. Lint (1992) suggests the use of options analysis in deciding whether to invest in advanced manufacturing technologies (AMT).

The new approach to cost-justifying CIM technologies is based on **integrated investment**. According to Raffish (1989), "Manufacturers were faced with a new challenge—integrated investment. These were investment opportunities that crossed product lines, functional department lines, and whose payback characteristics were different than the traditional investments." He adds, "Historically, strategic business planning has been driven with a Marketing and/or Finance emphasis. When the necessary forecasts were in place for annual sales, by product line or division, everyone else sort of marched to that tune. Today, we have learned some interesting lessons based on recent analysis. Planning need not always be driven from sales forecasts, manufacturing is not a necessary evil, and the tactical interpretation of the strategic plan must be integrated. This last point is not hard to understand, but is difficult to implement. (Figure 12.5 illustrates the general scenario.) The key to this method of planning is that all the technical and information requirements need to support the business plan, and they must be reconciled across the functional groups" (p. 11).

Most companies base their technology investment decisions on return-on-investment (ROI). One survey (McCallum and Vasilash, 1989) found that manufacturing companies investing capital in new machinery and equipment in 1989 expected, on

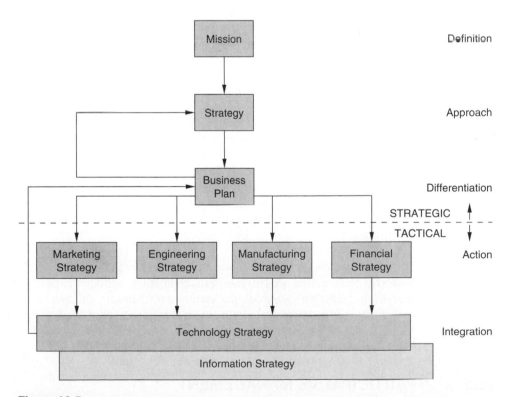

Figure 12.5
A new approach to planning that integrates technology and information strategies
Source: Reprinted with permission from "Justifying the Future" by N. Raffish, 1989, *ASKhorizons*, pp. 10–13.

the average, an ROI of 30%, though approximately half of the respondents were working with just 16%. Recognizing the importance of having productive equipment in the operations, one company was cited as working with an ROI of less than 10%.

12.4   EXPERT SYSTEMS

CIM decisions are more demanding than most decisions managers must make. Computers can help simplify decision making, however. Besides their conventional use in processing information, computers can, with the help of expert systems, serve as "advisors" to management.

An expert system is basically a computer program designed to emulate an expert—hence the name expert system. Expert systems apply facts stored in the computer and rules of thumb to help users solve decision-making problems. In its simplest form, an expert system consists of a **knowledge base** and an **inference engine**. The knowledge base is filled with facts and rules of thumb a human expert would use. The inference engine comprises the techniques of retrieving and using this knowledge.

Expert systems enable the users to capture the knowledge of experts in a series of statements, for example: "When condition P exists, I do Q." Once expertise has been stored in the computer, the system can help users solve problems in that area by suggesting likely outcomes or actions for a given set of conditions. The knowledge base can be created in two ways:

1. A set of IF-THEN-ELSE rules can be entered, in which case the expert system is called a **rule-based system**.
2. The program can contain a series of CONDITION-RESULT combinations. Such systems are useful in finding patterns in a set of data, for example in machine diagnostics to relate symptoms with causes. The knowledge is manipulated either by **backward chaining** or **forward chaining**. Also called **goal-driven**, backward chaining works with the given results to determine the likely initial conditions. The forward chaining starts with the given conditions to predict the possible results by going through the **decision tree**. A decision tree is a list of all possible options in making a decision; it resembles a tree with trunk and branches.

In most cases, expert systems are developed by the users themselves, resulting in powerful tools called **shells**. Like other software, modern expert systems are also user-friendly. As an example, one system can translate the user request "Show me the names and salaries of employees in the quality control department who got a raise in 1994" into a format it understands: SELECT name salary FROM employee WHERE department QC (quality control) AND raise 1994.

12.5   PARTICIPATIVE MANAGEMENT

Largely due to its success in Japan, companies have begun to encourage all employees to participate in managing the resources. New practices under this trend are collectively termed **participative management**. Participative management theoretically generates team spirit for the benefit of the company. Team spirit has been found to be effective only in smaller companies, however. Firms with more than 150 employees are too large to function like a team.

According to Beaumariage and Shunk (1991, p.2), a teamwork approach involves the following issues:

1. Defining Teamwork
   Definition of a Team
   Characteristics of Teams
   Notion of Empowerment
2. Organizational Considerations
   Company Issues
   Culture Shift
3. Employee Issues
   Management Input
   Team Participant Considerations
4. Mechanics of Teamwork
   Forming Teams
   The Rebel
   Rewarding Teamwork

Participative management and other personnel issues in CIM are discussed in the next chapter.

12.6   OUTLOOK

Developments in communications and the world's free markets have globalized manufacturing. The resulting climate provides opportunities for managers to learn from CIM experiences in other countries. Despite varying cultural factors and work ethics, CIM shares certain universal elements. Quality circles, just-in-time, and employee empowerment are some of the management tools that have broad CIM applications.

Besides applying practices proven both nationally and internationally, management may emphasize certain themes in their operations. This depends on the direction taken to achieve CIM. Boston University research (Box 12.4), for example, suggests that Japanese factories emphasize design, whereas U.S. companies focus on responsive organization and more efficient workers.

Managing manufacturing enterprises that are implementing CIM is challenging. The transition from managing the past environment to a CIM environment will take a decade or two. The 1990s may be remembered as the decade of this transition. According to Automation Research Corp., management challenges of the 1990s can be summarized as in Figure 12.6.

☐ BOX 12.4 *Future Factories: Three Flavors*

Three distinct kinds of factories are emerging in Japan, Europe, and the US, according to a Boston University (BU) survey of 500 successful manufacturing firms in these areas. One theme prevails in each region:

USA: The value factory focuses on responsive organization and more efficient workers.

Europe: The borderless factory surmounts cultural, linguistic, and ethnic differences to be pan-European.

Japan: The design factory introduces and individualizes products rapidly.

How does this affect managers' behavior? The survey asked them to pick the improvement program that was most important in 1990–92 (measured by investment of time, managerial effort, and money). The survey's ranking of the five programs in each region showed sharp differences in regional priorities.

United States
1. Linking manufacturing strategy to business strategy
2. Giving workers broader tasks and more responsibilities
3. Statistical process control
4. Worker and supervisor training
5. Interfunctional work teams

Europe
1. Linking manufacturing strategy to business strategy
2. Integrating information systems in manufacturing

3. Quality function deployment
4. Manager, worker, and supervisor training
5. Integrating information systems across functions

Japan
1. Integrating information function in manufacturing and across functions
2. Developing new processes for new products
3. Production and inventory control systems
4. Developing new processes for old products
5. Linking manufacturing strategy with business strategy

When four of the five items on Japan's priority list are absent from the other two, it's clear that the Japanese are moving along a different track.

US manufacturers came late to the quality revolution, and they may be late for the next one too, says Professor Jeffrey Miller of BU, who organized the survey. Because they're investing in people and organization rather than technology, they could well be left selling "good stuff cheap" while the Japanese sweep the market with "just-for-you" products, he says.

Boston University in the U.S., Waseda University in Tokyo, and INSEAD, the European Institute for Business Administration, in Fountainbleau, France, conduct the International Manufacturing Futures Survey each year. For more information, contact Boston University's Manufacturing Roundtable, 621 Commonwealth Ave., Boston, MA 02215; (617) 353-5077.

Source: From "Future Factories: Three Flavors," August 1991, *Manufacturing Engineering*, p. 16.

TRENDS

☐ Because of CIM, management style and culture are undergoing major changes to cope with emerging technologies and consumer demands.

☐ Conventional ways of managing manufacturing resources are being questioned.

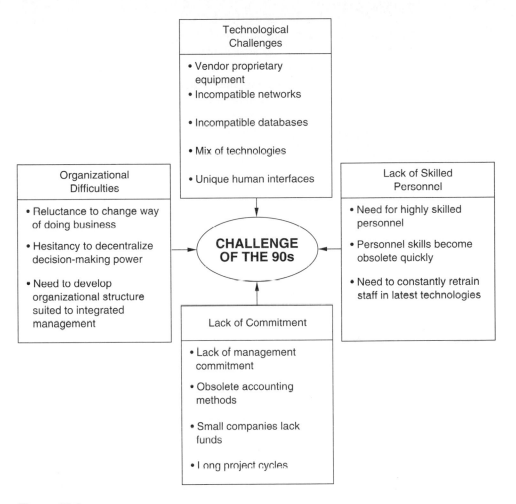

Figure 12.6
Management challenges of the 1990s
Source: From "Trend Is Less Islands of Automation and Towards Integration" by J. Murray and R. Ferrari, Mid-America 1990 Manufacturing Conference Proceedings. Reprinted with permission from SME.

❑ The basic concepts of Taylorism—including excessive division of labor—are inappropriate in today's modern factories, which are run by a highly educated and skilled workforce. Teamwork is found to be more effective; it is also compatible with CIM due to its integrating influence on personnel.

❑ Cost justification for CIM technologies must include intangible benefits such as enhanced product quality and company image.

❑ Modern management tools and techniques cross geographic boundaries. This encourages acceptance of management (and operations) practices proven effective in other countries.

❑ The use of expert systems in decision making continues to expand in CIM.

SUMMARY

Current issues of management that are appropriate for CIM were discussed in this chapter. The need for modern management techniques and practices was emphasized, as was the need for team efforts and "big-picture" planning and control. The necessity of considering intangible benefits in cost justifying CIM investments was pointed out. Challenges for management in the 1990s were also summarized.

KEY TERMS

Backward chaining

Decision tree

Forward chaining

Goal-driven

Inference engine

Integrated investment

Knowledge base

Lean production

Participative management

Rule-based system

Shells

EXERCISES

Note: Exercises marked * are projects.

12.1   Summarize Dr. Aaron's views on modern manufacturing wisdom in approximately 200 words.

12.2   Write a 250- to 300-word executive summary on DFM for senior management.

12.3   Explain the meaning of expert systems. How do they differ from conventional software?

12.4   List five major challenges facing management in the 1990s. Suggest how management can respond to them.

12.5*   Conduct a literature search for PC-based management systems available in the market. If you were the production manager of a midsized manufacturing company, which one would you prefer and why?

12.6*   If you were the president and CEO of a manufacturing company, what steps would you take to introduce CIM in your operations? (Imagine a manufacturing facility of your choice or refer to a company you know.)

12.7*   Visit a local company to learn about its management practices. Write a 200-word critique, pinpointing the deficiencies in these practices from a CIM viewpoint, and discuss it with the class or your work group.

SUGGESTED READINGS

Books

Bertain, L. (Ed.) (1991). *CIM implementation guide*. Dearborn, MI: CASA/SME (This guide comprises 18 chapters contributed by noted industry experts. It is useful in planning, organizing, and controlling CIM implementation. Benchmark case studies of successful CIM implementation are also included.)

Besterfield, D. H. (1994). *Quality control* (4th ed.). Englewood Cliffs, NJ: Prentice-Hall.

Harrington, J., Jr. (1984). *Understanding the manufacturing process: Key to successful CAD/CAM implementation*. New York: Marcel Dekker.

Savage, C. M. (1990). *5th generation management—Integrating enterprises through human networking*. Bedford, MA: Digital Press.

Veilleux, R., & Petro, L. (1988). *Tool and manufacturing engineers handbook: Vol. 5. Manufacturing management*. Dearborn, MI: SME.

Monographs and Reports

Beaumariage, K., & Shunk, D. (Eds.) (1991). Issues in migrating to teamwork. *SME Blue Book series*. Dearborn, MI: SME.

Chaintella, N. (1986). *Management guide for CIM*. Dearborn, MI: SME.

Kemelhor, R. E. (1988). *Intra/Inter-organizational system subtier environments*. Dearborn, MI: SME.

Murray, J., & Ferrari, R. (1990). *Trend is less islands of automation and towards integration*. Mid-America 1990 Manufacturing Conference, SME paper # MS 90-168.

Petrock, F. (1990). *Managing change for total quality: Stages of acceptance and resistance*. Mid-America 1990 Manufacturing Conference. SME paper # MS90-169.

Savage, C. M. (1986). *Fifth-generation management*. Dearborn, MI: SME.

Savage, C. M. (1987). *Fifth-generation management for fifth-generation technology*. Dearborn, MI: SME.

Journals and Periodicals

Journal of Applied Manufacturing Systems. St. Paul, MN: St. Thomas Technology Press.

Production. Cincinnati: Gardner Publications. (The articles in this monthly magazine are aimed at production managers of metalworking plants. The topics covered are current and presentations lucid.)

Management Memo. Published by the Management Technology Program and the Industrial Management Department, School of Industry and Technology, University of Wisconsin-Stout.

The Quality Circles Journal. A quarterly publication by the International Association of Quality Circles.

Articles

Horn, D. (1989). Expert systems emerge from their shells. *Mechanical Engineering. 111*(4), 64–67.

Lint, L. J. O. (1992, May-June). Real options analysis in advanced manufacturing technologies. *International Journal of Computer Integrated Manufacturing. 5*(3), 145–152.

McCallum E. D., Jr., & Vasilash, G. S. (1989). Capital spending 1989: America bounces back. *Production. 101*(1), 26–32.

Mehta, A. C. (1987, November). Integrating islands of automation is management, not technical problem. *Industrial Engineering*, pp. 42–49.

Nyquist, R. S. (1990). Paradigm shifts in manufacturing management. *Journal of Applied Manufacturing Systems. 3*(2), 1–8.

Raffish, N. (1989). Justifying the future. *ASKhorizons. 2*(4), 10–13.

Ramsey, M. (1989). Gaining proficiency in expert systems. *Mechanical Engineering. 111*(4), 73–78.

Sullivan, W. G. (1986, March). Models IEs can use to include strategic, nonmonetary factors in automation decisions. *Industrial Engineering*, p. 42.

CHAPTER 13
Personnel

"Technology, no matter how sophisticated, is useless if it outstrips the abilities of the people to understand, implement, and manage it."
—William E. Scollard, Ford Motor Co.

"Higher education and business are basically interdependent. One needs money to produce educated people. The other needs educated people to produce money."
—Milton Eisenhower

"Under CIM, jack-of-all-trades but master of none will give way to jack-of-all-trades and master of some."
—S. Kant Vajpayee, The University of Southern Mississippi

"We need more PhDs out there on the factory floor."
—Robert Hocken, University of North Carolina—
Charlotte, quoted in *ASEE Prism*, September 1991

"We find that 80% of the cost of implementation of a system is the education and implementation afterwards. Twenty percent is hardware and software acquisitions; the learning curve and associated costs are the lion's share."
—Tom Bogus in Enterprise Information Exchange (EIX) Issues in the
CIM Environment (a roundtable discussion), *SME blue book series*, 1991

"The factory of the future will not be characterized by 'expert' managers and (by implication) 'inexpert' workers. Instead, the manufacturing organization of the future will have a workforce that is actively participating in the continuous redesign and improvement of work."
—Carolyn Schuk, JIT segment manager, ASK Computer Systems

"The problem for many of us is that this CAD/CAM technology is moving so fast, it's difficult to keep up. New terms come and go before we understand what the old terms meant."

—Tom Beard

"Compared with spending on new plants or roads, investing in people would produce an even bigger payoff. High-tech jobs demand not the automation of yesterday's assembly lines but a new breed of skilled workers, contends former Labor Secretary Ray Marshall, who recently served on a commission examining workers' skills. New production techniques require decision making by even the lowliest worker, while modern quality control demands a knowledge of statistics common among Japanese high-school graduates but rare among many U.S. college students. "I've heard officials of foreign firms say, 'Don't quote me, but we have to simplify our machinery and dumb-down our training and orientation programs for workers in the U.S.,'" says Robert Reich, who teaches public policy at Harvard and is writing a book on competitiveness."

—*U.S. News & World Report*, July 16, 1990

"If a company is dedicated to being a world-class manufacturer, it has to have world-class manufacturing engineers. The idea that professional managers without engineering education can manage high-technology ventures that can compete globally is absurd."

—Jean Owen, associate editor, *Manufacturing Engineering*

"We are talking about some of the best paying jobs in manufacturing today, paying from $25,000 to $60,000 per year. Everyone talks about the demise of manufacturing, but it's a myth, and there are thousands of good jobs going unfilled to prove it."

—Bruce Baker, Tooling and Manufacturing Association

"Manufacturing people are tough, result-oriented, hands-on. They know that they make the product that makes the profit. They can take it. So who cares about image."

—Jean V. Owen, associate editor, *Manufacturing Engineering*

"To illustrate, in a recent article I wrote in collaboration with several academic colleagues on the subject of manufacturing curricula, we stated that the number of U.S. faculty equipped to teach a comprehensive course in Design for Manufacturability is probably less than 20."

—Philip H. Francis, vice president, Corporate Technology
Center, Square D Company, in *Proceedings of the ASME
1989 Mechanical Engineering Department Heads Conference*

"Seven years ago, I made a comment, as you may remember, there's no such thing as a computer system to control the manufacturing, or—if you will—substitute the word engineering environment. There are only computer assists to people systems. I don't care what we call them. Call it CAD/CAM, call it MRP, call it OPT, call it just-in-time, call it anything but late to lunch."

—Leo Roth Klein, Manufacturing Control Systems, in *Integrating
Process Planning and Production*, SME blue book series

"Just as the MBAs were the industry stars of the 1970s, the manufacturing engineering graduates will be the stars of the 1990s."

—Stuart M. Frey, Ford Motor Co.

"What is a manufacturing engineer? They are translators. They translate the output of the design engineers into the language of the shop floor."
> —Charles Stata, Digital Equipment Corp., in *Integrating Process
> Planning and Production*, SME blue book series

"I think we're in the same place now that we were 20 years ago with MRP. So we are struggling with the educational process, just like 20 years ago with MRP. That's why this time I'm spending a lot of time on training and education for my systems people."
> —Steve Covin, Compaq Computer Corp., in *ASKhorizons*, Summer 1989

"This software is not going to save you anything. The software's going to do nothing for you if you don't change the culture. Instill disciplines. Change the culture. Educate the people. Get the people to understand it. It's a people-driven system."
> —Joseph Vesey, Unisys Corp., in *Integrating Process
> Planning and Production*, SME blue book series

"Manufacturing engineering at Cambridge University (considered the best British undergraduate engineering school) is taught in close cooperation with various industries, with a significant industrial lecture content, six industrial projects, and modular concentrated short courses during the last year. They are so convinced of the importance of plant visits and plant projects that a large portion of the equipment funds is spent to buy buses instead of production equipment."
> —Eugene I. Rivin in *Journal of Applied Manufacturing Systems*, Spring 1990

Modern manufacturing depends heavily on people. Although CIM-oriented soft automation eventually reduces the number of people employed in a company, the educational and skill levels of those remaining are extremely high. In this chapter we discuss the personnel needs of CIM and the educational and training needs of such personnel.

13.1   IMPACT OF CIM ON PERSONNEL

Computer-integrated manufacturing and its building blocks such as CAD or CNC affect all company personnel, from operators to the CEO and president. Early predictions that only unskilled workers would be affected have been proven wrong. The restructuring and downsizing of a company reduce middle management positions as well. Employees in the 40-to-50 age group are being asked to retire, and those younger are being asked to retrain.

Harrington (1985) identified some of the changing skills of people working in CIM environments. For example, operators of NC machines need additional skills in part programming and CNC technology. Jobs of expediters are being eliminated altogether. Reading inspection instruments is less demanding since they have digital readouts and can print out the inspection results. As another example, cost estimating has been computerized to the point that anyone with keyboarding skills and some training can carry out this function.

Even areas that normally require higher skills have changed. For example, a designer's creativity and skills are challenged by CAD workstations that instantly test their ideas on design improvement. Knowledge-based software systems can even assess the quality of the designer's creativity by evaluating the manufacturability of an idea. Moreover, in CIM, designers need to know more about manufacturing, for example, part programs not just for machining but also for assembly, CMM inspection, or robotized packaging.

Most of all, management may need to undergo a cultural change. To begin, the president and CEO must believe in CIM. Their primary task is convincing other board members of CIM's leverage. Since CIM affects all three functions of management—planning, implementation, and control—change is required throughout the organization. Managers must switch from hard copy reports to electronic mail. The real-time environment under CIM demands faster turnarounds on decision making, which becomes a group activity. Meetings may be sudden, short, and highly focused, since the input information for the meeting will be clear, concise, and current.

CIM demands that specialists understand functions outside their areas. Specialists need to generalize more, and generalists need to specialize more. Under CIM, jack-of-all-trades but master of none will give way to jack-of-all-trades and master of some. This may initially be difficult, but knowledge-based computer systems smooth the transition by providing a helping hand.

Thus, the skills and practices of the past undergo profound change with CIM. As Harrington (1985) explains: "Indeed, it is safe to say that the impact of computer integrated manufacturing will be greater on the people involved than on the technology itself" (p. 280). The transition from conventional practices to those required under CIM benefits from training and retraining of the people affected.

13.2   ROLE OF MANUFACTURING ENGINEERS

In CIM environments, manufacturing engineers interact very closely with designers. They need to understand design, especially CAD, and the design process. CAD requires them to have insight into the principles of computer technology and the associated terminologies such as bits and bytes, RAMs and ROMs. The same is true for first-line supervisors or foremen who interact with operators, management, and plant equipment. Maintenance staff need to work more as a team with a common pool of expertise in areas as diverse as electronics, computers, hydraulics, pneumatics, and the usual mechanical and electrical systems.

Commissioned by the SME, A. T. Kearney Inc. conducted a survey to predict the job descriptions of manufacturing engineers by the year 2000. Entitled "Countdown to the Future: the Manufacturing Engineer in the 21st Century" and known as **Profile 21**, the survey results are based on the opinions of 7,500 manufacturing practitioners, a series of roundtable discussions, a Delphi study, a chief executive officer questionnaire, and an extensive literature search. It predicts that the environment in which future manufacturing engineers will operate will change due to:

1. increased product sophistication and variation
2. globalization of manufacturing
3. a multitude of social and economic factors

The major findings (Directory of Manufacturing Education, 1990) of Profile 21 are:

1. "Manufacturing engineers will function increasingly as **operations integrators**." Their main job will be that of integration engineers with the duties of coordinating people, information, and technology within the organization. An optimum balance between managerial and business skills on the one hand, and technical, scientific, and mathematical skills on the other, will be required.
2. "Given the growing overall importance of the manufacturing engineer's increasingly complex tools to everyday planning logistics and work flow, he or she must assume a greater role as a strategist." Future manufacturing engineers will possess some of the skills usually associated with people with a business administration background.
3. "Teamwork and people skills will play an increasingly important role in the work of manufacturing engineers." In the past, manufacturing engineers were concerned with technical matters only. Profile 21 predicts that increased sophistication of technology will render individual efforts less productive, and demand interaction with other personnel.

Future manufacturing engineers will need to be have both strong people skills and technology expertise, and to be effective team leaders. They will be aware of the company's global strategy, marketing and design priorities, and overall business plan. As Jean Owen (1989) states: "The ME of 2001 begins to sound like a camel—that horse designed by committee. Yet the camel is much-prized in its own domain, and for centuries wealth was counted in camels" (p. 49).

CIM Engineer and Technologist

With the diverse job functions, it may be appropriate to call manufacturing engineers of the 21st century **CIM engineers or technologists**. They will probably be expected to hold a master of science degree with a major in CIM or a similar discipline. The four-year bachelor of science program may be insufficient for the challenges of sophisticated plants. CIM engineers may also have one or two minors (secondary specializations) to supplement their learning in the major program.

Changes currently taking place in manufacturing will eventually result in three distinct roles for CIM engineers: operations integrator, manufacturing strategist, and technical specialist. According to the Profile 21 Executive Summary, they would have to be a three-in-one system. Their duties in these roles will cover various levels of breadth and depth of knowledge, as illustrated in Figure 13.1. The depth skills involve traditional technical knowledge, while breadth skills will be in areas such as personnel, communications, foreign language, finance, computers, and health and safety.

Figure 13.1
Three major roles of CIM
engineers and the associated
knowledge domains

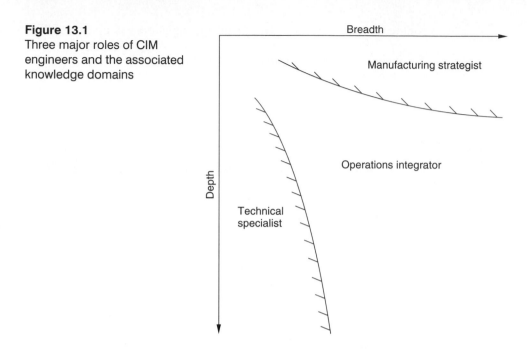

Stauffer (1989) quotes Donald G. Zook, 1986–87 SME president:

"Manufacturing is more complicated than in the days of stand-alone machines and large labor input. Today, we're dealing with flexible manufacturing cells and systems, the hierarchy of controls that tie them together, and the management information systems that feed them. CAD/CAM technology, with its computer graphics orientation, calls for using a single database, from the design of a component to its manufacture. The manufacturing engineer is the central figure in transforming the design into a product." (p. 64)

CIM Technicians

CIM engineers and technologists will need the support of CIM technicians, who may have two- or four-year degrees in CIM technology. The technicians will need significant training if their majors are in areas other than CIM.

Several junior or community colleges have already begun to develop and offer courses in CIM. Their curricula and subsequent job training program in the employing company should include the topics, as proposed by Amatrol Inc., shown in Figure 13.2. Amatrol calls the program an automation training pyramid with CIM at its apex. The learning modules comprise advanced systems technology for sophisticated tasking and flexible automation systems for integrated tasking. Comprehensive training in these modules would ensure effective operation of CIM plants.

13.3   ROLES OF INSTITUTIONS

CIM represents a change of drastic proportions. Manufacturers alone may not be able to cope with this change. It is essential that other institutions also participate in

AUTOMATION TRAINING PYRAMID

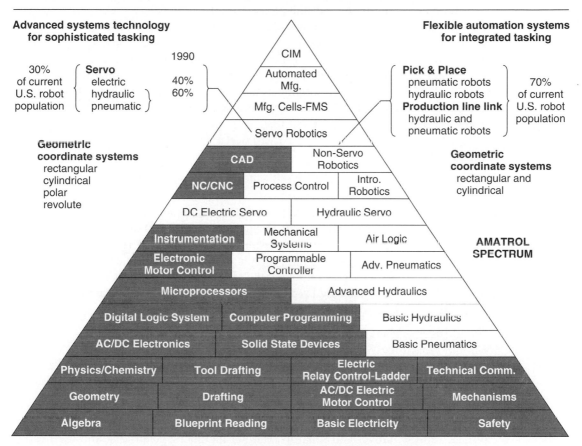

Advanced systems technology for sophisticated tasking

Flexible automation systems for integrated tasking

1990

30% of current U.S. robot population

Servo
electric
hydraulic
pneumatic

40%
60%

Pick & Place
pneumatic robots
hydraulic robots
Production line link
hydraulic and
pneumatic robots

70% of current U.S. robot population

Geometric coordinate systems
rectangular
cylindrical
polar
revolute

Geometric coordinate systems
rectangular and
cylindrical

AMATROL SPECTRUM

CIM

Automated Mfg.

Mfg. Cells-FMS

Servo Robotics

CAD — Non-Servo Robotics

NC/CNC — Process Control — Intro. Robotics

DC Electric Servo — Hydraulic Servo

Instrumentation — Mechanical Systems — Air Logic

Electronic Motor Control — Programmable Controller — Adv. Pneumatics

Microprocessors — Advanced Hydraulics

Digital Logic System — Computer Programming — Basic Hydraulics

AC/DC Electronics — Solid State Devices — Basic Pneumatics

Physics/Chemistry — Tool Drafting — Electric Relay Control-Ladder — Technical Comm.

Geometry — Drafting — AC/DC Electric Motor Control — Mechanisms

Algebra — Blueprint Reading — Basic Electricity — Safety

This interdisciplinary training approach gives in-demand
skills adaptable to changing technology

NC = NUMERICAL CONTROL   CAD = COMPUTER-AIDED DESIGN
CNC = COMPUTER NUMERICAL CONTROL   CIM = COMPUTER-INTEGRATED MANUFACTURING

Figure 13.2
Interdisciplinary training modules for CIM technicians
Source: Reprinted with permission from Amatrol, Inc.

bringing about CIM. Three major institutions—government, colleges and universities, and professional societies—have the most influence here.

Manufacturers

Manufacturers have the primary responsibility to develop CIM in their operations. Since the 1980s, industry has become more aware of the CIM concept and CIM's impact on the operational effectiveness to meet the market demand. The globalization of the market and consumer demands for personalized products are the major driving forces. High-tech sectors of the manufacturing industry such as computers, automotive, aerospace, and electronics have led the way in CIM innovations. Other manufacturers have implemented CIM mostly as proven technology.

The major limitation in advancing toward CIM is usually a lack of management vision, which fails to justify an investment because it is based on conventional monetary benefits. The other bottleneck is the lack of standardization due to the influence of free market forces, with little or no guidance from governments. General Motors' efforts toward MAP, for the benefit of CIM, are commendable in this respect. Providers of CIM technology usually undermine such efforts in favor of proprietary standards. But CIM's demands are such that they need to work together around some standard or understanding.

Government

Government at the federal and state levels should create a climate that encourages CIM implementation. Governments need to realize that investment in CIM technologies is costly and that the net effect of CIM will be fewer jobs in manufacturing. Favorable tax structures and support through national R&D laboratories are essential along with subsidies for worker retraining. Japan's MITI (Ministry of International Trade and Industry) is a good example of government's role as a facilitator.

The cover story of the January 1992 *Manufacturing Engineering* is entitled "Future View: Manufacturing Faces the Next Millennium." This issue includes articles on the following topics, all of which have a bearing on CIM:

Manufacturing Management in Crisis
Will the Workforce Work?
Who Needs Government?
A Game of Musical Factories
Tomorrow's Manufacturing Technologies

The issue is so strongly targeted at government that the publisher sent a copy to all U.S. representatives and senators, cabinet members, governors, and the head of every major federal agency. The magazine is lobbying for establishment of a cabinet-level secretary of manufacturing.

The poor performance of U.S. manufacturing is sometimes attributed to the disproportionately small population of engineers in this country compared with Japan. For every 10,000 people in Japan, there are 23 engineers but only 1 lawyer and 3 accountants. In the United States, the corresponding figures are 12 engineers, 20 lawyers, and 40 accountants. Thus, as a society Japan is more engineering-oriented while the U.S. is more litigious and finance-oriented. How can U.S. government correct this imbalance in favor of engineering and manufacturing? Here is a Herculean task for them, provided they agree with the hypothesis at the beginning of this paragraph!

Colleges and Universities

One basic problem for industries attempting to implement CIM is the dearth of skilled personnel in this area. CIM technologies change so rapidly that keeping pace with them is a never-ending task. Though recent graduates of colleges and universities bring in current skills, these skills may not be sufficiently multidimensional. Moreover, they need to be continually honed through retraining. For example, CNC programming skills on one machine and in one language may not be enough for CIM; their extension into other functions is essential.

Graduating students get an opportunity in colleges and universities to develop skills interaction through a senior project. But this experience may not be sufficient preparation for CIM. The interrelationships among various courses of the curricula need to be highlighted and reinforced adequately. Without it, education on and about CIM remains incomplete. According to Vajpayee and Jain (1991), manufacturing education could follow the example of law and medicine: The curricula could be case studies-oriented and team-taught around a typical manufactured product to reinforce the interrelationships among various functions of manufacturing.

According to Michael J. Kelly, director of computer integrated manufacturing and technology exchange at the New Jersey Institute of Technology (Stauffer, 1989), "Engineering graduates are usually able to apply their knowledge and analytical skills in solving specialized technical problems. Their education, however, has not provided them with the interdisciplinary background and systems orientation required to solve today's complex industrial problems. The theoretical science- and mathematics-based curricula encourages the analytical approach to problem solving, while system design, integration, and synthesis are what industry needs." Stuaffer adds, "But industry and the education community have been slow to adjust. Unlike other leading industrial countries, according to Zook, the U.S. does not recognize manufacturing engineering as a tangible engineering science like mechanical, electrical, industrial, and civil engineering" (p. 65). Administrators and faculty at colleges and universities need to make note of these comments and give manufacturing programs the attention they deserve. Box 13.1 discusses the link between education and the success of manufacturing.

Several educational institutions throughout the world offer programs entitled manufacturing engineering, engineering technology, or technology. SME's 1992 directory lists such programs in North America, some of which are accredited. Table 13.1 lists the **ABET**[1]-accredited manufacturing programs in the United States, while Figure 13.3 shows the growth in such programs since 1978.

Based on SME's 1992 *Directory of Manufacturing Engineering*, 971 two- and four-year institutions of higher learning in the United States were offering manufacturing engineering or engineering-related programs. Of these, only 55 (about 6%) were accredited in 1991.

Not all manufacturing programs offer a course in CIM. Of the 971 institutions, in 1990 about 25% offered at least one CIM course, mostly at the two-year junior or community colleges. Depending on the laboratory resources and the background of

---

[1]ABET (Accreditation Board for Engineering and Technology) is a nonprofit organization that monitors, evaluates, and certifies the quality of college-level engineering and engineering-related programs in the United States. It is governed by 21 member technical and professional engineering societies such as SME, ASME, and IEEE, representing more than 700,000 engineers.

◻ **BOX 13.1**   *The Power of Education and the Future of Manufacturing*

With "Synergy '92" as the theme of a recent joint meeting of the American Machine Tool Distributors' Association (AMTDA) and AMT—The Association for Manufacturing Technology, Harold J. Wagner declared that there is no more synergistic human enterprise than education. Mr. Wagner, who is serving as this year's Chairman of AMT, spoke to members about the importance of education to manufacturing. "An educated person is someone who can put more power back into the grid than he or she takes out," he said, as he outlined three key points.

First, the crisis in the workplace is an educational crisis. In the traditional workplace, he noted, work was divided into routines and managed hierarchically. In today's high-performance workplace, work is problem-oriented, flexible and managed by teams. "It requires workers who can think their way out of a problem," he said, warning that schools are not preparing young people for this environment.

Second, the crisis of competitiveness is an educational crisis. Mr. Wagner pointed out that in many other countries, schools are part of a joint effort with industry and government to improve the long-term competitiveness of manufacturing within their countries. He called for a similar long-term strategy in this country, with an emphasis on continuous improvement of educational techniques, using industry as a model. "It's time we in the private sector insist on applying total quality management principles to the schools," declared Mr. Wagner.

Finally, Mr. Wagner challenged the group to become part of the solution by adopting some basic principles: Produce the best people you can in your own business. Transform your workplace into a "learning place." And help your local schools. "You—one person, one business—can make a difference."

He concluded: "We belong to a capital-intensive industry. Let's put ourselves into a capital-intensive effort—into the capital that is going to give us the best possible return: our own employees and our own kids."

Source: "Final Comment," June 1992, *Modern Machine Shop*, p. 280.

---

instructors and supporting technical staff, current CIM courses are either management-oriented or technology-oriented. A balance is needed between management and technology orientations with multi-instructor teaching. In Appendix II such a balanced laboratory course (Vajpayee and Smith, 1991) is suggested.

A few colleges and universities have already taken the lead in developing an ideal CIM curriculum. For example, Figure 13.4 shows the modules of a 128-unit baccalaureate CIM program being developed by Professor Arthur Foston of the industrial technology department at California State University, Fresno. He has recently conducted a survey to determine the optimum number of units for each module. In a letter that accompanied his survey, Foston defines CIM as "a global closed-loop manufacturing system concept using computers to integrate and control information flow during all phases of manufacturing, including product inception, design, shipment and support." Foston's course structure is balanced, since it provides both a manufacturing foundation and general education, in addition to covering the technology and management of CIM.

Table 13.1
ABET-accredited manufacturing programs in the United States

Institution	Manufacturing Engineering		Manufacturing Engineering Technology	
	Bachelor Degree	Masters Degree	Associate Degree	Bachelor Degree
Arizona State Univ., Tempe, AZ				•
Arkansas, Univ. of, Little Rock, AR				•
Boston Univ., Boston, MA	•			
Bradley Univ., Peoria, IL	•			•
Brigham Young Univ., Provo, UT				•
Calif. Polytechnic State Univ., San Luis Obispo, CA				•
Calif. Polytechnic State Univ., Pomona, CA	•			
Central Connecticut State Univ., New Britain, CT				•
Central Piedmont Comm. College, Charlotte, NC			•	
Central State Univ., Wilberforce, OH	•			
Cincinnati, Univ. of, Cincinnati, OH			•	
East Tennessee State Univ., Johnson City, TN				•
Forsyth Tech. Comm. College, Winston-Salem, NC			•	
GMI Engineering & Mgmt. Inst., Flint, MI	•			
Gr. New Haven State Tech. Coll., North Haven, CT			•	
Hartford State Technical Coll., Hartford, CT			•	
Houston, University of, Houston, TX				•
Indiana Univ.-Purdue Univ., Fort Wayne, IN				•
Kansas State University, Manhattan, KS	•			
Mankato State University, Mankato, MN				•
Massachusetts, Univ. of, Amherst, MA		•		
Memphis State University, Memphis, TN				•
Miami University, Oxford, OH	•			
Midwestern State University, Wichita Falls, TX				•
Milwaukee School of Eng'g, Milwaukee, WI				•
Murray State University, Murray, KY				•
Nebraska, University of, Omaha, NE			•	•
New Hampshire Tech. Inst., Concord, NH			•	
New Jersey Inst. of Tech, Newark, NJ				•
New York, State University, Farmingdale, NY				•
Oklahoma State University, Stillwater, OK				•
Oregon Inst. of Technology, Klamath Falls, OR			•	•
Oregon State University, Corvallis, OR	•			
Owens Technical College, Perrysburg, OH			•	
Pellissippi St. Tech. Comm. Coll., Knoxville, TN			•	
Pittsburg State University, Pittsburg, KS				•
Rhode Island, University of, Kingston, RI		•		
Ricks College, Rexburg, ID			•	
Rochester Inst. of Technology, Rochester, NY				•

Table 13.1 *continued*

Institution	Manufacturing Engineering		Manufacturing Engineering Technology	
	Bachelor Degree	Masters Degree	Associate Degree	Bachelor Degree
St. Cloud State Univ., St. Cloud, MN				•
St. Thomas, Univ. of, St. Paul, MN		•		
Texas A&M Univ., College Station, TX				•
Thames Valley St. Tech. Coll., Norwich, CT			•	
Utah State Univ., Logan, UT	•			
Waterbury St. Tech. Coll., Waterbury, CT			•	
Weber State College, Ogden, UT				•
Wentworth Inst. of Technology, Boston, MA			•	•
Western Carolina Univ., Cullowhee, NC				•
Western Michigan Univ., Kalamazoo, MI				•
Western Washington Univ., Bellingham, WA				•
Worcester Polytechnic Inst., Worcester, MA	•			

Source: Reprinted with permission from "Degree Programs Continue Upswing," November 1991, *SME News*, p. 6.

Figure 13.3
The growth of ABET-accredited manufacturing programs in the United States

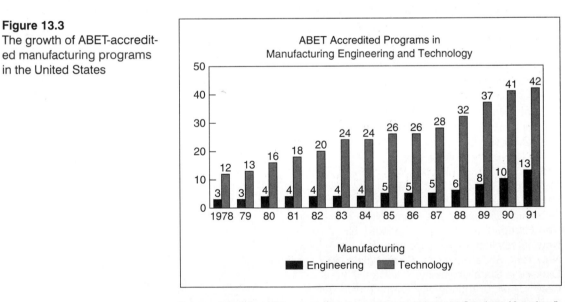

Source: Reprinted with permission from "Degree Programs Continue Upswing," November 1991, *SME News*, p. 6.

At his university, the author has been teaching a CIM course with an outline that follows the chapters of this text. In fact, the textbook is an outgrowth of this course, which has been taught since 1986.

Manufacturing curricula must strike a balance between an academic orientation and the needs of industry. Jay Zirbel of Texas A&M University has conducted

Figure 13.4
A proposed CIM curriculum

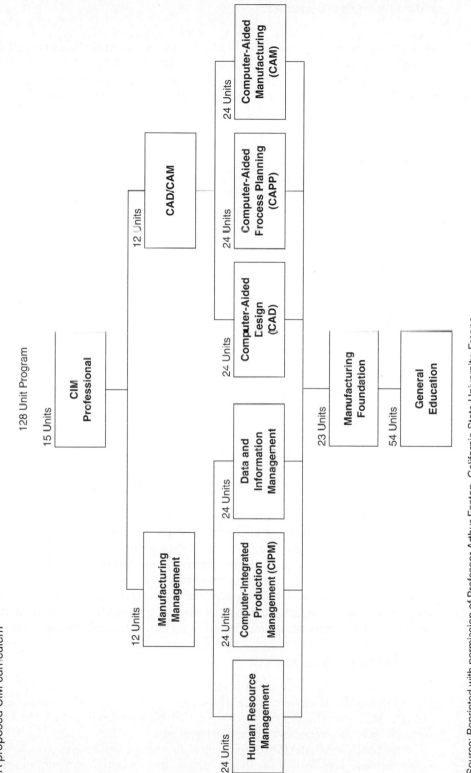

Source: Reprinted with permission of Professor Arthur Foston, California State University, Fresno.

research to determine the needs of industry. He has identified and ranked the tasks entry-level manufacturing engineering technologists will be required to carry out in the year 2000 and beyond. The research comprised both a survey and a Delphi study in which a panel of 14 experts participated. The findings of this work are as follows:

1. Manufacturing engineering technologists must exhibit strong work ethics, such as motivation, natural curiosity, and a sense of responsiveness without close supervision.
2. They must understand the importance of quality in all their actions and decisions.
3. Both oral and written communications will be extremely important.
4. The ability to work with colleagues in a team environment is of paramount importance.
5. Computer skills will be a must.
6. A sound background in the basics of manufacturing will continue to be required.
7. Skills on flexible automation will be necessary rather than desirable.
8. A knowledge of novel materials such as emerging composites and ceramics must supplement that of the conventional ones.
9. A working knowledge of health and safety considerations at the workplace will be important. (pp. 6–9)

Instructors need to maintain a balance between the educational and training needs of CIM. This balance is often lost during hands-on instruction with application programs meant to supplement the course. While it is tempting to teach the application packages extensively, fundamentals are at least as important as hands-on skills. Instructors at colleges and universities—where the primary purpose is education—should keep this in mind.

While covering computer systems in the various courses of a CIM curriculum, instructors should not spend an inordinate amount of time highlighting the computer's capabilities. They must point out that computers are not as capable as human beings at certain tasks. Table 13.2 compares the two on the basis of 14 criteria. A proper perspective on the strengths of computers versus human beings is essential, since dovetailing the two to maximize their outputs is the ultimate objective.

Lastly, it is essential to realize that most manufacturing engineers, like other engineers, are promoted to supervisory and management positions soon after entry-level jobs. In such positions, managerial skills become at least as important as technical skills. CIM curricula should therefore include sufficient management courses.

Professional Societies

Professional societies have an important role to play in advancing CIM. The SME and CASA/SME have been at the forefront in this role. Other societies such as ASME, IIE, and IEEE have also been active in disseminating CIM knowledge through interest groups and focused publications.

The SME and CASA/SME also help colleges and universities in developing curricula that are appropriate for the manufacturing industry. For example, under the

Table 13.2
A comparison between humans and computers

	Man	**Computer**
1. Method of logic and reasoning	Intuitive by experience, imagination, and judgement	Systematic and stylized
2. Level of intelligence	Learns rapidly but sequential. Unreliable intelligence	Little learning capability but reliable level of intelligence
3. Method of information input	Large amounts of input at one time by sight or hearing	Sequential stylized input
4. Method of information output	Slow sequential output by speech or manual actions	Rapid stylized sequential output by the equivalent of manual actions
5. Organization of information	Informal and intuitive	Formal and detailed
6. Effort involved in organizing information	Small	Large
7. Storage of detailed information	Small capacity, highly time dependent	Large capacity, time independent
8. Tolerance for repetitious and mundane work	Poor	Excellent
9. Ability to extract significant information	Good	Poor
10. Production of errors	Frequent	Rare
11. Tolerance for erroneous information	Good intuitive correction of errors	Highly intolerant
12. Method of error detection	Intuitive	Systematic
13. Method of editing information	Easy and instantaneous	Difficult and involved
14. Analysis capabilities	Good intuitive analysis, poor numerical analysis ability	No intuitive analysis, good numerical analysis ability

Source: Reprinted with permission from *Computer-Aided Design and Manufacture* (p. 16) by C. B. Besant, 1983, Chichester, West Sussex, England: C. B. Besant/Ellis Harwood Limited.

CASA/SME's Blue Book series, Kruppa and Gollajesse (1991) have identified the contents of CNC programming knowledge expected of four-year college graduates seeking entry-level jobs in manufacturing. The findings are based on a survey of U.S. manufacturing professionals whose population was represented by a sample of 500 SME senior members. The survey discovered that the demand for graduates with EIA-type CNC programming knowledge is substantially higher than for those with APT, Compact II, or conversational-type programming skills.

Manufacturing engineers and CIM personnel may find membership in SME and other professional societies beneficial. Addresses of societies associated with manufacturing are provided in Appendix I.

TRENDS

❑ The multidisciplinary nature of CIM's educational and training needs is increasingly understood by colleges and universities offering programs in manufacturing.

❑ Most manufacturing companies value the importance of education and training of their employees in CIM areas.

❑ CIM demands empowerment of employees—a difficult change in the style of personnel management at most companies.

❑ CIM engineers, technologists, and technicians are in high demand, a trend that is likely to continue for several years.

❑ The importance of manufacturing education is increasingly apparent to government and educational institutions. Manufacturing programs in colleges and universities have begun to offer more CIM curricula and courses.

SUMMARY

The educational and training needs of personnel working in CIM environments were emphasized in this chapter. Although CIM eventually reduces the number of employees in a company, the skills of those remaining are multidimensional, with design being the hub around which these skills revolve. Profile 21, which predicts what skills manufacturing engineers will need in the year 2000, was summarized in this chapter. The roles of industry, government, educational institutions, and professional societies were discussed. The current status of manufacturing education in the United States was reviewed.

KEY TERMS

Accreditation Board for Engineering and Technology (ABET)

CIM engineer or technologist

Operations integrator

Profile 21

EXERCISES

Note: Exercises marked * are projects.

13.1 Suppose you are the manufacturing manager of the XYZ company, which has decided to hire a CIM engineer and a CIM technician. Write their job specifications in 100–200 words each.

13.2   Prepare a summary of Profile 21 in 250–300 words (consult the original document in the library). Assume the summary will be presented to the academic council to seek its approval for a new four-year B.S. program in CIM.

13.3*   From the viewpoint of CIM discuss in 200–250 words the message conveyed by the article on page 136 in *Manufacturing Engineering*'s May 1991 issue.

13.4*   Study the manufacturing curriculum at your college or university and suggest how could this be modified to make it CIM-oriented.

13.5*   List the U.S. colleges and universities that offer at least one course in CIM. (Use SME's directory of manufacturing education or similar sources in the library.)

13.7*   Obtain an outline for a CIM course and study it. Suggest how it can be improved.

13.8*   Visit a local company to study and observe the duties of its manufacturing engineer. Discuss the educational and training needs if that individual were to be redesignated a CIM engineer.

SUGGESTED READINGS
Books

Harrington, J., Jr. (1985). *Computer integrated manufacturing*. Malabar, FL: Robert E. Krieger Publishing, p. 280.

Monographs and Reports

American Business Conference. (1987). *The challenge of global competitiveness*. Washington, D.C.: Author.

American Society for Engineering Education. (1987). *A national action agenda for engineering education*. Washington, D.C.: Author.

Directory of manufacturing education in colleges, universities and technical institutes. (1990). Dearborn, MI: SME

Directory of manufacturing education in colleges, universities and technical institutes. (1992). Dearborn, MI: SME

Hudson Institute, Inc. (1987). *Workforce 2000: Work and workers for the 21st Century*. Indianapolis: Author.

Kruppa, R. A., & Gollajesse, A. (1991). *The need for and content of CNC programming knowledge for four year, college prepared, entry level manufacturing engineers*. SME blue book series. Dearborn, MI: SME.

National Academy of Engineering. (1985). *Education for the manufacturing world of the future*. Washington, D.C.: National Academy Press.

Journals and Periodicals

Modern Machine Shop. Cincinnati: Gardner Publications. (The August 1991 issue of this monthly magazine deals with the educational and training needs of CIM. The seven articles discuss how manufacturers and educational institutions are responding to the task of training the workforce for the 1990s.)

Profile 21 Executive Summary. (1988). *Countdown to the Future: The Manufacturing Engineer in the 21st Century*. A.T. Kearney Inc. and SME.

Articles

Degree programs continue upswing. (1991, November). *SMENews*, p. 6.

Leybourne, A. E., & Vajpayee, S. K. (1989, summer/fall). Initiating a graduate program in manufacturing technology: One university's plan. *Journal of Epsilon Pi Tau. XV*(2), 24–29.

Owen, J. V., & Entorf, J. F. (1989, February). Where factory meets faculty. *Manufacturing Engineering. 102*(2), 48–55. (This article describes the various alliances that exist between manufacturers and U.S. universities to accelerate technology transfer. Various research centers, including NSF's engineering research centers, and other academic–industry collaborations are listed.)

Owen, J. V. (1989, November). Images of the manufacturing engineer. *Manufacturing Engineering*, pp. 56–62.

Stauffer, R. N. (1989, September). Getting manufacturing education up to speed. *Manufacturing Engineering*, pp. 63–66.

Vajpayee, S., & Jain, A. (1991). Teaching manufacturing as law/medicine. *International Journal of Applied Engineering Education*. 7(4), 258–263.

Vajpayee, S., & Smith, G. A. (1991). Design of a lab course on computer-integrated manufacturing. *International Journal of Applied Engineering Education*, 7(3), 207–211.

Wiebe, H. A., & Scott, C. (1988). The integration of multiple disciplines in graduate manufacturing education. In L. P. Grayson & J. M. Biedenbach (Eds.), *Proceedings of the 1988 ASEE Conference* (pp. 288–290). Washington, D.C.: American Society for Engineering Education.

PART V

Epilogue

This concluding portion of the text comprises the last two chapters. Chapter 14 presents some of the emerging technologies and management practices that directly or indirectly influence CIM. In Chapter 15, we look at CIM's ramifications.

CHAPTER 14
Emerging Technologies

"The point is, we have the best machine tools we can get with today's technology. Now, we're going to add something to them that will not try to make the machines better but will try to make the parts better."

—Jack McCabe, Mechanical Technology, Inc. (on the role of real-time in-process gaging on machining)

"Ken Garfano, vice president of manufacturing for Itran Corporation, believes that the merger between vision and CMMs eventually will raise inspection throughput by 500 to 600%."

—*Manufacturing Engineering*, November 1989

"Voyager's pictures are an example of image processing. Machine vision occurs when a computer is used to interpret the image."

—Bob Dewar, Perceptron

"Initially, when we first saw vision, it was seldom robust. Now, the equipment is mature; it stands up well."

—Glenn Jimmerson, Ford Motor Co.

"Traditional computer-integrated engineering/manufacturing techniques can't keep pace with the task of synchronizing the vast quantity of rapidly changing and interrelated data about parts, products, and their manufacturing processes—from their conceptualization to the customer. Salzman writes that the evolving technological environment requires new CIM systems that incorporate data handling and control capability in powerful networked desktop computers; new optical storage media that can mingle alphanumeric data with graphics or pure images; and object-oriented CAD/CAM technology that can show the real-world characteristics of parts, products, and processes."

—*Manufacturing Engineering*, November 1989

"The first thing you have to do to achieve this is buy the philosophy that multimedia communication is the backbone of your management structure."

—Paul Mace, president, Paul Mace Software in an interview with *InfoWorld*, January 11, 1993

"Although concurrent engineering carries the word 'engineering' in its title, it is really not an engineering or technology discipline. It is truly an organizational, human resources, and communications discipline between various organizations."

—Dale DeLorge in SME blue book *Product
Design and Concurrent Engineering*, 1992

CIM is now considered feasible by most manufacturing companies. The doubts of the 1980s have disappeared. CIM is the culmination of several technologies and emerging management practices. Chapter 14 presents some of the upcoming technologies and practices that are likely contributors to CIM.

In 1991, the U.S. White House identified 22 areas of technological development that are critical to national prosperity and security (*World Almanac*, 1992). Flexible computer-integrated manufacturing is one of them. Other areas included on the list were surface transportation, environment, material processing, electronic materials, ceramics, composites, high-performance metals and alloys, intelligent processing equipment, micro- and nano-fabrication, system management, software, microelectronics and optoelectronics, high-performance computing/networking, high-definition imaging and displays, sensors and signal processing, data storage, computer simulation, applied molecular biology, medicine, aeronautics, and energy. Since several of these areas directly or indirectly affect CIM, their development will undoubtedly help manufacturers achieve CIM.

14.1   EXPERT SYSTEMS

Expert systems are essentially software that helps users achieve what the services of an expert would provide. It is a branch of computer-based artificial intelligence. At the core of an expert system is a knowledge base and an inference engine that operates on the knowledge base to develop a solution or response. The system may be interfaced to the end user or to an array of sensors and effectors to communicate with the plant and processes.

The major advantages of expert systems are:

1. Captures the expertise of employees who may not be there tomorrow due to job change, retirement, or death.
2. Synergetic effect with the knowledge of several experts.
3. Resulting decisions are consistent.
4. Knowledge can be updated, revised, and improved.
5. Knowledge can be shared and used when an expert is busy or not available.

Box 14.1 describes an injection molding expert system.

What role can expert systems play in a CIM environment? Since CIM involves decision making at various levels and an expert system aids in decision making, CIM has a definite need for expert systems. CIM managers find expert systems helpful in comprehending the interactions among the large number of factors CIM involves and in decision making.

<table>
<tr><td>⬜</td><td>**BOX 14.1**</td><td>*Homegrown Experts*</td></tr>
</table>

Joe Bryant, technical services scientist for a major manufacturer in Royston (Georgia) cites several distinct reasons for using expert systems technology. At one time, his company's injection blow-molding process created dimensional tolerance problems and other product defects. While staff members could turn to the company's experts with questions about many aspects of the process, Bryant says no one person could "become the expert in everything. We needed a way to capture the expertise of many people."

Bryant set out to capture this expertise with an expert system. With it, he stored information from experts at the company and from technical manuals for the injection-molding equipment. Bryant also added graphics so that the system could display pictures showing how to correct for various problems. The information-gathering process was easier than some had feared. "We went from no knowledge to a system in one week," Bryant reports.

One advantage of using expert system technology for his injection-molding application, Bryant says, was that it allowed him to combine expertise and retain it, even after people left the company. "The expertise is now available to the crew on the third shift, or even when the expert is home sick in bed," he says. Indeed, the system demonstrated its value shortly after it was introduced by suggesting a solution to a problem that would have taken days of downtime to research and isolate. "An expert system is like the fire sprinkler system—you don't really get a return until you have a problem," he notes. "Now everyone here understands the value."

Source: From "Gaining Proficiency in Expert Systems" by M. Ramsey, April 1989, *Mechanical Engineering*, p. 77.

Experts systems are being put to use in the following areas of CIM:

1. *Management*. Expert systems help managers make better decisions. For example, a manager can determine why a product is losing market share and can seek suggestions and recommendations (from the expert system) on how to improve the situation.

2. *Maintenance*. The cost of lost production arising from the failure of major CIM equipment and devices is usually high, since they are expensive. To minimize the chance of breakdowns, predictive maintenance is desirable. Modern predictive maintenance relies on **signature analysis**, which involves on-line capture and analysis of signals, such as vibration or noise, that reflect equipment "health." Expert systems are effective in such analyses and in troubleshooting the failed equipment. They can handle on-line—sometimes in real-time—incoming data from sensors and make decisions to help keep the plant operational.

3. *Design*. In product designs requiring input from various disciplines, for example in the development of IC packages, expert systems offer a distinct advantage.

4. *CAPP*. The sophisticated task of generating the "best" process plan for a given part or product benefits from expert systems. Some common CAPP expert systems (Chang, Wysk, and Wang, 1991) are: GARI, TOM, SIPP, SIPS, and DCLASS.

Until recently, expert systems had required mainframes or minis as platforms. Today, micro-based packages are also available. The use of LISP—the programming language for expert systems—creates difficulty in integrating LISP machines (hardware especially designed for AI work) with the conventional computers used in manufacturing. This is avoided by programming in C language, which also allows linking expert systems with existing databases. It is possible to store expert system information in relational databases to facilitate information sharing.

A recent development is a programmable expert system computer configured especially for the tough environment of the factory floor. The system consists of plug-in modules that may be interconnected via an optional rack-mounted enclosure or over a LAN. After developing the expert system application program on a PC, the user can download the program to a microprocessor that interfaces with alarms, PLCs, or PID controllers. Up to 15,000 production rules and 15,000 logical variables, enough for an average-size manufacturing operation, can be contained.

An example of a real-time expert system for a CIM environment is G2 from Gensym Corp. It is designed for large applications involving hundreds of variables, such as process control, CIM, financial trading, network monitoring, and automatic testing. An increase in average uptime of transfer machines from 48% to 90% is reported in automotive manufacturing from another expert system, System 90 from Septor Electronics.

Rather than a complete expert system, the trend is to buy development tools that help build an expert system. NASA has evaluated 20 commercially available expert-system-building tools, which are basically software. Such tools reduce the time of developing expert systems by an order of magnitude, compared to that required with traditional languages such as LISP or C. In one application (Romero, 1991), mainstream developers found the reduction in development time from 180 person-months with C language to 30 person-months with an expert system language, a savings of 83%.

One of the successful development tools is ART (Automated Reasoning Tool). It has helped, for example, International Harvester of Chicago design and build an expert system, called Wire Harness Cost Estimator, for use in producing cost-effective wire harnesses for trucks.

As with any software, expert system development comprises problem recognition, development team selection, project management, feasibility study, analysis, design, implementation, testing, technology transfer, maintenance, documentation, and training (Romero, 1991).

14.2   COMPUTER VISION

Computer or machine vision is basically the reconstruction of explicit and meaningful descriptions of physical objects from images. As a technology it covers a wide range of applications. For example, in the military, it is used to locate and identify enemy naval vessels from aerial photography. In medicine, it is used in diagnosis based on computer-aided tomographic (CAT) scans. In manufacturing, computer vision is used in a variety of tasks, such as location and orientation of parts on conveyer belts, assembly, and robotic scanning of mechanical parts. Today, it is feasible to develop PC-based vision systems that can handle moderately complex tasks, for example sorting out nonconforming washers (Lim and Vajpayee, 1987). Figure 14.1 shows the four subsystems that a typical robotic vision system comprises.

All vision systems can be classified into one of three basic types. The simplest one is the classic two-dimensional system with a fixed third dimension. Such systems attempt to find something on or about the X-Y plane. They are similar to the imaging on an optical comparator except that a computer rather than an operator handles the interpretation of the image. The second type is **stereo vision**, based on two cameras. Such systems permit measurement of depth as well, provided the scenes are simple. Both cameras look at the same image, each generating close to 100,000 pixels. In the third type, based on **structured light**, a known light—preferably from a laser—is projected for viewing by the camera(s). Changes in the pattern

Figure 14.1
A robotic vision system comprises four subsystems

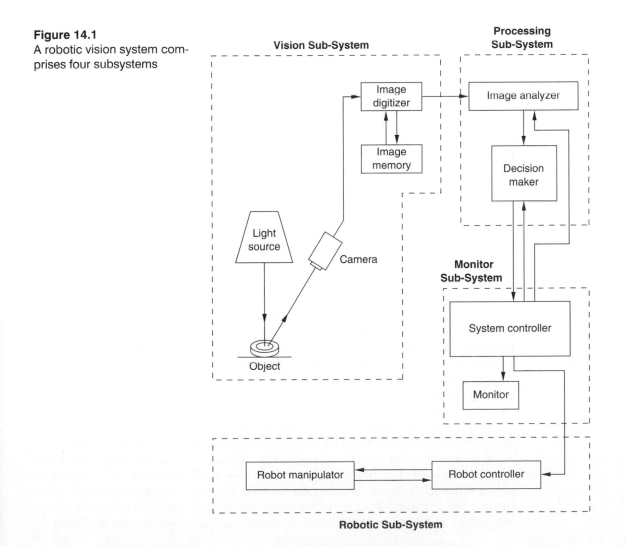

as it moves along the surface are analyzed to measure the third dimension and related information.

Digital image processing is a major task of computer vision. It is a technique to process the view that the vision system captures. Initially developed during the 1960s to enhance the quality of images transmitted from space, its application has extended to manufacturing, primarily due to the advent of PCs. Two major areas of application are robotics and inspection for quality. Depending on the needs of the application, the level of analysis can vary widely.

For inspection and guidance applications of machine vision, image processing involves three steps:

image acquisition
low-level processing to generate a rudimentary image
high-level processing for feature extraction

Manufacturing requiring feature classification and detection represents low-end, cost-effective technology. A typical image-processing system requires additional hardware and software along with a charge-coupled device (CCD) or vidicon and an analog-to-digital converter (ADC). The selection of digitizer is based on the speed requirement of the application.

14.3   LASERS IN MANUFACTURING

LASER is an acronym for light amplification by stimulated emission of radiation. It evolved as a process around 1960 and is now a powerful tool with several manufacturing applications. A laser may concentrate an enormous amount of energy onto a small area, causing the target material to heat up and even vaporize. The light emitted from the laser differs from other natural and artificial light sources: It is more monochromatic, coherent, nondivergent, and brighter. Because of the broad range of tasks it can perform in a wide variety of applications, the laser holds a unique place among nontraditional processes.

The primary reason behind lasers' use in manufacturing is their ability to process materials faster and/or better than conventional processes. For example, replacing a multipass arc welding process with single-pass laser welding reduces the number of passes and saves machine time and labor cost. As another example, a laser can cut sheet metal faster with more precision and less distortion than conventional techniques. In many cases the parts are ready for the next operation immediately after laser cutting, since burrs are absent.

Some of the machining lasers deliver as much as 25 kW of power. High-power lasers process faster and maintain the desirable metallurgical properties of the workpiece. A laser-based process is compatible with soft automation, and hence with FMC and FMS technologies. This gives rise to laser cells which, in conjunction with robot gantry and work handling modules, are effective processing centers. Lasers are more suitable for producing limited volume and/or prototype manufacture than for mass production. They are used in two areas of manufacturing: machining and metrology.

Machining

By controlling its power density, a laser can perform a wide variety of machining tasks, from ablating to welding. Based on the power densities required to melt or vaporize, materials can be grouped into three classes (Engel, 1989), as shown in Table 14.1. The last column gives some examples of materials in each class. Note that (a) the power required for vaporization is 100 times greater than that for melting, and (b) class I materials require 100 times more power than those in class II, and 10,000 times more than those in class III.

Lasers are widely used to drill in aircraft parts, automotive components, and hard ceramics. They produce small holes using single or multiple pulses from a stationary laser beam. Round holes with diameters ranging from 0.005 to 0.050 inch (0.127 to 1.27 mm) are possible at length-to-diameter ratios up to 100 (Benedict, 1987).

Welding with lasers is most effective when low-distortion, high-speed autogenous welds are desired. Parts to be welded must be held tight at a gap less than five percent of the smaller thickness. Shielding gases such as argon or helium are used locally to protect the weld puddle from oxidation and contamination.

Laser heat treating is a transformation hardening process used for nearly distortion-free, localized surface hardening to depths up to 2 mm. For this type of hardening to be effective, the material must contain sufficient carbon (greater than 0.3%) to produce a martensitic phase. This process is generally performed by CO_2 lasers, because they offer enough power to be cost-effective.

Metrology

Besides machining, lasers are used in noncontact measurement and inspection. Although lasers' metrological applications are not new, their linking with a variety of techniques and equipment as an alternative to traditional methods of gaging, part inspection, and machine tool calibration and alignment is.

Table 14.1
Classification of materials

| Class | Power density, W/in^2 | | Examples of materials |
	Melt	Vaporize	
I	10^7	10^9	Good thermal/electrical conductors such as Cu, Al, Au; Refractory metals such as Va, Ta, Mo
II	10^5	10^7	Ferrous metals such as carbon steel, stainless steel; Alloys such as hastalloy, inconel, monel
III	10^3	10^5	Organic/plastic materials such as wood, plastics; Metals such as Pb, Zn, Cd

Source: From "Classification of Materials" by S. L. Engel, February 1989, *Manufacturing Engineering*, p. 43.

It is the unique property of monochromaticity that permits laser applications in metrology; this property enhances the visibility of interference fringes. Lasers typically used in metrological applications are low-power, helium-neon systems. These produce light, at a wavelength of 6328 angstroms (1 angstrom = 0.6 micrometer), which is coherent in phase and a thousand times more intense than any other monochromatic light source.

Laser applications in metrology include interferometry, scanning, and holography. One of the earliest metrological uses of low-power lasers was in interferometry, which determines distance or thickness by measuring wavelengths. The interferometer splits the original laser into measurement and reference beams. The measurement beam travels to a reflector mounted on the part whose distance is to be measured, while the reference beam is directed at a fixed reflector. On reflection, these beams are recombined into a single beam that travels to a photo detector. In-phase waves produce a series of bright bands, while out-of-phase waves produce dark bands. The number of the alternating bright and dark bands, also called **interference fringes**, determines the distance (Farnum, 1990).

One recent application of lasers in metrology is machine tool calibration and alignment testing. Alignment by laser is superior to the conventional method based on levels, squares, ordinary lightwaves, and so on. The cost advantage of laser-based alignment testing results not only from the accuracy of the process but also from its speed.

In-process gaging based on laser scanning is a promising inspection technique. In laser scanning, the object is placed between the laser beam and a receiver containing a photo-diode, and a microprocessor computes its dimensions based on the shadow. The noncontact nature of laser scanning makes it suited for in-process measurement as in inspection of hot rolled or extruded material. Its comparative simplicity has led to the development of highly portable scanning systems. Laser-based scanning systems compete with the more sophisticated technology of machine vision, the choice depending on the cost and difficulty in transferring the image into usable data.

Holography is a method of taking three-dimensional pictures without a lens. Ordinary photography records only the amplitude and frequency (or wavelength) of the light waves; holography preserves the phase as well. Though this technology was developed long ago in the late 1940s, its commercialization has become attractive only recently when monochromatic, coherent light sources such as lasers became available.

Holography has not been wholeheartedly adopted as a laser-based inspection method. Called holographic interferometry, the technique involves superimposing a hologram of the stressed object onto a second hologram of the unstressed object, or on the object itself. Commonly used sources for stressing the object are heat or sound. Depending on the extent of stressing, the wavefronts are at variance and yield a specific pattern of interference fringes. The analysis of the resulting pattern reveals defects that may not be visible to the naked eye and yields data on an object's stress or vibration severity.

Holography has been used in engineering analysis of a wide range of parts, such as turbine blades, gun barrels, and computer components. The technique is commonly used in the automotive industry. Major automobile manufacturers now routinely use holographic interferometry to analyze the parts likely to contribute to noise and vibration. Fully assembled automobiles are sometimes examined this way to study the vibration characteristics of large flexural components such as door panels (Farnum, 1990).

Laser Meets CNC

Lasers' distinct advantages in machining and metrology have led to several recent developments. Coupling a laser with CNC and a display screen can produce a versatile system, for example, to automatically cut parts along complex contours (Vaccari, 1991). To capitalize on the advantages of both the laser and the CNC, lasers are used on multiaxis CNC machines in lieu of conventional machines. Only a wisp of burr needs to be removed after machining with such a setup; as in other CNC processes, switching from one part program to another takes less time.

Recently, lasers have joined FMSs as a new tool for rapid prototyping and model making. The tedious, and often lengthy, process of making prototype parts has been shortened by using **laser stereolithography**. This new process begins with a normal 3D CAD model generated on PCs. The stereolithography system then translates and processes the design data.

14.4   CONCURRENT ENGINEERING

Concurrent engineering is a newly emphasized approach that considers the design and development of a product concurrently with the other related functions of manufacture and support. Such an approach encourages consideration of all the factors at the outset, such as quality, cost, maintenance, and repair. In other words, product design engineering occurs simultaneously with process design engineering. The new approach yields best results for both the product developers and consumers. The terms *simultaneous engineering* and *concurrent engineering* are synonymous.

Is concurrent engineering a new concept? In my opinion, it's not. Manufacturing has always required concurrent considerations of all departments and functions. But the concept is getting closer attention because of stringent demands for product quality and shorter times to market. The emphasis makes sense, since computers (especially in CIM environments) can rapidly assemble all relevant information so product design and manufacturing issues can be considered simultaneously. Thus, as with several other "new movements" in manufacturing, CIM has made concurrent engineering old wine in a new bottle.

According to DeLorge (1992), "The term 'concurrent engineering' can have somewhat different scope and meaning in different industries. The scope and involvement of outside suppliers is generally where the largest variances occur. The first step in concurrent engineering may be entirely within one company, utilizing only the internal employee organization. The ultimate level of concurrent engineering has internal and external partners working as a unified team, from conception through completion of the project." (p. 1)

Based on a four-year project by the Harvard Business School on large-scale concurrent engineering:

1. One company reported that most initiatives investigated by the top management using a top-down implementation had failed. Directives, training programs, reorganizations, slogans, and pep talks were ineffective given the magnitude of the change needed to implement concurrent engineering.
2. Second, the consensus was that to implement concurrent engineering most effectively, a company should first have a business problem/challenge that everyone believes must be solved. Many organizations implemented concurrent engineering only when the rewards were survival, not just "the right thing to do." (DeLorge, 1992, p. 3)

14.5 MULTIMEDIA COMMUNICATIONS

Multimedia communications represent a technology of the 1990s and beyond. Many people refer to it simply as **multimedia**. Multimedia integrates information in any form—live videos, taped digitized video sources, sound, text, and numerical data—on a PC and communicates them together. The decreasing cost and enormously improved capabilities of the PC technology is behind the development of multimedia. Since CIM is based on integration via computers and manufacturing cannot do without information, multimedia technology is extremely compatible with the concept of CIM. In an interview, Paul Mace, an expert on multimedia, emphasized (Box 14.2) its impact on the information management. As with most emerging technologies, multimedia's benefits to CIM will depend very much on how boldly management implements multimedia in the operations.

14.6 BEST PEOPLE PRACTICES

CIM's integrating influence brings together not only the technologies but people as well. Strong interactions among company personnel are essential for success with CIM.

In 1992, under the editorship of Victor Muglia, CASA/SME conducted a roundtable discussion by a group of experts on "people practices." **People practices** mean the day-to-day activities that are done to, for, around, and with employees. They exclude the tasks carried out by employees for the business, such as making the product or providing the service. Thus, people practices involve only people and not machines. "Johnston (a participant in the roundtable) envisions best people practices as those practices that are 'the best a business can possibly do for its employees that results in the best that individuals can do for their business'" (Muglia, 1992, p. 6).

The results of the discussion have been published as SME's blue book entitled *1992 Best People Practices*. The preface of this document makes a strong case for the importance of personnel: "Successful CIM systems consist of well-engineered and well-integrated technology, business processes, and people practices. Many unsuccessful systems have ignored the need for business re-engineering and quality people practices. Others have implemented re-engineering and people practices as a last resort to prevent business from failing. A basic understanding of people practices by the engineer, planner, manager, and implementor is critical to achieve effective solutions. Benchmarking provides the means for measuring the effectiveness of people practices. This document is intended to provide a basic introduction to people practices" (Muglia, 1992, p. 1).

<table>
<tr><td>■</td><td>**BOX 14.2**</td><td>*Mace Sees Multimedia Altering Corporate Structure*</td></tr>
</table>

InfoWorld: Multimedia is a very broad subject. Where would you like to begin?

Mace: The first thing we need to recognize is that multimedia is an adjective. We've been grappling for two or three years to find out what person, place, or thing the adjective modifies. "Multimedia what?" And the answer is, it always means multimedia communications. It's about integrating information in all forms onto the now ubiquitous personal computer.

Also, distributed computing had to happen and, clearly, the PC has won. There is not merely a terminal on every desk today. There's a full-blown computer there and, now, everyone agrees they should connect them together.

InfoWorld: How will this change corporate management structures?

Mace: A management structure is a conduit of information. Managers pass commands, orders, requests, and requirements down from the top to the people who actually make the products or provide the service to the customers. Management translates information about how that manufacturing or service end of the business is doing and filters that back up to the top.

Most companies don't recognize the overlap between that very legitimate management need and the information processing function. There's little reason for information management by the managerial team. Those people don't need to be involved in the massage of data. Data is usually only input at one level in a corporation, and that is where the money goes in and out—where the service is performed or the product is manufactured. All other corporate data is an abstraction of that fundamental data.

Companies failed to recognize that. They do information gathering at multiple levels throughout the company, and therefore require huge staffs of people. But, as the management structure changes, the rolling white-collar layoff that we've seen in this recession is not going to roll back. Those information carriers are being elimi-

nated because computers can carry that information without all the flappers in between.

InfoWorld: But at same time you end up with too much information.

Mace: Corporations are information rich, typically, and communication poor. They possess information in many forms—in paper and digital form, as live video conferences, and in telecommunications.

The problem is that the traditional idea of data and information exchange from bottom to top and back down by a computer is still in the form of numbers. Principally, what gets translated are spreadsheets and numerical analyses. It's data, not knowledge. People are not informed by it because it never makes it past their eyeballs. That information can be presented in more usable forms, but the technology for presenting it is only now coming to us.

Using multimedia, the information can be processed once and presented in many forms so that anyone can, depending upon their needs and sophistication, be informed. Information also becomes less redundant. A company is merely spinning its wheels when people in 27 different places are all duplicating effort.

InfoWorld: What's it going to take to make multimedia communications ubiquitous?

Mace: The obstacle to having a multimedia information network is the company's concept of how it manages itself. It takes a rather bold decision on the part of the executive directors of the company to decide that they will democratize the workplace—that they will make information generally available through a corporatewide PC network.

When more people from my generation move into the top management positions, then they'll be able to enable the people who are 10 years younger to actually pursue this. From the person who sweeps the factory floors to the person who sweeps the old members off the executive board, there's a need to understand the company's purpose and what it's doing.

InfoWorld: Then this is a generational change as companies move into the next century?

Mace: Right. People would like to feel that they're heard, that they have some input into how business is done, how a product is manufactured, how the company is managed—that they can go to their terminals and register how they feel about it.

But there has to be a willingness on the part of the corporation to allow its management structure to be fundamentally changed. One-way information exchange in a corporation tells you that this is merely a conduit for the president or CEO to tell everybody else what to do. When you make it a two-way communication street, where it's interactive multimedia information, now you don't just inform the work force. You also let them give back to you at the highest level what they're capable of expressing—not only about themselves, but about the company.

InfoWorld: So you see multimedia as a major tool for enabling change?

Mace: Absolutely. The flexibility that is allowed when you can integrate live videos, taped digitized video sources, sound, text, and numerical data all into the same place is that people will find their own proper level of expression.

If you make the fundamental information available to all the people who might need to know about it, then the chances are that they might do something imaginative with it and say something back to senior management that might help improve what the company is doing.

The first thing you have to do to achieve this is buy the philosophy that multimedia communication is the backbone of your management structure. When you accept that and build upon that concept, then the amount that you have to spend for hardware to implement it decreases significantly and you might actually see an improvement in the bottom line.

But if you don't understand it, then merely throwing hardware and software at the problem won't help. The key to this is "interactive." Merely using it to send orders down does not do you any good.

Source: "Mace Sees Multimedia Altering Corporate Structure," January 1993, *InfoWorld*, p. 17. Copyright January 11, 1993, by *InfoWorld* Publishing Corp., a subsidiary of IDG Communications, Inc. Reprinted from *InfoWorld*, 155 Bovet Road, San Mateo, CA 94402. Further reproduction is prohibited.

Muglia's report attempts to answer why the best people practices are essential to CIM, especially its management. The report groups various practices of manufacturing and management into four categories. The first category is traditional practices; the last three are modern practices grouped under "good," "better," and "best." Manufacturing tasks that benefit the most from modern practices have been divided into seven classes:

1. organizational values and beliefs
2. job design and job analysis
3. individual and organizational development
4. working and work environment
5. performance measurement and development
6. employee communications and information
7. employee involvement and participation

The major conclusions for the traditional and the "best" practices are summarized in Table 14.2.

Table 14.2
Traditional and best people practices.

Item	Traditional	Best
Values	Inconsistent and ad hoc values of executives; Noncommunicated	Defined and broadly accepted values based on trust, respect, dignity, cooperation, mutual benefits, and partnership
Orientation	Business-only oriented values	People-oriented values integrated with business values
Organizational style	Decentralization; Independent departments with formal boundaries organized by functions	Smaller P&L units. Interdependent cross-functional teams with overlapping boundaries organized by end product
Information, power, rewards	Management controlled	Workers, teams, and management shared
Decision control	Upper management	Lower point of control
Focus	Management centered	Employee-centered with business needs
Management style	Autocratic (vertical)	Coach and leader with cross-functional responsibilities
Hierarchy	Steep pyramid	Flat pyramid and highly networked
Layers	7 to 10	2*
Span of control	1:1 to 6:1	26:1 to 60:1†
Job design	None	Systematic job design including enlargement, enrichment, and rotation
Employee selection basis	Experience and education	Proven skills ability after passing Situation or Task, Action, and Results (STAR) interviews
Selection decision maker	Supervisor	Peer team members and subordinates consensus
Learning	Learning stops when goal is met	Continuous learning
Education	Minimum education to meet minimum job requirements	Continuous education in business and interpersonal areas
Education and training expense/year	Less than 0.5% of salary expense	More than 5% of salary‡ expense
Employee development	None to company driven	Individual driven and company supported with rotation through multiple jobs each for 1–3 years

* 5 or 6 under "good" and 3 or 4 under "better" practice

† 7:1 to 12:1 under "good" and 13:1 to 25:1 under "better" practices. Also, the appropriate term in the case of "better" or "best" practices is *span of influence*, not *span of control.*

‡ 0.5 to 2.5% in companies that follow "good" and 2.6 to 5.0% in those that follow "better" people practices.

Table 14.2 *continued*

Item	Traditional	Best
Individual responsibility	Completion of assigned tasks	Be involved and accountable
Differentiators (Inhibitors)	Separate parking, eating, clothing, privileges, benefits	No differentiators
Worker's role	Limited to assignment	Involved participation
Supervision style	Autocrat/administrator; Supervisors plan, direct, control, and monitor processes	Leader, coach, enabler; workers and leaders jointly plan, direct, and control processes
Approvals	Many approvals needed; 3 to 5 signatures for expenditures up to $2,000	Few to no approvals; no signature required for expenditures up to $2,000
Information reported	Daily/weekly/monthly/ quarterly/annual standard reports	Demand only exception reports
Knowledge focus	Minimum to meet job need	Continuous knowledge expansion
Knowledge base	Minimum to complete task	Knowledge of business, interpersonal skills, decision making, etc.
Communication	Tell them only what they "need to know"	Let people know what, how, and why of change early
Performance input	Supervisor	Peers, subordinates, employee, and others
Hourly worker promotion	Based on seniority	Based on merit and performance based on lateral knowledge/skills
Compensation basis	Payment for time, inflation and compatability level	Payment for output, quality, skill, knowledge, and performance
Compensation percentage at risk	None	Greater[§] than 25%
Satisfying customers	Meet minimum "get by" requirements	Driven to exceed customer requirements

[§]0 to 10% in "good" and 10 to 25% in "better" people practices companies.

Source: Adapted with permission from the CASA/SME Blue Book, "1992 Best People Practices," reprinted with permission of the Computer and Automated Systems Association of the Society of Manufacturing Engineers, Copyright 1992.

The roundtable participants believed the following companies had the best people practices in 1992:

Amoco	Apple Computer	AT&T
Boeing	Caterpillar	Chrysler
Computervision	Corning Glass	Cummins Engine
Dana Corporation	Delco Electronics	Digital Equipment
Eastman Kodak	Federal Express	Ford
General Electric	General Foods	GM, Cadillac
Harley Davidson	Hewlett Packard	Honeywell
IBM	Los Angeles Times	Mazak
Motorola	Packard Electric	Proctor & Gamble
Rockwell Int'l	Saturn Corporation	Texas Instruments
TRW	Volvo	Xerox
3M		

TRENDS

❑ CIM continues to be an enabler. It encourages concurrent engineering and facilitates simultaneous consideration of all the factors involved in product design and manufacture.

❑ The increasing use of expert systems in manufacturing supports CIM development.

❑ Computer vision applications in manufacturing continue to expand.

❑ Laser technology is widely practiced in manufacturing; it is no longer an unconventional process. The ability of lasers to adapt quickly and easily to various performance levels makes them ideal for a wide range of machining and metrology applications.

❑ Laser and CNC technologies have begun to merge into a system, resulting, for example, in five-axis CNC laser machines. Such equipment can machine complex geometries, say of aircraft parts, more effectively.

❑ Current developments in multimedia communication will assist the expansion of CIM.

❑ The emerging best people practices are compatible with the concept of CIM. They cater to the personnel needs of CIM by integrating employees.

❑ Box 14.3 summarizes some recent trends in CIM technologies observed at the 1990 International Manufacturing Technology Show (IMTS-90) exhibition.

| BOX 14.3 | *A Dozen Trends to Watch at IMTS-90* |

With the every-two-year IMTS (a show devoted to machine tools and other manufacturing technology) cycle, it is only natural to raise the question, "What possibly could be new?" In conversation with leading observers, we came up with twelve trends:

1. Turn/Mill Combination.   This is not a new concept. Kearney and Trecker created a combination more than 50 years ago. Industries around the world, however, have made the discovery that quantum productivity improvement can be made if setups and queuing for secondary operations are eliminated. The credo now is: "Do all machining operations in a single setup." The turn/mill concept allows single-setup operation for parts that have a primarily round geometry. The concept is now found on turning equipment from the smallest screw machine to the largest lathe, whether single- or multiple-spindle.

2. Combination Drive and C-axis Motor.   Milling flats, drilling cross holes, or machining cam slots on a turned workpiece usually requires a positive reference on the part. This requires a C-axis drive that divides a single rotation of the held workpiece into as fine a resolution as 0.001 degree. Such a precise rotation is normally obtained by disengaging the spindle drive motor and activating a separate C-axis drive. Several lathe builders are now offering a single motor that is a combination drive and C-axis. This development will simplify both the programming and execution of mill/turn CNC programs.

3. Five-Axis or 3/2.   The concept of doing it in a single setup is not confined to turned parts. The advanced CNC machining centers always had a five-axis capability. This meant a tilt and swivel of the head or a special tilt and swivel of the table. An emerging concept utilizes a head that can be oriented in either the vertical or horizontal mode. With the normal X, Y, and Z machine axes under programmed control, the ability to swivel the head

to either vertical or horizontal often eliminates a second setup to complete machining operations on the workpiece. A more sophisticated version of the 3/2 concept is a head that can be programmed-directed to any orientation between vertical and horizontal. In this instance, it is a true five-axis unit. The advantage of the 3/2 head is a lot of added capability for a modest increase in price.

4. High-Speed Spindles.   Isolated examples of high-speed spindles have been displayed for the past ten years. Often they were add-on units to be used for special applications. The big problem with any spindle running at speeds from 15,000 to as high as 100,000 RPM has been heat generation and the resultant dimensional growth that is the enemy of precision. Builders of such spindles are now solving those problems with magnetic bearings, ceramic balls for bearings, a cooling jacket surrounding the bearings, or a combination.

5. Probes and Feedback.   Having the right information before an unacceptable condition is reached is one of the most powerful quality assurance tools now available. Various arrays of probes, sensors, and other feedback devices now detect trends so that corrective action can take place before a critical point is reached.

6. The Ubiquitous Laser.   Forget the laser as a star wars development. It has become a working shop-floor resource. It cuts, heat treats, welds, measures, and inspects. Perhaps the most significant development is the downsizing of the laser gun itself so that it no longer sticks out fifteen feet behind the machine tool. Powerful cutting lasers now fit into a body that has been reduced to approximately three feet. The laser is carving out some very significant niches where it is the best technology available. The laser can be finely controlled and focused. This is its strength.

7. Breaking Down the Design/Manufacturing Wall.   Tradition holds that designers worked on one side of the wall and manufacturing on the other. When the design was completed, it was pitched over that wall and manufacturing was left to program the design for NC and then produce it. The modern CAD/CAM systems and other computer software are breaking down that wall. Computer-generated design data is being used more and more as the basis for NC programming. In its advanced stages, it is also being used for total part processing, including inspection. Software development is the fastest growing and most rapidly changing phenomena of all the manufacturing resources.

8. Electronic and Software Invasion.   There is a tendency to think of control and software advances in terms of machine controls or complete CAD/CAM systems. However, electronics have been proved on simple gaging equipment. Not only are digital gages easier to read, but many of them can even generate statistical distribution curves and calculate standard deviation for SPC.

9. 32-bit Now Standard.   The conversion from the old hard-wired NC units to the new CNC was confirmed in 1976. In 1990, the 32-bit control will be standard. It simply means that data is processed faster and in greater quantities. This allows split memories, better feedback and communication, and the elimination of data starvation that often caused dwell marks, overshoot, or undershoot in the execution of an NC part program.

10. Cutting Tools March Ahead.   Developments in cutting tools are keeping pace with machine, control, software, and spindle advancements. High positive-rake-angle face milling cutters that require substantially less horsepower than the old negative-rake type will be shown by most of the major cutting tool manufacturers. The high-axial-rake cutters are made possible by fine-grain carbides and physical-vapor-deposition coatings that produce a sharp edge able to withstand the action of a sustained milling operation. In addition to advances in coatings and geometries, the major tooling suppliers are also developing complete tool control and offset systems so that all tools are properly inventoried, identified, and offset controlled.

11. FMS and Cells.   Flexible machining systems captured the manufacturing community's imagination about ten years ago. Since that time the FMS has been in a holding pattern. One or two examples may be shown, but the emphasis definitely will be on cells that offer most of the advantages of an FMS but are far less complex to install and operate, and certainly less expensive to buy. With present software and communication capability, it is possible to network cells so that, in effect, an FMS is created in manageable steps. Look for two- and three-machine cells with simple robots or pick-and-place units as the idea of the hour.

12. Accessories.   The final links in the machining process are the accessories and workholding fixtures. Many advances have been made in this area. Programmable boring heads, modular tooling, vises, ceramic fixturing, and many other concepts have come into their own.

Behind these developments is the crying need for better capital equipment utilization, faster throughput, and reduced work-in-process. If these three factors are maintained under assured control, productivity and quality will improve.

Source: "A Dozen Trends to Watch at IMTS-90" by K. M. Gettelman, July 1990, *Modern Machine Shop*, pp. 102–107.

SUMMARY

Some of the emerging technologies that have begun to positively impact CIM were discussed in this chapter. Expert systems, computer vision, lasers, and concurrent engineering are among the technologies helping CIM in various ways. When lasers are integrated with CAD/CAM systems and manufacturing processes, prototyping takes less time and expense. Although initial investment in laser systems is high, it can be justified in some types of manufacturing applications.

A brief introduction to multimedia was presented, since its integrating capabilities benefit CIM. Finally, the changing roles of management and other personnel, including best people practices, were covered.

KEY TERMS

Concurrent engineering

Interference fringes

Laser stereolithography

Multimedia

People practices

Signature analysis

Stereo vision

Structured light

EXERCISES

Note: Exercises marked * are projects.

14.1   What are expert systems and how do they support CIM?

14.2   Explain concurrent engineering in 200–250 words.

14.3   Write a 400-word essay on computer vision. How does this technology help CIM?

14.4   Describe any case study of laser use in manufacturing.

14.5   What is meant by multimedia? Discuss how it can help CIM.

14.6*   Visit a local manufacturing facility and study its people practices. How do they compare with the "best" practices discussed in Chapter 14?

14.7   Circle T for true or F for false or fill in the blanks.

    a.   Expert systems are sophisticated NC equipment.   T/F

    b.   Concurrent engineering is the same as simultaneous engineering.   T/F

    c.   "Best people practices" mean

        Practices by the smartest persons

        Most effective practices by people

        Practices by employees to their best ability

        None of the above

    d.   By multimedia, we mean coaxial and fiber-optic cables and other similar cabling systems.   T/F

    e.   Signature analysis is _____

SUGGESTED READINGS

Books

Ballard, D. H., & Brown, C. M. (1982). *Computer vision*. Englewood Cliffs, NJ: Prentice-Hall.

Benedict, G. F. (1987). *Nontraditional manufacturing processes*. New York: Marcel Dekker, pp. 299–320.

Chang, T., Wysk, R. A., & Wang, H. (1991). *Computer-aided manufacturing*. Englewood Cliffs, NJ: Prentice-Hall.

John, R. F. (1978). *Industrial applications of lasers*. Academic Press, pp. 274–275.

Miller, R. K. (1985). *Artificial intelligence applications for manufacturing*. Madison, GA: SEAI Technical Publications.

The World Almanac and Book of Facts. (1992). New York: Pharos Books, p. 200.

Monographs and Reports

DeLorge, D. (1992). *Product design and concurrent engineering* (SME blue book series). Dearborn, MI: SME.

Liu, D. (1990). *Expert systems* (SME blue book series). Dearborn: MI: SME.

Muglia, V. O. (1992). *1992 best people practices* (SME blue book series). Dearborn, MI: SME.

The nature and evaluation of commercial expert systems building tools (rev. 1, NASA TM-88331 (NS87-28281/NSP). Washington, D.C.: NASA.

Romero, J. E. (Ed.) (1991). *Expert systems: How to get started* (SME blue book series). Dearborn, MI: SME.

Articles

Blattenbauer, J. A., & Kim, Y. (1989, July). Bringing image processing into focus. *Mechanical Engineering*, pp. 54–56.

CIM Update. (1990, April). *Manufacturing Engineering*, p. 26.

Engel, S. L. (1989, February). Classification of materials. *Manufacturing Engineering*, pp. 43–46.

Farnum, G. T. (1990, January). Measuring with lasers. *Manufacturing Engineering*, pp. 45–46.

Gettelman, K. (1990, June). The German march toward advanced manufacturing. *Modern Machine Shop*, p. 95.

Gettelman, K. (1990, July). A dozen trends to watch at IMTS-90. *Modern Machine Shop*, pp. 102–107.

Lim, W. S., & Vajpayee, S. (1987). Development of a vision-based inspection system on a micro-computer. *Computers in Industrial Engineering*. 12(4), 315–24.

Parks, M. W. (1987, January). Expert systems: Fill in the missing link in paperless aircraft assembly. *Industrial Engineering*, pp. 37–45.

Schoonmaker, P. M. (1989, December). Troubleshooting an expert system. *Mechanical Engineering*, pp. 56–58.

Tompkins, J. A. (1990, winter). Where is manufacturing headed? Dynamic consistency. *Journal of Applied Manufacturing Systems*. 3(2), 43–46.

Vaccari, J. A. (1991, April). Precision laser cuts improve quality. *American Machinist*, pp. 45–46.

Vasilash, G. S. (1989, November). Vision: It's back—and sticking around. *Production*, pp. 38–46.

Vasilash, G. S. (1990, May). Some things managers might want to know about. *Production*, pp. 42–47.

White, K. W. (1989). Is there a future for vision with neural nets? *Vision* (MVA/SME quarterly). *3*(3).

CHAPTER 15

Ramifications of CIM

"An economy can only be as strong as its manufacturing base."
—Akio Morito, chairman, Sony Corp., in *Harvard Business Review*, May-June 1992

"By the 21st century, a vast system of electronic 'highways' could link research centers, manufacturing enterprises and educational institutions throughout the country. Eventually, every home could have access to two-way multimedia data, sound and video."
—*U.S. News and World Report*, November 16, 1992

"Our factories are more than places to produce things. They are providers of education, formulators of a sense of purpose, catalysts of a practical orientation, and promulgators of the discipline society needs. Factories are interwoven with our culture and our way of life. If our industrial situation changes, the rest of our economy is profoundly affected."
—Frederick M. Zimmerman and Marilyn Magee-Powell, *Journal of Applied Manufacturing Systems*

"CIM is pushing manufacturing into the footsteps of farming. Like the farmers of today, fewer manufacturers employing still fewer people will be able to produce in the future all that society would need."
—S. Kant Vajpayee, professor, The University of Southern Mississippi

"As a percentage of total manufacturing costs, direct labor in many American and Japanese plants is well under ten percent. At the new ultra-modern Mazak facility opened in 1987 in Worcester, England, the figure is less than five percent."
—Ken M. Gettelman, *Modern Machine Shop*

In the previous 14 chapters we discussed the various technologies and management concepts appropriate to CIM. It is obvious from the discussions that CIM is a broad discipline. Current developments in CIM suggest that it is no longer an innovation but a conventional practice. In this closing chapter, the outlook for CIM and its impact on manufacturing are presented.

15.1   DOES CIM EXIST?

Does a CIM facility exist today? The answer to this question is both yes and no, depending on how much you expect of a CIM facility. It also takes us back to the first chapter, which began with the question: What is meant by CIM? If CIM means a high level of integration in which computers and flexible automation perform tasks cost-effectively, with little human intervention, then yes, CIM plants do exist. On the other hand, if CIM signifies a utopian concept, a lights-off and entirely automated operation, a **factory of the future**, then no such plant exists. Will such a plant ever exist? Probably.

The current status of CIM falls somewhere between these two expectations. What exists in a few selected manufacturing enterprises in the U.S., Europe, and Japan is a **pseudo-CIM** facility. Boxes 15.1 through 15.4 describe some of these facilities and developments to illustrate that CIM has already begun to attain perfection. However, we must not expect anything revolutionary out there in the real world, where profit is the motive behind all actions—including exploitation of CIM. Companies implement CIM only when, and if, it makes business sense.

 BOX 15.1   *Mazak Dedicates CIM Facility*

Mazak Corporation has completed the expansion and reorganization of its Florence, Kentucky, machine tool factory and dedicated the plant as a "CIM facility." An IBM host computer and a series of VAX computers link the management functions with production departments. Production scheduling simulation guides purchasing decisions and, aided by a new automatic storage and retrieval system, minimizes work-in-process inventory. Raw materials are transported by AGVs. Production takes place in three FMSs for machining and, for sheet metal work, a flexible fabrication system. The facility is expected to produce 105 machine tools per month, with an eventual goal of 120 per month.

Source: From "Mazak Dedicates CIM Facility," July 1990, *Modern Machine Shop*, p. 44.

BOX 15.2   *Ingersoll Represents World Class Manufacturing*

Ingersoll is an international company based at Rockford, Illinois. It makes special machines and equipment. One of the LEAD Award winners from CASA/SME for excellence in manufacturing, its Rockford facility approximates a CIM plant. The 959,000 sq ft plant is air-conditioned, not for the comfort of the employees but for the health of machines so that they can produce the tolerances desired.

The shop is divided into two sections: the heavy machining area processes workpieces over one cubic meter in volume, the light machining area those below this size. The latter contains 15 machining centers. An FMS is the attraction of the light machining area. The FMS, comprising 5 machining centers, has reduced the number of machines (17) used earlier, and at the same time increased the throughput. The plant can produce a lot size of one; in a particular year, for example, FMS handled more than 7,000 prismatic parts, of which 30 percent were in lots of one.

Like most manufacturing companies, the accounting department's computer was bigger than that of the manufacturing department. It was soon realized that in a CIM environment it should be the other way round. Today, the manufacturing department's computer at Ingersoll is bigger than that of accounting's.

Ingersoll started the computerization of its manufacturing in the 1970s with NC machines. They moved to CAD in 1975 and to CAM in 1979. According to George Hess, vice president of Systems and Planning, "By the end of the '70s, almost everything was automated. But not talking." There were 1,300 different application programs and 225 master files. The move to CIM came when it was decided to roll the master files into a single integrated database, not an easy task. It took two years to achieve that.

The hub of the hardware is a National Advanced Systems NAS-XL/70 central processor that can process 31 mips (million instructions per second). The central database is accessed 8 million times per day for tasks, such as CAD, CAM, NC, and all types of information for the business, which include marketing reports, personnel, purchasing, BOM, shop scheduling, word processing, electronic mail, etc. It is done using 500 alphanumeric video display terminals and 200 vector graphics video display terminals. The central processor's slave machines are a NAS-6600 computer, five VAXs and MicroVAXs for solids modeling and FMS development/control, and one hundred PCs.

Besides the various manufacturing software packages, Ingersoll has CIMPLEX, a process planning and generative numerical control (PPGNC) package from Automation Technology Products (ATP, Campbell, CA) for automatic NC part programming. CIMPLEX is like an expert system that is used to design a 'features enriched' solids model of a part, and then to generate an optimized NC program automatically for producing the part.

Source: From "Impressions of Ingersoll" by G. Vasilash, February 1989, *Production*, pp. 40–46.

BOX 15.3   *Allen Bradley's World Contactor Assembly Facility*

Within Allen Bradley's (A-B) headquarters complex at Milwaukee, Wisconsin, there is a 45,000 square foot CIM plant. It is a model of the "factory of the future." Built in just two years during 1982–84 at a cost of $15 million, the plant makes motor contactors and control relays for the world market.

How did it come about? Based on market research for motor starters and other business factors, the company decided back in 1975 to expand globally. Today its international business outside North America is 25% in comparison to 3% then. A task force under Larry Yost, vice president of Operations for the Industrial Control Division, was set up in 1982. The task force was a multidisciplinary team with members from cost and finance, marketing, product development, quality assurance, manufacturing, machine building, and management information systems. The composition of the task force clearly illustrates that A-B management knew what CIM was! J. Tracy O'Rourke, president and chief executive officer of A-B, admits, "We wouldn't have been successful without a multidisciplinary team approach." The primary goal of the project at the planning stage of product design and facility construction was lowering production costs by reducing the cost of materials, direct and indirect labor, and scrap and rework. Their goal was not any different; this is what all manufacturing companies strive for.

What are the special features of A-B's CIM plant? The 300-foot by 150-foot plant has only two doors. Raw materials (brass, steel, silver, coils and springs, molding powder) enter through one door, and the finished product exits the other. The facility is grouped in four key areas: a control room, a plastic molding cell, a contact fabrication cell, and the assembly line. An employee is checked in when the magnetic ID badge is inserted in the reader.

The production cycle starts when A-B distributors or field sales offices enter an order. Most of A-B's worldwide distributors and all the U.S. field sales office are connected to the company's mainframe computer at Milwaukee via telecommunications. Every day at 5 a.m. this computer transfers all the orders received by the previous day to another computer, Vista 2000—an A-B product, located in the control room of the plant. The plant computer breaks these orders down into production requirements and electronically transmits these to the cell level PLC (programmable logic controller), A-B's own PLC-3 that works as a master controller. The PLC-3 is in contact with 26 smaller programmable logic controllers, PLC-2/30s, on the factory floor. These control the assembly stations. Two data highways, again A-B products, connect the PLC-3 with the PLC-2/30s.

"With the operation we now have, which essentially uses zero labor, we are successfully achieving a 60 percent cost advantage," claims O'Rouke.

During the first two years of operation, over 5,000 A-B customers have toured the plant. A-B's facility closely approximates a "paperless factory." It is the Mecca for anyone interested in manufacturing.

Source: From "Allen Bradley's World Contactor Assembly Facility" by J. Pierce (Ed.), December 1988, *Management Memo* (a publication of the management technology program and the Industrial Management Department, University of Wisconsin-Stout), pp. 9–14.

BOX 15.4   *An Off-the-Shelf Approach to CIM*

The company has just scrapped its minicomputer-based CAD/CAM stations for a faster system which has been tied into the existing DNC network. They are now in the process of integrating the design, process engineering, scheduling and quality computer systems via a relational database management system. The CNC coordinate measuring machine (CMM) automatically feeds back SPC data, and fourteen additional APC analysis processors will soon be added to the shop floor to streamline system data-gathering capabilities.

Who does this sound like? John Deere? IBM? Ford?

Hardly. It's K&G Manufacturing, an 85-person job shop in Faribault, Minnesota. Although K&G's degree of integration is enough to turn a head, what is most impressive about this seemingly complex setup is that it has been achieved with standard, off-the-shelf software products all running on a local area network (LAN) utilizing personal computers (PCs).

To be sure, this is not a typical job shop, at least in its use of technology. It is, however, an example of what almost any shop can achieve by applying widely available software on relatively inexpensive hardware platforms.

Standard Connections

There are at least three relatively recent technological developments that make K&G-style integration possible.

i.   First is the evolution of personal computers. Newer, more powerful PCs are taking on jobs previously reserved for considerably more expensive minicomputers.

ii. There has been an explosion of PC-based software developed for manufacturing applications. Many of these products have been tailored specifically for made-to-order environments.

iii. Companies are going about the business of tying all these computers and programs together with standard integration products.

Although PCs deliver a wide array of computerized benefits, it is this last development that makes information sharing practical for the small manufacturer that is wary of the expense, complexity and inflexibility of custom-written software.

Along with CAD/CAM, K&G also installed a DNC network to transport part programs to the shop's CNC machine tools. The network was supplied by Intercim and included a generic post-processor. This program management system eliminated paper tape and spared K&G the expense of buying individual post-processors for the shop's diverse combination of machine tools and controls.

A major computer manufacturer refers to this concept as value-added data flow. The idea is that the first person to work with the information on a part creates a data file in a way that is useful to downstream users. Thereafter, each successive user of the file adds only that data which is specific to his or her function. Thus, as the file is used it becomes more complete, but never requires any user to recreate previously-established information.

CIM can provide these tools and information, not just to operators, but to everyone involved with the throughput and quality of a production operation. And all the while, it can improve the quality of the work environment. There is an old adage that knowledge is freedom. In manufacturing, knowledge enables people to do the right things on their own, without being told. That kind of freedom is good for the individual and good for the organization.

Source: From "An Off-the-Shelf Approach to CIM" by T. Beard, May 1989, *Modern Machine Shop*, pp. 54–64.

CIM's "forward march" is often held back by practices we have used since the dawn of the 20th century. For example, most operations are based on **Taylorism**, which divides work into subtasks for productivity gains in labor-intensive industries. But, as Box 15.5 explains, CIM is the antithesis of Taylorism for managing human resources. CIM's impact on current manufacturing practices is enormous and, therefore, change will be slow.

The major technical hurdle to the realization of full-blown CIM facilities is the shortage of fully integrated comprehensive software. Such software is monumentally difficult to develop, since full-blown CIM operations are very complex, even for a medium-sized company. A modular approach (Box 15.6) is the only practical way of developing integrated CIM software. Since the hardware and networking technologies required for CIM implementation are already available, CIM software remains the only missing link.

We are only beginning to imagine the factory of the future under a fully implemented perfect CIM. We know that the factory of the future will be expensive, not only the initial investment but also on a recurring cost basis. The technologies

---

BOX 15.5   *Control Is Still Essential*

Instead of controlling people as a utility resource, we are moving to an era where people on the shop floor are important resource managers. To better understand where we are going in this trend, it helps to know where we came from.

Our industrial heritage is based on the Taylor model. Its guru was Frederick W. Taylor who wrote and lectured as the 19th century was going out and the 20th was arriving. Skilled labor was in short supply and hordes of unskilled laborers with a primitive agricultural background were emigrating from southern and eastern Europe to the United States.

Taylor knew that utilization of this labor resource, deficient in both language and advanced educational skills, involved breaking down manufacturing operations into very simple repetitive tasks that could be taught very quickly and learned by anyone. Thus, the unskilled laborer became the adjunct to his machine or process and was regarded as an object to be closely controlled. The individual was not to use any discretion or provide any feedback. He was to do his simple task—period. Control was absolute and complete. It was used very effectively in creating the mass production industries. Henry Ford took the Taylor model and put the country on wheels. Although it worked less well in job lot

industries, Taylorism became the credo for the American plant floor. It was the best available at the time, but it had three great weaknesses.

First, there was no means to accommodate whatever creative impulses the individual worker might have had. Second, a worker doing nothing but simple repetitive tasks has a wide variability in how these tasks are executed. This is not good for building quality into a product. Thirdly, the Taylor model could not have been better designed to smash the self-esteem of the individual worker.

Cheap and compliant labor eventually disappeared. Automation did much to eliminate the repetitive tasks and make them more consistent. This also eliminated the need for vast quantities of unskilled labor. The computer has brought a significant degree of control to more complex operations. The result has been a shift away from direct labor as the primary manufacturing cost factor. It now rests in capital equipment, work in process, and general overhead.

Control is needed as much today as it was a century ago. But the shift is away from the labor to capital.

Source: From "Control Is Still Essential" by K. Gettelman, June 1990, *Modern Machine Shop*, pp. 6–8.

BOX 15.6   *Is This a CIM Software?*

Mac-Pac from Anderson Consulting is an integrated, on-line manufacturing, and resource planning and control system. Each Mac-Pac system is divided into 12 modules:

Design engineering
Manufacturing engineering

Inventory control
Requirements planning
Master scheduling
Shop floor control
CONBON (card order notice/bring out notice for JIT planning)
Capacity planning

Purchasing
Product costing
Inventory accounting, and
Distributed processing control.

Mac-Pac/D is a version designed for aerospace

industry and defense contractors, and is COBOL language-based.

Source: From "Is This a CIM Software?" *Modern Machine Shop*, p. 180.

involved in CIM are not static, but require frequent updates, which will make CIM look like a mirage. For example, Mazak Corp., a major world builder of flexible manufacturing systems (FMSs), recently found it financially more attractive to replace rather than upgrade two FMSs in their operations in Japan and one in the United States. The replaced FMSs were only eight and five years old. Thus, factories of the future will have faster turnover of their machinery. This demands a new approach to financial management and personnel training needs, as discussed in Chapter 14.

15.2 SCENARIOS

CIM-based operations possess two special features:

1. Individual computers are interconnected to form a distributed information system. Such a system handles all data collection and information processing. These include design data, manufacturing data, customer orders, inventory data, maintenance, and quality assurance data such as process capabilities.
2. The entire operation will comprise modular subsystems, each controlled by a computer, which carries out the following major tasks:

    Product Design. This task is carried out interactively using a computer that operates on stored product models. Optimization of the design to maximize profit is the main criterion.

    Production Planning. A computer-based system generates optimized production and process plans based on the output from the product design module. It takes into account scheduling conflicts, inventory position, resource utilization, line balancing, and other considerations pertaining to planning.

    Production. Parts are formed, machined, and assembled if required, at various workstations comprising FMCs or FMSs with minimum human intervention. Sufficient on-line data collection and real-time analysis take place to aid the unmanned production. A variety of sensors are used and adaptive control of the processes is practiced. Material handling is computer-controlled using AGVs, stacker cranes, robots, and other machines. They store, retrieve, find, transport, load, and unload parts and fixtures throughout the plant. Inspection of parts, subassemblies, and assemblies is computer-controlled and linked with the sensors downstream and with the design function upstream.

The scope of CIM now includes consumers as the most vital link within the overall integration. Consider the following scenario:

> John is excited that he will be passing his driving test on the very day he is legally entitled to drive. He was awake late last night choosing from the menu on his personal computer as to what type of car he would like to drive. All his selections in regard to color, sunroof, type of engine, stereo system, etc., have been saved in a file. After passing the test he will rush back home to press the enter key of his computer, which will place his order directly to the nearest manufacturing plant. He will have made his decision, however, based on price quotes and delivery dates from several manufacturers. The only thing left to do will be to go to the nearest dealer and pick up his personalized car on the promised day.

The integration efforts fueled by CIM are closing the gap between the drawing board and the factory floor. *U.S. News and World Report* magazine (July 20, 1992, p. 56) predicts the scenario illustrated in Figure 15.1. According to this scenario, you should be able to order a part for your car the way you order a cup of coffee and donuts at a restaurant.

According to a third futuristic scenario, CIM-based job shops will be operated like a laundromat. Vajpayee (1984) described the following scenario:

> Imagine that you are the owner of a factory—miles away from your home. You have watched late last night a film on the television in your bedroom. It is a miserable morning; you are enjoying your bed-tea and do not wish to get out of bed to go to the factory to start the morning shift. A moment later you remember, however, the need to produce goods that morning to honor a promised delivery.
>
> Supposing that the above relates to the year 2000 and that you have kept pace with the manufacturing practices of the day; this is what you are likely to do.
>
> Switch on the terminal—the same screen on which you watched the film last night—from the remote control pad under your pillow. Press a few keys on this pad and there on the screen is the morning view of your factory gates and the surroundings. See how faithfully your recently-hired nightguard—the new robot BAHADUR, is keeping watch on everything. He is as fresh as you left him last evening. You can also see that the materials for use in today's production have already been delivered at the gates against your order placed via the same terminal, which incidently you use also to credit the supplier's account. And you decide to proceed with the morning production shift without getting out of bed.
>
> Enter a few secret code numbers in the remote pad and BAHADUR gets started. It talks to one of its friends (equivalent to a forklift truck and its operator in today's terms), who starts fetching the materials in. In the meantime you instruct (in fact you speak in plain language as you would do to your manager in 1984) your central computer via the terminal to produce the required goods, pack them, and arrange to deliver them to the customer. And that's all. The rest of the day is yours to enjoy with the family.
>
> In 1994 the above does sound like a manufacturing fiction, especially if it relates to the batch-type production, but it could become a reality by the turn of the century.

Figure 15.1

Computerized spare parts: The required part for car repair is ordered for delivery within minutes.

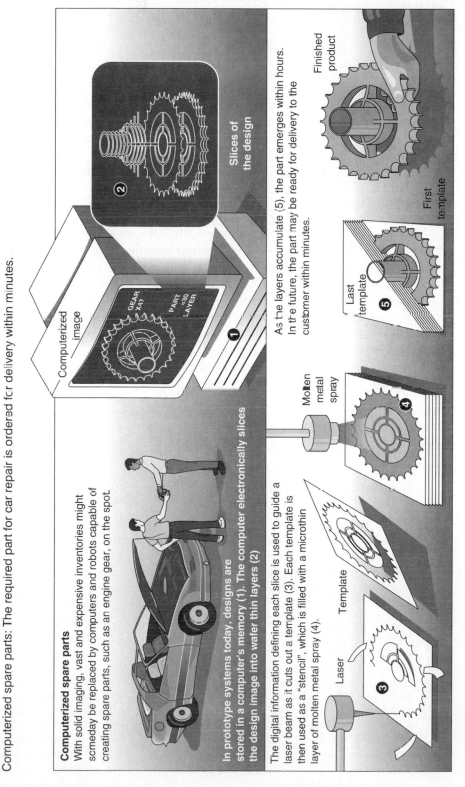

Computerized spare parts

With solid imaging, vast and expensive inventories might someday be replaced by computers and robots capable of creating spare parts, such as an engine gear, on the spot.

In prototype systems today, designs are stored in a computer's memory (1). The computer electronically slices the design image into wafer thin layers (2)

The digital information defining each slice is used to guide a laser beam as it cuts out a template (3). Each template is then used as a "stencil", which is filled with a microthin layer of molten metal spray (4).

As the layers accumulate (5), the part emerges within hours. In the future, the part may be ready for delivery to the customer within minutes.

Computerized image

Slices of the design

Laser   Template

Molten metal spray

Last template

First template

Finished product

Source: From *U.S. News and World Report*, July 20, 1992, p 56. Copyright, July 20, 1992, U.S News & World Report.

15.3 CONCLUDING REMARKS

Prior to industrialization, the wealth of nations was measured only in terms of the farm output. Today, it includes the output of the manufacturing and service sector as well. The Industrial Revolution propelled, some two centuries ago, creation of today's world of manufacturing. Industrialized nations under their free market economies today produce an enormous amount of manufactured goods, so much so that environmental pollution has become a problem.

The current trend is toward a service-industry-based economy and society, and CIM seems to be accelerating that trend. Manufacturing has begun to follow the footsteps of farming. Today, less than 3% of the U.S. population produces all the food we need, along with enough surplus to donate or sell to other countries. It is reasonable to expect that manufacturing will in the near future (probably by the year 2025) be able to produce, with fewer people, all the manufactured goods we will need, with surplus for other countries. CIM is already making it happen.

How will such a change in manufacturing affect people? Imagine as few plants and production workers in the year 2025 as there are farms and farm workers today. This gigantic migration of the workforce from manufacturing to the service sector will be slow, but steady. How will the service- and CIM-based economy affect people's lives and lifestyles? These are the questions we leave for social scientists to ponder.

TRENDS

❑ The pace of CIM development continues unabated. All the right "ingredients" are available; the only thing holding it back is time.

❑ The move toward downsized but efficient manufacturing continues.

❑ The loss of manufacturing jobs in developed countries will continue. The manufacturing industry in such nations has two options: to move the operations overseas or to implement CIM; and both mean fewer manufacturing jobs in these countries.

❑ As long as computers and communications continue to improve—a definite trend—CIM will continue to influence manufacturing. The influence of computers and communications on manufacturing is nothing special; they affect our lifestyles, too. As *U.S. News and World Report* (November 16, 1992, p. 92) wrote:

Technology proposal is to move the nation's infrastructure into the 21st century with advanced electronic communications highways—an idea that Japan has already poured billions into. The goal is to enhance productivity by moving vast quantities of digital data seamlessly over high-capacity networks as sound, graphics and video between universities, corporations, industrial research centers, health-care facilities—and ultimately everyone's home. With such a capability, corporations could send computer-generated drawings across the nation in an instant to cooperate in the design and manufacture of new products. Remote communities could send X-rays over the wire for expert diagnosis at regional health-care centers. Telecommuters could work at home by tapping into a corporations's computer systems. High-speed data highways "will have as much impact on how and where people live as the interstate highways of the 1950s."

❑ CIM began in the 1970s as a concept. Its evolution since then into a technology has been fueled by the unbelievable developments in computers and communications. Meanwhile, its scope has expanded to include the management tasks as well. Today, in the 1990s, CIM is "doable," as illustrated by the various boxes in this chapter and elsewhere; it is now a conventional practice.

SUMMARY

This concluding chapter reviewed the current status of CIM development. The boxes illustrated how a few existing manufacturing facilities approximate the CIM concept. A lack of comprehensive CIM software remains the major obstacle to full-blown CIM; the required hardware and networking technologies are already available.

Three futuristic scenarios of buying a new car, ordering a replacement part, and operating a laundromat-style job shop illustrated where CIM is leading manufacturing. Manufacturing in the future is predicted to resemble contemporary farming—fewer people employed in fewer factories producing all the manufactured goods we will need. The major constraint is likely to be the capacity of the environment to provide refuge to all the used, unused, misused, and abused goods.

KEY TERMS

Factory of the future

Pseudo-CIM

Taylorism

EXERCISES

Note: Exercises marked * are projects.

15.1   What difficulties do companies interested in implementing CIM face?

15.2   Do you agree with the prediction that manufacturing in the future will be like today's agribusiness? Support your position with some relevant data and/or logical arguments.

15.3* Write a three-page essay on an actual CIM facility. (Contact an appropriate company for relevant information.)

15.4* Visit a local manufacturing company to find what it is doing to implement CIM in its operations. Following the visit, prepare a five-year plan for the company to achieve CIM. Discuss this plan with appropriate company personnel.

15.5   Circle T for true or F for false.

    a.   Taylorism is compatible with CIM.   T/F

    b.   The major hurdle to CIM implementation is software.   T/F

c. In companies operating in CIM environments, plant computers
are more critical than business computers.   T/F

d. Manufacturing jobs will decrease as CIM advances.   T/F

e. CIM is an appropriate technology only for industrialized nations.   T/F

SUGGESTED READINGS

Books

Black, J. (1991). *The design of the factory with a future*. New York: McGraw-Hill.

Dauch, R. (1993). *A passion for manufacturing*. Dearborn, MI: SME.

Gunn, T. (1992). *21st century manufacturing: Creating winning business performance*. Dearborn, MI: SME.

Hirano, H. (1989). *JIT factory revolution: A pictorial guide to factory design of the future*. Dearborn, MI: SME.

Journals and Periodicals

Journal of Applied Manufacturing Systems. St. Paul, MN: St. Thomas Technology Press.

Articles

Beard, T. (1989, May). An off-the-shelf approach to CIM. *Modern Machine Shop*, pp. 54–64.

Salas, R. (1989, August). The future impact of CIM. *Production*. 101(8), 78–79.

Sheridan, J. H. (1991, April 20). The CIM evolution. *Industry Week*. 241(8), 29–51.

Vajpayee, S. (1984, February). Factory of the future. *Proceedings of Conference on Machine Tools Design*. Department of Mechanical Engineering, Institute of Technology, Banaras Hindu University, Varanasi, India. pp. 30–40.

Vajpayee, S., & Reiden, C. E. (1991, fall). Computer-integrated manufacturing and its ramifications. *Journal of Industrial Technology*. 7(4), 31–35.

APPENDIX A
Other Resources

Besides the various books, monographs, reports, journals, periodicals, magazines, and articles cited in the text, and the selected bibliography that follows Appendix B, several other resources can help in learning about CIM. The primary resources fall into six categories.

1. PROFESSIONAL SOCIETIES

Several professional societies actively disseminate information about CIM. They publish and distribute books, videotapes, journals, and magazines exclusively on CIM or in areas closely related to CIM. These organizations also sponsor conferences, symposia, meetings, and workshops for exchange of ideas and CIM case studies.

The major professional societies active in CIM include:

Society of Manufacturing Engineers (SME) and Computer and Automated Systems Association of SME (CASA/SME)
> One SME Drive, P. O. Box 930, Dearborn, MI 48121, USA
> Telephone: (313) 271-1500
> SME is fully devoted to the cause of manufacturing. The society was chartered in 1932 as the American Society of Tool Engineers with headquarters in Detroit. With current membership of over 80,000 drawn from 70 countries, SME is internationally recognized as the premier organization for manufacturing professionals.

International Institution for Production Engineering Research (CIRP)
> 10 Rue Mansart, 75009 Paris, France
> CIRP is an international organization of manufacturing "pundits." It is in fact a coveted club with very selective membership.

The Association for Manufacturing Technology (AMT)
> Founded in 1902 as the National Association of Machine Builders, it changed its name in 1988 to NMTBA-The Association for Manufacturing Technology. Effective January 1992, it became known simply as AMT. Based in McLean, VA, it publishes the *Economic Handbook of the Machine Tool Industry*, which includes domestic and international data.

American Production & Inventory Control Society (APICS)
> Founded in 1957, APICS disseminates knowledge in production planning and control and conducts certification examinations in this field.

As associations of manufacturing professionals, SME and CIRP devote their resources exclusively to manufacturing engineering and technology. While SME targets its activities at practicing professionals, CIRP concerns itself with dissemination of cutting-edge research findings and long-term forecasts on manufacturing trends.

In addition to these mainstream societies, other professional organizations are active in certain aspects of manufacturing. By its very nature, CIM permeates them all. Some major organizations marginally related to CIM include:

American Society of Mechanical Engineers (ASME)
    345 E. 47th Street, New York, NY 10017
    Telephone: (212) 705-7785

Institution of Industrial Engineers (IIE)
    25 Technology Park, Norcross, GA 30092
    Telephone: (404) 449-0460

Association for Integrated Manufacturing Technology, formerly
    Numerical Control Society (AIM Tech)
    5411 E. State Street, Rockford, IL 61108

American Society for Quality Control (ASQC)
    310 West Wisconsin Avenue, Milwaukee, WI 53203

American Society for Engineering Education (ASEE)
    11 Dupont Circle, Suite 200, Washington, D.C. 20036

National Association of Industrial Technology (NAIT)
    3157 Packard Road, Suite A,
    Ann Arbor, MI 48108-1900
    Telephone: (317) 677-0720

2. TRADE EXHIBITIONS/SHOWS

Industrialized nations each year organize numerous exhibitions and trade shows involving professional societies, promoters, vendors, and users of CIM equipment and technologies. Visits to these exhibitions and shows are highly educational, especially to keep up to date on new developments. Some of the major trade shows are:

AUTOFACT

The term AUTOFACT is a shortened form of AUTOmatic FACTory. Beginning in 1977 and regularly sponsored by the SME, AUTOFACT is an annual exhibition and conference that takes place in the fall in Detroit or Chicago, hubs of mechanical manufacturing in the U.S. It is the largest showcase of CIM hardware, software, and manufacturing services in the western hemisphere. An increasing number of CIM products and related technologies have been exhibited at AUTOFACT.

IMTS

The International Manufacturing Technology Show (IMTS, formerly known as the International Machine Tool Show) is the premier event of the industrial world and the largest manufacturing show. The 1992 IMTS held in Chicago attracted 1,500 exhibitors from 30 nations and covered 1 million square feet of manufacturing machinery up and running. The show encompasses new products, processes, systems, and software relating to all areas of manufacturing, such as CAD/CAM, robots, CIM, machining centers, FMS, and lasers.

WESTEC

The Western Metal and Tool Exposition and Conference (WESTEC) is held at the Los Angles convention center. More than 500 companies display their latest products relating to CIM and other areas of manufacturing. Sponsored by SME, American Society for Metals International (ASM), and American Machine Tool Distributors' Association (AMTDA), WESTEC has been visited by 1 million people since its inception in 1964. An additional sponsor for WESTEC '90 was The Association for Manufacturing Technology (AMT).

EASTEC

Except the venue, which is the Eastern States Exposition Center in Springfield, MA, EASTEC is in many respects similar to WESTEC. First staged by the SME in 1979 as the Hartford Tool & Manufacturing Engineering Exposition, it has been held yearly since 1981.

EMO

The European Machinery Organization (EMO) Machine Tool Exposition is held in Hannover, Germany, every four years. EMO is the largest manufacturing equipment show in the world. A recent show attracted 2,100 exhibitors from 36 countries and 235,000 visitors.

Japan Machine Tool Show

With Japan's reputation for quality consumer products and its significant share of the international market for machine tools, the Japan Machine Tool Show attracts visitors from around the world. The show is a window on Japanese CIM technologies.

JIMTOF

Held annually, the Japan International Machine Tool Fair (JIMTOF) is one of the three largest international machine tool shows, the other two being the EMO and IMTS.

3. VIDEOS

A number of videos about CIM are available. Most are targeted at management and contain successful case studies. Some deal with technical aspects as well. Videotapes on CIM include the following:

Computer Integrated Manufacturing. 1989. $170.
> The University of Iowa, Audiovisual Center/Department J
> C215 Seashore Hall, Iowa City, IA 52242
> Telephone: (319) 335-2539

Implementing CIM: A Strategic Challenge. 1987. $95.
> SME, One SME Drive, P.O. Box 930
> Dearborn, MI 48121
> Telephone: (313) 271-1500

CIM: Focus on Small and Medium-Sized Companies. 1986. $200. SME.

Manufacturing Insights. SME produces videotapes on current issues under the title *Manufacturing Insights.* The tapes highlight state-of-the-art manufacturing systems and processes and are popular with manufacturing professionals. The format includes an introduction, interviews with industry experts, case studies, and a summary. Cost varies from $125 to $200, and there are more than 50 tapes to choose from.

4. VISITS TO MODERN MANUFACTURING FACILITIES

Several facilities in the U.S. and other industrialized nations, especially in Japan and Western Europe, have operations that approximate a CIM environment. A visit to such facilities is worth the effort. Boxes 15.1 through 15.4 in Chapter 15 described four such operations. Some of the U.S. facilities have won Leadership & Excellence in the Application & Development of CIM (LEAD) awards.

LEAD Awards

Through its LEAD award, CASA/SME recognizes manufacturing companies for their efforts in implementing CIM. Since 1981, a LEAD has been awarded each year to one company for excellence in CIM implementation, and since 1985 to one college or university for excellence in CIM education and research.

The following companies have been LEAD award recipients:

1993	The Foxboro Co., Intelligent Automation Division, Foxboro, MA
1992	Tektronix's Circuit Board Division, Forest Grove, OR
1991	Dana Corp., Spicer Universal Joint Division, Lima, OH
1990	Corning Asahi Video Products Co., State College, PA
1989	Digital Equipment Corporation, Colorado Springs, CO
1988	Allen Bradley, a Rockwell International company, Milwaukee
1987	Texas Instruments, Defense Systems & Electronics Group, Sherman, TX
1986	Martin Marietta Energy Systems, Inc., Y-12 Plant, Oak Ridge, TN
1985	Cone Drive Division of Textron (formerly Cone Drive Operations of Ex-Cell-O Corporation), Traverse City, MI
1984	General Electric Co., Steam Turbine-Generator Operations, Schenectady, NY
1983	AT&T Technologies, Inc. (formerly Western Electric), Printed Wiring Board Facility, Richmond, VA
1982	Ingersoll Milling Machine Company, Rockford, IL
1981	Deere & Co. (formerly John Deere), Waterloo Tractor Works, Waterloo, IA

The following colleges and universities have been recognized with LEAD awards for their contribution to CIM:

1993	University of Illinois at Urbana-Champaign, IL
1992	Nanyang Technological University (NTU), Singapore
1991	Cranfield Institute of Technology, Bedford, England
1990	Arizona State University, Tempe
1989	North Carolina State University, Raleigh
1988	University of Wisconsin, Madison
1987	Rensselaer Polytechnic Institute, Troy, NY
1986	Georgia Institute of Technology, Atlanta
1985	Lehigh University, Bethlehem, PA

Information on LEAD awards and participation can be obtained from:

CASA/SME, LEAD Award Program
    One SME Drive, P.O. Box 930
    Dearborn, MI 48121
    Telephone: (313) 271-1500

5. GOVERNMENT AND PRIVATE AGENCIES

Several government agencies and not-for-profit organizations promote CIM. Federal agencies such as the National Bureau of Standards (NBS) and various laboratories of the U.S. Department of Energy have been active in areas relating to development and transfer of CIM technologies. Some major organizations are the following.

NIST (National Institute of Standards and Technology)

Formerly called the National Bureau of Standards (NBS), NIST has a legislative mandate to assist in transferring advanced manufacturing technologies to small and medium-sized businesses. Toward this objective, it has developed an automated manufacturing program. As a part of this program, NIST has built a sophisticated system called Advanced Manufacturing Research Facility (AMRF) at its site in Gaithersburg, Maryland. AMRF is a prototype CIM system designed to support research on standards for the factory of the future. The areas of current research are:

1. software and hardware interfaces
2. information representation, management, and communication
3. process control and measurements
4. factory reference models

One of the main aims of AMRF is to encourage the growth of CIM.

The Association for Manufacturing Technology (AMT)

The AMT is a trade organization. It offers membership to companies that manufacture machinery and related equipment as well as products and software used in the production of discrete durable goods. It was founded in 1902 as the National Machine Tool Builders' Association (NMBTA) when metalworking machine tools were virtually the only machines used in manufacturing. The organization realized in the 1980s that modern manufacturing is more than machine tools—it deals with issues of design, material handling, assembly, quality assurance, factory communications, and more. Moreover, modern materials such as ceramics and composites require specialized processing. To acknowledge these changes, NMTBA broadened its scope to serve manufacturing systems industry and changed its name in 1988 to The Association of Manufacturing Technology, but retained the initials NMTBA for continuity. In January 1992, it changed to AMT.

National Center for Manufacturing Sciences (NCMS)

NCMS is a nonprofit organization formed in 1986 to help member companies develop new manufacturing technologies. Headquartered in Ann Arbor, MI, it is a consortium of some 100 U.S. manufacturing companies. Supported by those members and the Pentagon, NCMS's mission is to find ways to help U.S. manufacturing regain the lead. Its address is:

National Center for Manufacturing Sciences
900 Victors Way
Ann Arbor, MI 48108
Telephone: (313) 995-0300

NCMS is divided into the following six areas, each under the direction of a Technical Program Committee.

1. manufacturing processes and materials
2. production equipment design, analysis, and testing and control
3. manufacturing operations
4. manufacturing data and factory control
5. information and technology transfer
6. strategic issues

Computer-Aided Manufacturing International, Inc. (CAM-I)

A nonprofit industrial research organization, CAM-I is engaged in advanced manufacturing research and development. For example, in 1976 it contracted McDonnell-Douglas Automation Company to develop for its sponsoring members a variant-type process-planning software called CAM-I CAPP. More than 400 systems of this software have been sold through CAM-I or CASA/SME.

MANTECH Programs

The U.S. Air Force Manufacturing Technology (MANTECH) program was set up in 1947 to improve productivity, enhance quality, and reduce the life cycle cost of weapon systems. The program funds application-oriented projects that generate and disseminate technical information and knowledge. Recently, a directorate was set up at Wright-Patterson Air Force Base to transfer the military technology for civilian use. Manufacturing companies interested in finding useful CIM technologies under the MANTECH program should contact:

MANTECH Technology Transfer Center
WL/MTX
Wright-Patterson AFB, OH 45433-6533
Telephone: (513) 256-0194

6. BULLETIN BOARDS

Although information on CIM is available in books, journals, conference proceedings, databases, and commercial software, accessing not-for-profit software through bulletin board systems (BBS) is another alternative. Computerized bulletin boards started in the late 1970s when users could connect their eight-bit microcomputers via modems to swap programs and question each other. The basic bulletin board consists of a modem, a microcomputer, phone line, and software. The BBS software guides the caller with a series of menus to its various functions. Thousands of BBSs exist in the world. The major benefit of BBS is that you can use someone else's software, within legal restrictions, and allow others to use yours. Several bulletin boards are available for accessing CIM-related information, including software. The four major ones are the following:

SME-ON-LINE. SME launched its first electronic-bulletin-board information service, SME-ON-LINE, in 1993. Aimed at manufacturing professionals, the service is available 24 hours a day, seven days a week. SME-ON-LINE is divided into three areas: bulletins, messages, and files. Bulletins contain updated information on news, books, videos, coming events, and so forth. Messages connect users with experts in specialty areas. The files contain thousands of application programs available for downloading. SME-ON-LINE can be reached at (313) 271-3424.

Information on Technology in Manufacturing Engineering (INTIME). This is a computerized record of more than 13,000 articles, books, and technical papers published by SME. The INTIME databank includes the title and abstract, which may be obtained as computer printouts. The INTIME data is also available on magnetic tapes for loading into the user's mainframe or minicomputer. Call (313) 271-1500 and ask for INTIME.

Computers in Mechanical Engineering—Information and Software Exchange (CIME-ISE). CIME-ISE is a free bulletin board service offered by the ASME publications directorate. It began operating in 1987 and has since compiled more than 1,000 programs. Files have been categorized by functions. Several software tools are available as shareware. Open 24 hours a day, it can be reached at (608) 233-3378. All necessary help and information are available on line.

Computer Software Management and Information Center(COSMIC). Operated by the U.S. National Aeronautics and Space Administration (NASA), COSMIC distributes software developed with NASA funding to industry, academia, and other government agencies. It is a part of NASA's Technology Utilization Network. Though most of the software is aimed at basic research, some may be applicable to advanced manufacturing. Contact John A. Gibson, The University of Georgia, at (404) 542-3265.

APPENDIX B
A Laboratory Course

As with any engineering or engineering technology course, a CIM course[1] can be more effective if supplemented with hands-on practice. However, a CIM lab course is expensive to set up and takes a significant amount of time and effort to plan and implement. An important question is: What type of experiments should a CIM course offer? An answer to this question is provided in this appendix. The experiments suggested here have been designed to encompass all facets of CIM. Depending on the strength of existing resources such as faculty specializations and interests, and the type of manufacturing laboratories already available, the suggested experiments can be modified, or new ones added, to develop the lab course appropriate to a specific curriculum.

The lab is suitable as a one-semester course for students majoring in manufacturing, mechanical, or industrial engineering, or in engineering technology. The special features of the course are: (a) the experiments simulate a real-world situation, (b) they highlight interactions of a CIM environment, and (c) miniprojects allowing further study of the experiments are required.

1. COURSE STRUCTURE

The experiments revolve around a hypothetical manufacturing company called Ultra Modern Manufacturing Incorporated (UMMI). Figure B.1 describes UMMI in terms of product variety, volume, manufacturing operations, financial turnover, workforce, and so on. The company profile has been designed to represent typical discrete manufacturing with international operations. It is assumed that UMMI's philosophy is to keep up to date with technology and people practices.

2. CONTENTS

The experiments have been organized so the course can be offered over a semester with three or four contact hours per week. Table B.1 lists the experiments and their purpose. A detailed description of the experiments, their objectives, and mode of interaction with each other are presented in the next section. The experiments have been designed to interact with as many

---

[1]Vajpayee, S. K., & Money, E. (1989). Laboratory-based teaching of computer-integrated manufacturing. In J. M. Biedenbach (Ed.), *1989 ASEE Southeastern Section Proceedings*. Washington, DC: American Society for Engineering Education.

Figure B.1
Profile of Ultra Modern
Manufacturing Inc.

UMMI is interested in evolving into a CIM operation. The company aims
    to practice and implement the latest in manufacturing technologies,
    such as CAD/CAM, robotics, FMC and FMS, and automated process
    control.

Name : Ultra Modern Manufacturing Inc.

Goal : Maintain and enlarge market through quality

Strategy : Move toward the next generation of automation with
    CIM in mind

Headquarters: Hattiesburg, MS

Products : Gas-powered ride-on and walk-behind (both manual
    and gas/electric-powered) lawn mowers

Turnover : $1.35 billion annually

Plants : Hattiesburg (HB), MS
    Dearborn (DB), MI
    Derby (DY), England
    Patna (PT), India
    The Hattiesburg plant makes engines, motors, and gear-
    boxes for all the operations. Each plant makes or buys
    components locally and assembles them in lawn mowers
    for their respective markets.

Markets : The two U.S. plants serve the American market, the
    Derby plant caters to the European market, and the
    Patna plant to the Asian and Australasian market.

other functions as occur in the real world. This interaction is illustrated in Figure B.2, where small circles represent the experiments described in Table B.1.

With their decreasing cost and increasing power, microcomputers play a major role in CIM. That is why most experiments aim at hands-on experience around microcomputers. A full-fledged lab for the experiments suggested in this course will require several resources (see Table B.2).

3. EXPERIMENTS: OBJECTIVES AND INTERACTIONS

0. Course

In the first meeting, the course is introduced and its objective explained together with CIM's role in modern manufacturing. The structure of the course and an outline of the experiments are presented. The interrelationships among the various experiments are highlighted with the help of Figure B.2. The class is divided into groups, if required, and a schedule is finalized. Students are shown the experiment sites and equipment, and safety procedures are explained.

1. Business

Business is an integral part of manufacturing. UMMI's marketing department has established that there is a need for high-quality lawn mowers. UMMI would like to determine the selling price the market could bear, and compare the cost and risks of market failure with the profit and probability of success. Students are expected to determine whether a test market should be conducted using a Bayesian or similar approach. To calculate the profit, the gearbox of the lawn mower is considered as a candidate. The various cost elements of R&D, manufacturing, inventory, personnel, and so on are considered in determining its cost.

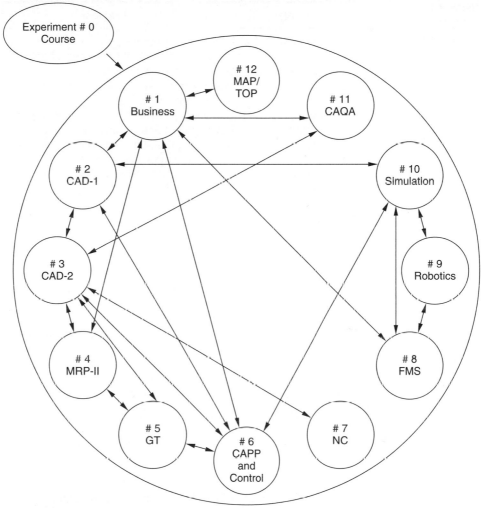

Figure B.2
Interaction among the experiments in the suggested CIM lab course

This experiment interacts with experiments 2, 4, 6, 8, 11, and 12. Experiment 4 produces the inventory carrying cost, and the manufacturing cost is determined from experiments 6 and 11.

2. *CAD-1*

The concept of design for manufacturing is explained in this experiment. To bring out quality products, UMMI needs to modify the design of the existing gearbox. Stress analysis of a gear and its shaft is carried out using a finite element package. The end result of the experiment is a design specification of these components.

This experiment interacts with experiments 3, 6, and 10. The design data is used in drafting (experiment 3) and is also the basis for the process plan in experiment 6. This interaction ensures that the process plan (experiment 6) does not create any difficulty at the production stage. A dynamic simulation of the design is carried out in experiment 10 as a supplementary exercise.

Table B.1
List of experiments

#	Title	Purpose
0	Course	Introduction to the experiments
1	Business	Management and other business functions
2	CAD-1	Design and analysis of a part
3	CAD-2	Drafting and drawing using AutoCAD
4	MRP-II	Computerized manufacturing resources planning
5	GT	Classification systems and coding of parts
6	CAPP	Development of a process plan
7	NC	NC code generation using CAM software
8	FMC	Design and operation of a FMC
9	Robotics	Feasibility and justification for a robot
10	Simulation	Simulation of a process plan
11	TQM	Design and implementation of a TQM program
12	MAP	MAP's history and current status

3. CAD-2

This experiment involves drafting the parts designed in experiment 2. The objective is to draw the gear reduction box of the lawn mower using a commercial drafting package such as AutoCAD. Production drawings of the gear and its shaft are made along with the assembly drawing of the gearbox. The drawings must contain all the information necessary for production.

This experiment interacts with experiments 4, 5, 6, 7, and 11. The drawing of the gear-shaft is stored in a database created by AutoCAD, and is used in experiment 7 to generate the NC code. The bill of materials for the gearbox is prepared in experiment 4. There are several gearboxes in the design database and the individual components of all these are examined later in experiment 5 to develop a GT framework concept. Quality characteristic information from the drawings is used in experiment 11 for inspection purposes.

4. MRP II

In addition to planning the material requirements, MRP II involves the planning of other resources such as labor hours, machine hours, floor space, and capital. The importance of MRP II in a CIM environment is emphasized in this experiment by planning for such resources as gearbox production. A software package such as STORM is used to carry out MRP II tasks.

This experiment interacts with experiments 1, 3, and 5. The inventory carrying cost is used to determine the unit cost in experiment 1. From the assembly drawing (experiment 3) of the gearbox, the bill of materials is prepared. Given a set of gearboxes with various output requirements, and the bills of materials, the GT concept is applied in experiment 5 to implement cellular manufacturing of similar products.

5. GT

UMMI makes gearboxes of varying output characteristics, resulting in components of different sizes. The company needs to classify them and other similar components to manufacture them in a cell and thus realize the benefits of GT. In this experiment, a manufacturing cell is designed (layout) for the gear-shaft family.

This experiment interacts with experiments 3, 4, and 6. Components similar to those in the bill of materials (experiment 4) are identified by classification codes. Along with those designed in experiment 3, these parts are examined for GT. The experiment highlights the fact that not all the parts a company produces may fit into the GT concept; production of some parts is outsourced wherever cost-effective.

6. CAPP

Before the newly designed gear-shaft can be manufactured, its process plan must be developed. This is done using the variant system of process planning, since a similar design already exists.

This experiment interacts with experiments 1, 2, 3, 5, and 10. The experiment is used to calculate the manufacturing cost (for experiment 1). From the design drawings of experiment 3 and the physical dimensions of the part, a process plan is developed. If there is any conflict with the processes, design specifications are reexamined and modified to match the process capabilities of the available resources. The process plan is later simulated in experiment 10 for optimization.

7. NC

Though the gear-shaft can be machined by an individual operator, the (completed) process plan and the corresponding resources plan streamline its manufacture along with the other components. This experiment consists of generating the NC code using a software package such as SmartCAM, and downloading it to a machining center.

This experiment interacts with experiment 3. The design drawing of experiment 3 for the gear-shaft is used to generate the NC codes.

8. FMC

UMMI is expecting an increase in demand for their lawn mowers over the next five years. Moreover, to offer product variety the company is planning to develop a manufacturing cell. In the future, other cells may be developed and integrated into an FMS. The objective of the experiment is justification analysis for a FMC, given the technical and economic data.

This experiment interacts directly with experiments 9 and 10, and tangentially with the business function (experiment 1).

9. Robotics

UMMI is considering a pick-and-place robot to replace the manual operation of unloading the gear-shaft from the machining center to the conveyor. Given the product shape and its physical characteristics, students develop a technical specification for the robot along with its cost justification.

This experiment interacts with experiments 8 and 10. In experiment 8, students enhance the FMC by incorporating a robot. In experiment 10, the robot and its working are simulated.

10. Simulation

Being a powerful tool for testing new product models or changes in existing ones, simulation is used to investigate possible improvements in production facilities. For UMMI, a process line is computer-modeled using a software with animation and "what if" capabilities.

This experiment interacts with experiments 2, 6, 8, and 9. It is carried out for the gear-shaft designed in experiment 2 and the process plan prepared in experiment 6. The designed component is simulated dynamically. A simulation of the FMC station (experiment 8) incorporating the robot (experiment 9) may also be carried out here.

11. TQM

To emphasize the company goal of maintaining its market through quality, this experiment involves development of a TQM program as well as use of SPC for the gear-shaft production line. The experiment is in two parts. The first one involves developing a quality manual for

UMMI; in the second part, students learn to develop control charts for gear-shaft production using a software package such as SQCpack.

This experiment interacts with experiments 1 and 3. The drawing of the gear-shaft in experiment 3 provides the values.

12. MAP

This experiment involves a library search on MAP's history and its current status. The preparation of an "executive brief" for the top management is also required. The actions UMMI should take now to ensure that future automation will integrate easily in a MAP/TOP environment are pointed out. For example, should the company set up a standing task force for the purpose?

Experimental Procedure

The course consists of 13 experiments conducted over a 15–16 week semester. A handout of the scheduled experiments should be distributed to the students in the first meeting. Each experiment is described under the following headings.

Title	:	
Objective	:	
Assignment	:	What is to be done in the experiment?
Procedure	:	A brief outline of the procedure to be followed.
Results	:	Tabulate any results.
Discussions	:	Discuss the results.
Report	:	To be submitted before the next meeting. Besides describing what has been learned, the report should also contain a critical discussion on how the experiment interacts with other experiments within the CIM concept.
Comments	:	Students should be encouraged to make observations or comments that might be helpful in enhancing the experiment in future.

4. MINIPROJECT

In addition to the basic experiments, each student carries out an individual project. The project is based on one experiment of their choice, subject to the approval of the instructor, and requires an in-depth study. The project's goals are well-thought-out and suggestions researched on how the experiment could be improved, and the associated cost. A written report and oral presentation to the class are required on each project.

5. EVALUATION

Each experiment and the corresponding report is graded for 50 points, making a total of 650 points for the experiments. The miniproject carries 350 points. The points are converted to letter grades according to: $A \geq 900 > B \geq 800 > C \geq 700 > D \geq 600 > F$.

6. FEATURES

The need to supplement classroom learning with a laboratory course is desirable for effective teaching of CIM. The experiments of Table B.1 are typical topics under CIM, which could be modified depending on the facilities available and as the course matures.

Table B.2

Facilities required

The basic facilities and costs required for the experiments are listed here. Depending on the availability of funds, course enrollment, and what already exists, the inventory of items may differ from what is suggested below.

Expt. #	Title	Facilities	Cost ($)
0	Course	none	none
1	Business	Software such as Lotus 123 or dBASE III	675
2	CAD-1	An FEA package	500
3	CAD-2	Software such as AutoCAD	2,500
4	MRP-II	Software such as STORM	2,000
5	GT	A GT package	1,000
6	CAPP	A CAPP package	2,000
7	NC	Software such as Smart CAM	2,500
8	FMC	A flexible processing system such as the Fischer-Technik line	5,000
9	Robotics	"Minimover" educational robot	1,500
10	Simulation	Software such as SIMFACTORY	2,500
11	TQM	Software such as SQCpack	500
12	MAP	A copy of MAP 3.1	700

With enrollment of 20 students, two students together conduct an experiment in each meeting. The class meets twice a week, requiring hardware resources for 10 students. With these assumptions, the following hardware will suffice.

Five 486 machines with 120-Mb hard disk, math coprocessor, and enhanced graphics, one plotter, and one printer.

The course has been designed to benefit both engineering and engineering technology curricula. Some of its important features are:

1. CIM is taught in a laboratory environment where students learn about its various constituents and their interactions by conducting experiments.

2. An attempt has been made to approximate the real-world situation by directing the experiments around a discrete-manufacturing company interested in implementing CIM.

3. The course provides hands-on practice on PCs and relevant software.

4. The miniproject encourages library research.

5. Despite the built-in interdependence, each experiment has been designed to be self-contained.

6. The lab course can be combined with a lecture course on CIM.

7. A closed-loop approach, with feedback from students, has been suggested.

Glossary

This glossary defines or explains the key terms listed at chapter-ends and boldfaced in the text. It also contains some additional terms associated with CIM.

ABET (Accreditation Board for Engineering and Technology) A government-recognized agency in the United States that accredits educational programs leading to associate, baccalaureate, and master degrees in engineering or engineering technology.

Access path The way data is related, both logically and physically, to the data structure.

Acoustic coupler An electronic device that facilitates transmission of digital data through a standard telephone headset.

Adaptive control A technique by which control parameters are automatically adjusted in response to measured process data to achieve near-optimum performance.

Address A group of numbers that uniquely identifies a station in a local area network or a location in computer memory; similar to the address of a house on a street.

ADC (analog-to-digital converter) It converts analog signals in digital data for processing by computers or other digital devices.

AGV (automated guided vehicle) AGVs are used on plant floors to move materials, tools, and other items. They are self-controlled or under the control of a computer and follow a specified path.

AI (artificial intelligence) Programmed intelligence that can be stored in computers to imitate the capabilities of human beings, such as self-learning, adaptation, and reasoning.

AIAG (Automotive Industry Action Group) A group of automotive manufacturers that develop and implement appropriate technologies in their operations.

Alphanumeric characters A group of characters comprising the letters of the English alphabet, decimal digits, common symbols such as %, $, @, and some control characters.

ALU (arithmetic logic unit) The portion of the hardware that performs arithmetic and logical operations.

AML (a manufacturing language) A language developed by IBM for programming robots.

ANSI (American National Standards Institute) A U.S. association charged by industry and the federal government with the task of generating and disseminating standards. As the U.S. standards organization, it acts as a liaison with the ISO.

APT (automatically programmed tools) One of the main languages for programming numerically controlled machine tools and other NC/CNC equipment.

Application layer   The highest level in the ISO's layered OSI model of data communications protocol. This layer is closest to the user.

Application program or software   A computer program that carries out a task specific to the users' needs.

ARPA   (Advanced Research Projects Agency) An agency of the U.S. defense department, currently referred to as DARPA (Defense Advanced Research Projects Agency).

ARPANET   Sponsored by ARPA, it is a network of computers located at U.S. universities (some non-U.S. as well) engaged in defense research. In use since the 1960s, it has been the platform for many advances in data communications.

ASCII   (American Standard Code for Information Interchange) An industry-standard character code widely used for information interchange among data processing, communication, and associated manufacturing systems. It is a seven-bit code system prevalent in microcomputer hardware.

ASME   (American Society of Mechanical Engineers) The ASME is partially engaged in disseminating manufacturing knowledge. It sponsors Manufacturing International, a conference held every two years, and publishes CIM-related articles in its transactions and *Mechanical Engineering* magazine.

AS/RS   (automated storage and retrieval system) A high-density rack storage system for rapid storage and retrieval of parts and tools; may be computer-controlled.

Assembler   A program that translates the mnemonics entered on an ASCII keyboard into binary codes of the microprocessor.

Asynchronous transmission   A data transfer technique in which the time interval between the stop bit of the last character transmitted and the start bit of the next character is variable. (See also synchronous transmission.)

ATC   (automatic tool changer) A mechanism, usually under the command of part programs, that automatically changes the tool on NC machines.

Attribute   A characteristic of an entity or object that identifies or describes it. For example, a part's attributes may be part number, material, weight, and so on.

AUTOFACT   (Automatic Factory) An annual exhibition of, and conference on, advances in manufacturing.

Backward chaining   A process of finding the causes by beginning with the effect (final state or condition) and tracing the path backwards.

Ball bar test   A test for assessing the quality of CMMs, in which a bar having balls at its two ends is used.

Bandwidth   A measure of a transmission line's capability, usually expressed in Hz. It is the difference between the highest and lowest frequencies at which transmitted signals do not lose more than a specified amount (usually 50%) of power. For example, if measurements over the line show that all signals under 500 Hz and over 3.5 kHz lose half the power, but any signal within this range retains at least 50% power, then the bandwidth of the line is 3000 (3500 – 500) Hz.

Baseband   Transmission system in which only a single set of signals is present on the transmission medium at a time (contrast with broadband).

BASIC   (Beginner's All-purpose Symbolic Instruction Code) A popular high-level computer language for general use.

Batch production   The type of manufacture in which products are made in batches. Batch production of discrete parts and their assembly into end products benefits the most from CIM.

Baud rate   A measure of data transmission speed. Sometimes baud is used instead of baud rate. In most cases, baud is a rate of one bit per second.

Baudot code   A five-bit code used in the 1950s on paper tape and punched cards.

BCD (Binary Coded Decimal) A coding system for representing decimal numbers in which each digit is designated by a four-bit binary code. For example, decimal number 25 is 00100101 in BCD (0010 represents 2 and 0101 represents 5). The most common BCD code is the 8-4-2-1.

Bezier curve A method of approximating the workpiece geometry in CAD, or tool paths in CAM, through special curves.

Binary A condition or system that relates to two possible values or states.

Binary system A numbering system based on the digits 0 and 1. These two digits are called bits (binary digits). Computers work and communicate based on this system.

Bits The two binary digits 0 and 1. A bit is the smallest unit of digital data.

Blank A semifinished piece of metal or nonmetal, to be formed and/or machined into its final shape and size.

Block A complete set of information that gives one instruction to an NC machine. Like a sentence, a block makes complete sense and can stand on its own.

BOM (bill of materials) A list of parts and material data referring to the end product.

BOMP (BOM processor) Software that processes BOM files.

bps An acronym for bits per second, bps is a measure of transmission speed.

B-rep (boundary representation) One of the several techniques by which part geometry is defined in CAD systems. The part is described as composed of explicit boundaries, which, in turn, are considered composed of curves.

Broadband In a broadband system, different sets of signals can coexist on the transmission line at the same time; this is achieved by translating each set to noninterfering frequencies. A broadband system can thus handle a variety of signals, such as audio, video, and others. (See also baseband.)

B-splines A technique of generating curves and surfaces for approximating (by computer) the actual tool path or machined surface.

Buffer A temporary area for holding something. In computer systems, the buffer is the portion of the memory that has been set aside for holding the data temporarily until it reaches its destination. On the shop floor, a buffer is an area or space for storing items (blanks, semifinished or finished products, tools, etc.) until the next machine or material-handling device such as a robot is ready to fetch it.

Bus A collection of circuits that act as a communication path for two or more devices. In bus topology, all stations and devices use the same single transmission medium to which they are attached.

Business cycle The ups and downs in the business climate.

Bypass relays Relays in a ring network that permit message traffic between two nodes normally not adjacent. These allow the removal of any node in such a network for servicing and other activities.

Byte A set of eight bits, processed together as a unit; it can be a subset of a computer word. A byte represents one alphanumeric character in ASCII. It is also used to express a computer's memory size.

C A popular programming language. The UNIX operating system and most UNIX application programs are written in C. The portability associated with UNIX is primarily due to the fact that C, unlike other high-level languages, allows programmers to write system-level codes that will work on any computer with a standard C compiler.

CACE (computer-aided cost estimating) The process of using computer systems in cost estimation.

CAD (computer-aided design) This term is broadly used to denote the use of computer systems in designing components and end products. It includes all activities associated with product design, such as modeling, drafting, and analysis. In the beginning, it meant drafting only, but nowadays the letter D stands for design, which includes drafting as well. Modern CAD systems also help the designer carry out analysis for stress, vibration and noise, thermal deformation, and more to optimize the design.

CADD   (computer-aided drafting and design) In the early stages of CAD development, this term was coined to emphasize the design capabilities of software packages beyond drafting. Now, since most packages include both design and drafting, the term CADD has become superfluous. CAD is a better terminology and is preferred to CADD. However, CADD is used sometimes to emphasize the drafting part of design.

CAD/CAM   (computer-aided design and computer-aided manufacturing) This term was coined to highlight the link between CAD and CAM. In the 1970s, computer use in design and in NC programming developed separately. When it was realized that the data generated to represent a part in CAD could also be used for NC programming (i.e. in CAM), the term CAD/CAM was coined to emphasize this link. Later versions of both CAD and CAM systems began to offer the capability of transporting CAD data into the CAM arena. Such a capability is a standard feature of today's CAD/CAM products.

CADCAM   Same as CAD/CAM; the absence of the slash is meant to reinforce the desired strong link between CAD and CAM.

CAE   (computer-aided engineering) The use of computer systems in functions that are essential to engineering a product. Examples of such functions are DFM considerations, modal analysis, material selection, and finite element analysis. CAE ensures that the product will carry out its intended function during normal operation.

CAM   (computer-aided manufacturing) This term was coined to denote the second stage of computers' use in manufacturing, the first being CAD. In the early days of CAM, the term was limited to mean computer use in NC programming tasks only. Later its scope broadened to include other areas of production as well. Today, CAM represents all uses of computers in management, control, and operation of a manufacturing plant, except those in design and business functions. It includes tasks such as process planning, part programming, robotics, material handling, tool management, fixturing, inspection using CMM, and more. In fact, CAM is as broad as CIM, except that the latter emphasizes integration. CAM connotes piecemeal applications that may lead to *islands of automation*—an undesirable result for CIM.

CAMAC   (computer-assisted measurement and control) The use of computer systems in measurement and control of processes.

CAM-I   (Computer-Aided Manufacturing—International) A private U.S. research consortium that promotes the use of computers in manufacturing by funding projects in all areas of CAM or CIM. It was set up before the CIM concept came into being.

Canned cycle   Used in the context of NC machining to mean a preset sequence of operations initiated by a single command.

CAPP   (computer-aided process planning) Use of computer systems in process planning, which involves selecting processes and their parameters for embodying the design into a tangible product. CAPP bridges the gap between CAD and CAM.

Capital productivity   A measure of how well the invested capital is used to generate the desired output (which may be profit).

Carrier signal   A continuous sinusoidal signal, applied to a communications medium, that conveys information only if altered in some way such as amplitude-modulated or frequency-modulated. It is the alteration that contains information. The detection of a carrier signal means that the line is in use (data is being transmitted).

CASA/SME   Formed in 1975 as the Computer and Automated Systems Association interest group within SME, its objective is to provide comprehensive and integrated coverage of modern computer-based technologies and management techniques for the advancement of manufacturing.

CD-I   (computer disc-interacitve) A computer disc containing data or information with which the user can interact.

CD-ROM   A computer disc containing an enormous amount of data or information that can only be read (not written onto). One such disc can store every word of a pile of books ten feet high.

Cell   A production unit consisting of two or more workstations with the material handling units, storage buffers, and other associated devices necessary for their logical connection. The cell is an independent facility that can process most effectively a family of parts or models. When it is flexible, the term FMC (flexible machining or manufacturing cell) is used to denote the facility.

Cellular manufacturing   The concept and practice of cell-based production.

CIE   (computer-integrated engineering) An integrated use of computer systems in engineering a product.

CIM   (computer-integrated manufacturing) Integrated use of computer systems in manufacturing. In its broadest sense, as in this book, CIM includes all the functions of operating a manufacturing business. It covers CAD, CAE, CAM and all other uses of computers in managing the entire enterprise.

CIM engineer or technologist   A job title for manufacturing professionals whose duties demand education and skills in CIM technologies and management.

CIM I   (computer-interfaced manufacturing) To highlight the lack of true integration, I is appended to the term CIM. In this sense, the middle term in CIM stands for *interfacing*. CIM I denotes efforts aimed at interfacing resources.

CIM II   (computer-integrated manufacturing) To emphasize that the real need of CIM is true integration, II is appended to the term. CIM II represents the advanced level of integration, rather than the efforts that merely interface the resources. Thus, CIM II denotes true CIM.

CIM wheel   A conceptual wheel model developed by CASA/SME to portray the meaning of CIM.

CL file   (cutter location file) A computer file that contains data on cutter locations for processing a part.

Clock   An inside-the-hardware pulse generator that generates basic timing signals for synchronizing the computer operations.

Clock rate   A measure of a computer processor's speed in MHz.

CMM   (coordinate measuring machine) Inspection equipment, basically an NC device, that measures the coordinates of part surfaces at desired points.

CMOS   (complementary metal-oxide semiconductor) Semiconductors of a special type, used in developing integrated speeds.

CNC   (computer numerical control) The technology and devices in which a computer numerically controls the process or equipment such as a machine tool. The use of computers makes NC systems more flexible since changes are made more easily.

Coaxial cable   One type of electrical cable in which the conducting wire is covered by an insulation that is surrounded by a metallic tube whose axis coincides with that of the wire (hence the term coaxial).

COBOL   (common business-oriented language) A high-level language widely used in business applications.

Code-39   A bar coding system.

Collision   Destruction of data when simultaneous transmission is attempted by two or more nodes over a single medium.

Communication hardware   Hardware especially designed or tailored for the needs of communication.

COMPACT II   A source language used in programming NC machines.

Compiler   A program, also called a language processor, that translates the statements of a high-level language into machine codes.

Computer-integrated enterprise   A broader view of CIM, integrating the whole enterprise.

Computer language   Language used in programming computers.

Computer program   A set of commands a computer can understand; also called software.

Computer virus   A program intentionally developed by unauthorized persons to erase or contaminate useful computer data. The virus is spread through networks.

Computer word   A set of bits processed together by the computer. The word size in modern PCs is 32 bits.

COPICS   (communications-oriented production information and control system) A computer system developed by IBM for managing the various production functions in an integrated way.

Concurrent engineering   The concept and practice of various functions or departments working together, from the beginning, to engineer a product.

Constant data   Items such as the headings of a report that do not change from report to report.

Control charts   A technique of investigating the performance of (manufacturing) processes from a product quality viewpoint.

Controller   A device, usually computer-based, used to control a process, equipment, machine tool, robot, and so forth.

CPU   (central processing unit) The hardware responsible for interpretation and execution of instructions given to a computer.

C-post   Customized postprocessor that prepares NC data for a specific machine tool.

CRC   (cyclic redundancy check) A method for detecting errors in a message by performing calculations on the bits and appending the result of the calculation to the message. The receiving station repeats the calculation at the other end and compares its result with the appended one to determine whether transmission was error-free.

CRP   (capacity requirements planning) Planning of requirements for production capacity. If capacity is found to be insufficient, alternatives such as overtime, extra shifts, or subcontracting are considered.

CRT   (cathode ray tube) An important component inside monitors that display text and graphic data. The CRT creates images by focusing an electron beam on the screen.

Crosstalk   The unintentional induction of signals from one communication channel onto another.

CSG   (constructive solid geometry) A technique of describing an object's geometry, using a set of solid primitives such as boxes, spheres, or cylinders to construct the model. CSG is one of the techniques of representing solids in a CAD system. CSG views a part as composed of solid primitives and their combinations. The boundaries of the part are implicit since they are the by-product of CSG construction, rather than defined explicitly. The CSG representation is unambiguous and enables calculations of part attributes such as weight and center of gravity and their associativity with the part.

CSMA/CD   (carrier sense multiple access with collision detection) A technique that enables multiple stations or nodes to access one transmission medium.

Custom macro   A block of program codes developed by, or for, the user that replaces a block of keystrokes with a single entry key.

Data   Values physically recorded in the database.

Database   A collection of interrelated data or information stored on a data storage device.

Data dictionary   Also called data directory, it is the database's table of contents. It defines the database and identifies the data items used by application and program modules. A necessary tool for the DBA.

Data model   A conceptual method of structuring data. The three common models are hierarchical, network, and relational—the latter two being more common.

DBA   (database administrator or administration) The person or department in charge of data administration in an organization. Duties include the purchase and development of appropriate DBMS, design and control of the database or databases, responsibility for backup and recovery, and so on.

DBMS   (database management system) A set of system software that facilitates the management and control of the database.

DEC   Digital Equipment Corporation.

Dedicated automation   Automation that is dedicated to a specific product, process, or task, and is difficult to modify for processing a different product. Also called hard automation.

Demodulation   The process of separating the data signal from the carrier signal.

Desktop computer   A computer small enough in size and weight for an office desk.

Detroit-type automation   Detroit is the hub of the U.S. automotive industry. The type of automation practiced in such facilities before the advent of programmable or soft automation is refereed to as Detroit-type automation. Because of its inflexibility, this is also called hard automation.

DFA   (design for assembly) The concept and practice of taking all elements into account at the product design stage to ensure ease of assembling.

DFM   (design for manufacturability or manufacturing or manufacture) This term represents "old wine in a new bottle." That the design must be adapted to match the capabilities of the available production resources is, and always was, practiced by good designers. With the use of computers as CAD/CAM systems, design's shortcomings in terms of manufacturability become apparent. The practice of DFM minimizes these shortcomings; it leads to a set of rules which, if followed at the product design stage, facilitate production.

DFMA   (design for manufacturing and assembly) Consideration of all the relevant factors at the product design stage itself to achieve ease of both manufacture and assembly.

Discrete manufacturing   The type of manufacture in which discrete components, subassemblies, and/or assemblies are the end products. The term is used to differentiate this type of operation from process-type manufacturing, in which production is continuous.

Discrete-parts manufacturing   See discrete manufacturing.

DMIS   (Dimensional Measuring Interface Specification) A specification for exchange of inspection data among several devices.

DNA   (Digital Network Architecture) DEC's layered data communications protocol.

DNC   (distributed numerical control) A system of interconnected NC equipment that allows data or information exchange to take place. DNC once was an acronym for direct numerical control, where it meant the control of a machine tool by downloading the part program from a mainframe or minicomputer.

DOS   (disk operating system) A PC software system that oversees the interaction between the CPU and the magnetic disk system for storing information. It does the "housekeeping" to manage the PC by, among other things, preparing the hardware to run application programs.

Domain   An attribute or characteristic of an entity in a relational database, for example the column in a table (or relation). It also means the data values an attribute can assume, for example the domain of all the part numbers.

DSS   (decision support system) AI-based computer system especially developed for users to aid them in decision making.

Duct system   Tubular enclosures that hold cables so the cables can be passed through the floors and walls of the communication site.

Duplex   A system in which signals transmit in both directions.

DXF   (data exchange format) A specification for exchange of data between two CAD systems, developed by Autodesk, Inc.

Dynamic routing   A technique for directing messages through a network in which the decision about the transmission path is made "on the fly" as traffic congestion dictates.

EBCDIC   (extended binary coded decimal interchange code) A code used by most third-generation computers in which eight bits represent a character. An eight-position field allows 256 different bit configurations, which are sufficient to represent any data of the human world into that of the computer's.

EIA   (Electronic Industries Association) A professional organization of the electronics industry, it also develops standards for character coding in the NC area and continues to develop standards as needed.

Ed(s)   (editors) Person or software that carries out editing.

Encoding   The description of analog signals in digital form. For example, an analog signal's instantaneous amplitudes can be assigned eight-bit binary numbers, thus yielding 256 digital voltage levels.

End effector   A gripper or actuator at the end of a manipulator or robot arm.

Entity   An object or thing about which data is stored.

Ethernet   A baseband protocol invented by Xerox Corp., Ethernet is commonly used as a LAN for electronically connecting the various devices and components of CIM. It is a CSMA/CD-based network system that runs at 10 MHz and uses coaxial cable.

Even parity   A system in which the total number of 1s in a transmittable character is even. If required, an additional 1 is added at the parity position to achieve it.

External memory   Memory other than the internal memory of a computer; external memory is essential for storing large amounts of data and long programs.

Factory of the future   A conceptual, hypothetical automated factory likely to be developed in the near future. Being futuristic, such a factory is like a mirage that moves further away as you reach it.

FDM   (frequency division multiplexing) Sharing of a communication line by allocating specific frequency bands for the devices.

FEA   (finite element analysis) A technique of analyzing structures for design integrity. The structure is assumed to be composed of small elements whose properties and boundary conditions determine its behavior under load.

Fiber-optic cable   A data transmission medium consisting of glass or similar fibers. Light emitting diodes (LEDs) introduce light, representing the signal, into the cable at one end. After bouncing off the inside of the fiber surface, light reaches the other end, where it is converted back into an electric signal by a detector.

Field   A specified area in a record to hold data.

File   The storage space for a set of data.

File manager   Software that allows the user to work only on one file at a time.

Firmware   Software implemented in user-modifiable hardware, which is a microprocessor having ROM. This allows frequently used programs and routines to be invoked by a single command.

Flag This term serves the same purpose in computer communication as the tangible flag does in rural mailboxes. It is a program-readable piece of information signifying that data are available, space is available, or an operation has been completed, so that the affected device can act.

Flat file A file with rows and columns for displaying the data.

Flexibility Capability of modification or adaptation.

Flexible automation The type of automation that can be modified easily, primarily through software, to accommodate design changes in the product within the constraints of a production facility. Also called soft automation. (See Detroit-type or hard automation.)

Flexible fixturing The process of designing, developing, or using a fixture that can be modified easily for other parts.

Floppy disk A thin, flexible disk that stores information magnetically. Also called a diskette, it is commonly used with PCs.

Flow control A hardware and/or software-controlled mechanism that stops transmission when the receiving station is unable to store the data received.

FMC (flexible machining or manufacturing cell) A flexible production facility to process a "family" of parts.

FMS (flexible machining or manufacturing system) A highly sophisticated production facility, costing millions of dollars, that can manufacture any product within a family. It comprises several FMCs with appropriate material-handling systems serving them. Modifications in FMS to cope with product variations are easy to implement since the system is primarily software-driven.

Forming Manufacturing processes in which part shape and size are formed in a cavity or otherwise, rather than machined off a blank.

FORTRAN (Formula Translator) A high-level programming language invented by John Backus for use by scientists and engineers. One of the most commonly used languages.

Forward chaining A process of finding the effect (final state or condition) by tracing the path forward, beginning with the known causes or given conditions.

Frame A group of bits that are communicated together. A frame's first few bits comprising the address and control information are called a header. The header is followed by the actual data, and then the trailer—the last few bits for error detection. Framing is the process of dividing the data to be transmitted into groups of bits, and then adding the header and trailer.

Full duplex A communication system's capability of transmitting in both directions simultaneously.

4GL (fourth-generation language) Very high-level language used in decision support systems, DBMS, and so on, and in developing application programs.

G (Giga) A prefix in SI units to mean billion.

Gateway A computer system, including software, that allows two LANs of different protocols to communicate. It translates all protocol levels from the physical through the application layer, and thus can interconnect any two networks.

GDP (gross domestic product) The monetary value of all the goods and services produced by a nation within its borders. GDP excludes income from overseas operations, and is therefore less than GNP.

G-post Postprocessor of a generic type.

GM (General Motors) GM is the registered trademark of General Motors Corp., a giant U.S. company. One of the three leading U.S. automobile manufacturers, the company was instrumental in developing MAP.

GNP   (gross national product) The value of all the goods and services produced by a nation in a specific year. Sometimes, GDP is used in place of GNP; GDP excludes the income from overseas operations.

Gouging   Error introduced in the workpiece due to finite size of the tool or cutter.

Graphics   Storage, processing, and display of geometric data by the computer. Development in graphics since the 1970s has rendered CAD/CAM very useful in CIM.

Gray code   A variation of the binary code so that only one bit position is affected when the value changes to the adjacent one.

GT   (group technology) A concept in which parts' similarities are identified, and used to group the parts for the benefit of design and manufacture.

Hard automation   Same as Detroit-type automation or dedicated automation.

Hard disk   A rigid, sealed, metallic disk used for data storage. Also called fixed or Winchester disk.

Hard gage   Conventional physical gages used in inspection. (See soft gage.)

Hardware   The physical items of a computer system.

Hex   Abbreviated form of hexadecimal system.

Hexadecimal system   A numbering system based on 16 digits—the 10 decimal digits, 0 to 9, and the first six letters of the English alphabet, A to F.

High-level language   A computer language, such as BASIC or FORTRAN, that has been developed for the convenience of human beings.

Higher nibble   Left four bits of a byte.

Host computer   A networked computer through which communications among the rest of the computers and other devices take place.

Hybrid network   A local area network that employs two or more topologies and access methods, such as token ring and CSMA/CD bus.

Hz   (Hertz) A unit of frequency in cycles per second.

IBM   International Business Machines Corp.; known worldwide as a leading computer company.

IC   (integrated circuit) An electronic circuit with components etched on silicon wafers. ICs are the building blocks of microprocessors and various digital devices including computers.

ICAM   (Integrated Computer-Aided Manufacturing) A project of the U.S. Air Force to model the manufacturing function and its communication needs.

Icons   Symbols used to indicate commands as user input.

Idling signal   A signal applied to a transmission line to indicate that no data is being transmitted. With this signal on, the line is said to be "idling," indicating that the line is operational.

IEEE   (Institute of Electrical and Electronics Engineers) A professional society of electrical and electronic engineers; active in developing standards relating to computers and communications.

IEEE 802   A standard that specifies physical and link layers.

IEEE 802.2   A standard for logical link and medium access control; used in IEEE 802.3,4,5 standards.

IEEE 802.3   A standard that specifies CSMA/CD (for Ethernet).

IEEE 802.4   A standard that specifies token bus (for MAP).

IEEE 802.5   A standard that specifies token ring.

IGES   (initial graphics exchange specification) A standard developed by private companies and the U.S. National Bureau of Standards; initially released in 1981 and revised several times. IGES aids in the exchange of design data between CAD systems of different makes.

IIE (Institution of Industrial Engineers) As a premier professional society of industrial engineers, IIE disseminates CIM knowledge pertaining to productivity and analyses of manufacturing as a system.

Inference engine Expert system module responsible for drawing conclusions based on rules and data.

Information Meaning of the data as conveyed to the user.

Information technology The technology in which data and information are the products.

Input device A device through which a user can communicate with computers.

I/O (input or output) **device** A device that can be used to send input or receive output from a computer.

Interface Physical or logical link between two devices or systems.

Internal memory The space within the processing unit of a computer for temporary storage of data and programs being worked with.

Interpreter A language processor that translates the source codes, line by line, into machine codes during the execution of a program.

ISO (International Standards Organization) A United Nations agency charged with the task of developing and implementing standards worldwide.

Islands of automation Flexible automation trend of the 1970s in which individual operations or stations were automated, mostly computer-based, one at a time with little or no regard to similar efforts elsewhere in the company. It was thought that the individually automated workstations or processes, called islands of automation, could easily be interconnected later when needed. But this proved to be difficult, forcing early thinking about CIM.

JIT (just-in-time) A new approach to processing or producing items just before their use. JIT reduces the investment in inventory.

Join A basic operation that combines tuples from different relations having common domain. The other two basic operations in relational DBMS are selection and projection; *select* extracts certain rows from the data, whereas *project* extracts the specified column. Select and project are operations on a single table (relation) while join involves two or more tables based on common domains.

K A prefix in computer technology that represents 1,024.

k (kilo) A prefix in SI units that represents 1,000.

KBS (knowledge-based systems) See knowledge-based systems.

Kermit A software system that enables file transfer from mainframes to PCs and vice versa.

Kernel A major component of the UNIX operating system, kernel responds to user requests and interacts directly with the hardware.

Key A value or item used to identify specific data in the database.

Knowledge base The stored rules and data that are used for inference purposes.

Knowledge-based systems Computer systems, especially software, that allow development, storage, and execution of inference rules in addition to the usual data storage and manipulation. Also called expert systems.

Labor productivity An indicator of how well available human resources are being used. Measured as output per employee hour.

Ladder diagram A diagram useful in programming PLCs.

LAN (local area network) A facility, usually a combination of wiring, transducers, adaptor boards, and software protocols, that interconnects workstations, computers, machine tools, robots, and other digital devices in a CIM environment within a plant or several plants in close vicinity. LAN is a data communication network covering limited distance of a mile or two. It is proprietary and provides high-bandwidth communication over inexpensive media (usually coaxial cable or twisted pairs) as well as switching capability.

Language processor   Software that translates or interprets computer languages.

Laptops   PCs that are light, small, and portable.

Laser stereolithography   A new process that allows rapid development of prototypes.

LEAD award   (leadership and excellence in the application and development of CIM award) Presented annually by CASA/SME to one manufacturing company and one college or university for achieving excellence in CIM.

Lean production   The concept and practice of improving productivity by reducing or eliminating unnecessary resources.

LED   (light emitting diode) Basic electronic element used in displays.

LISP   (list processing) A language designed for artificial intelligence work and noted for its ability to process procedures (list) in the way other languages process data.

LOGMARS   (logistics applications of automatic marking and reading systems) A bar code specification the U.S. Department of Defense requires suppliers to adopt.

Loss function   A mathematical function conceived by Taguchi to emphasize the loss to society from poor-quality products.

Low-level language   Computer language in a format that is easily understood by microprocessors, but difficult for humans.

Lower nibble   The right four bits of a byte.

LSB   (least significant bit) The rightmost bit in a group of bits.

LSI   (large scale integration) The technology by which a large number of electronic components are assembled on a single chip or board.

M   (Mega) A prefix in SI units for a million.

Machine control unit   The microprocessor-based control unit of a modern machine tool.

Machine language   The language of digital computers based on 0 and 1.

Machine tool   A machine such as a lathe that can be used to produce another of its own kind.

Machining center   A modern machine tool with milling as the primary operation.

Mainframe   Shortened form of mainframe computer.

Main memory   Memory within the processor unit of computers.

MAP   (manufacturing automation protocol) A communication standard specifically tailored for the needs of automation in manufacturing.

Market research   Efforts made to collect consumer feedback on products.

Mass storage   External memory for storing large amounts of data and/or programs.

MCD   (machine control data) Data that controls the movement of tools and the workpiece on NC machines. Also called "G" codes.

MCU   (machine control unit) See machine control unit.

Medium   A person, mechanism, electronic pathway, or any other means of conveying information from one point to another.

MFLOPS   (million floating point operations per second) A unit for measuring the processing speeds of CPUs.

Microcomputer   A small-size computer with its CPU on a single microprocessor chip.

Microprocessor   A miniature CPU on a single integrated-circuit chip.

Micros   Abbreviated form for microcomputers.

Minicomputer   A computer with capabilities in between those of the mainframe and the PC.

Minis   Clipped form for minicomputers.

MIPS   (million instructions per second) A unit for measuring the processing speeds of CPUs.

MIS   (management information system) Software programs especially designed to produce information in formats suitable for managers.

MIT   Massachusetts Institute of Technology.

MODEM   An acronym combining the terms modulator and demodulator. Modems use digital data to alter a signal so the signal can be transmitted or received over an analog medium such as a telephone line.

Modula-2   A programming language.

Modular handling system   A material handling system that can be modified easily to suit the needs of other products or tasks.

Modulation   A process of converting digital signals representing 0s and 1s into analog signals of sound so that they can be transmitted over telephone lines.

MRP   (materials requirement planning) Planning for the materials required to manufacture a product. MRP is mostly computer-aided.

MRP II   (manufacturing resources planning) Denotes the planning of all the resources, such as materials, machines, and operators, required for the manufacture of a product.

MSB   (most significant bit) The leftmost bit in a group of bits.

MTR   (maximum transmission rate) The theoretical rate beyond which no transmission is possible through the medium.

MTBF   (mean time between failures) A measure of product reliability.

Multimedia   Integrated information that may include live videos, taped digitized video sources, sound, text, or numerical data.

Multiplexing   The process of combining and sending different signals over a single transmission medium and separating them at the receiving end.

Multitasking   A computer system's capability that allows two or more computing tasks, such as interactive editing and complex numerical calculations, at the same time.

NASA   (National Aeronautics and Space Administration) A U.S. government agency charged with the responsibility of space exploration.

National income   The value of goods and services produced by a nation after the cost of capital and taxes are subtracted from GDP.

NBS   (National Bureau of Standards) A U.S. government agency for standardization; now called NIST.

NC   (numerical control) See numerical control.

NCMS   (National Center for Manufacturing Sciences) A consortium of U.S. and Canadian corporations for developing and implementing next-generation manufacturing technologies. NCMS brings together industry, governments, academia, philanthropic organizations, and others to coordinate efforts in this direction.

NEMA   (National Electrical Manufacturing Association) A trade organization established for the benefit of electrical manufacturers.

Network topology   The pattern of connection between points in a network. The common types of network are mesh (each point connected to all others), star, bus, and ring.

Neutral language   A language that allows data exchange between two devices seamlessly (i.e., without needing translation).

NGC   (next-generation controller) A project to develop the next generation of controllers. Financed by the U.S. Air Force's ManTech program, NGC is being developed by Martin Marietta at the National Center for Manufacturing Sciences (NCMS).

Nibble   A group of four bits.

NIST   (National Institute of Standards and Technology) Formerly NBS.

NML   (neutral manufacturing language) A conceptual language that would allow data exchange seamlessly within all functions of manufacturing.

Nodes   Service points in a network at which service is rendered or a communication channel is interconnected.

NUBS   (Nonuniform B-splines) A technique for generating CAD surfaces. It combines the merits of the Coon's patch and Bezier methods.

Numerical control   A technique for effecting control through numbers and symbols. Initially, NC technology developed on its own, but later merged with computer technology. This resulted in rapid growth of computer use in manufacturing.

NURBS   (Nonuniform rational B-splines) The most advanced technique of (mathematically) generating a surface in CAD.

Object-oriented DBMS   A database management software that organizes information using models that represent real-world entities as objects rather than records.

Object-oriented programming   A programming technique in which sections of the program code and data are represented, edited, and used in the form of objects, such as graphic elements and window components, rather than as strict computer codes. Such an approach allows development of toolkits that facilitate programming.

Object program   The compiled version of a high-level language program as translated into machine language.

OCR   See optical character recognition.

Octal system   A numbering system based on the eight digits 0 to 7.

Odd parity   An error-checking system in which the total number of 1s in a transmittable character is odd. If required, the parity position carries a 1 to achieve it.

Off-line programming   The process of developing a program on a system that physically simulates the actual system on which the program will be run. This approach frees the actual system to continue to operate normally, thereby maintaining high utilization.

Operating system   The group of software that handles a computer's housekeeping chores while using the hardware for applications.

Operations integrator   A new terminology to denote that modern manufacturing needs technical personnel who are skilled in several areas. CIM engineers may be called operations integrators.

Optical character recognition   A computer-based optics-oriented system that can recognize characters.

Optobus   A communication bus based on light signals rather than electrical signals.

ORACLE   A relational database management system, developed by Oracle Corp., for various operating systems including UNIX.

OS/2   (operating system/2) The operating system used with IBM's Personal System/2 line of computers. It was designed to optimize the performance of Intel's 80286 microprocessor.

OSI   (open systems interconnection) An architecture of a layered collection of standards developed under the aegis of ISO for interconnection of heterogenous computer and communication networks. It facilitates data communication between devices of different makes.

Output device   A device such as a printer that receives the output of a computer's calculations, analyses, and the like.

Override   A process in automated equipment, such as NC machine tools, whereby the operator can manually change the value of a parameter from what is in the program.

Packaged software   Ready-made software.

Packet   A group of data and control bits that are switched and transmitted together. The bits are arranged in a specific order, and the control bits include source and destination addresses.

Packet switching   The transmission of data in small, discrete switching packets rather than in streams, done to optimize use of physical data channels.

Parallel interface   A digital interface with multiple data lines, with each line transmitting one bit of data so that the bits travel together. (Contrast this with a serial interface, in which only one line is available and the bits follow each other.)

Parallel processing A computing strategy designed into the hardware and software, in which a single large task is divided into parts, each carried out in parallel on separate processors.

Parallel transmission The technique of transmitting a group of bits in parallel (i.e., together). (See also serial transmission.)

Parametric design An approach to designing in a GT environment where the part belongs to a family. Rather than designing the exact parts, a generic part is designed with its dimensions as variables. The designer or the application program simply specifics the variables' values (parameters) to develop a specific part design from the generic model.

Parametric programming A method of part programming in which the part belongs to a family. Rather than program for a particular part, a generic part is programmed with variables as its dimensions. The programmer simply specifies the values (parameters) for the new part to create the specific program. The concept relies on GT principles.

Parity The number of 1s in an eight-bit character or other groups of bits, which may be odd or even. Parity check is a technique of ensuring error-free transmission. It can verify against one (or odd numbers of) error only.

Parity bit Bit 0 or 1 is used at the leftmost position in an eight-bit character to render the sum total of all 1s either even or odd, depending on whether the system is an even- or odd-parity type.

Part A component that is either an end product on its own or is assembled into an end product. Also called workpiece.

Part program compatibility The condition in which a part program written for processing on one machine is suitable for another machine with little or no modification.

Part programming The process of writing appropriate codes or programs for processing or machining the part on NC machinery.

Participative management The concept and practice of managing the human resources in a way that all employees participate in the decision-making process.

Pascal A computer programming language.

Payback period An economic analysis method for deciding whether to fund a project. Payback period is the time in which invested capital pays for itself through cost savings effected by the investment.

Payload Maximum weight that can be carried at normal speed of operation; used as one of the specifications for robots and other machines.

PC See personal computer.

PCB See printed circuit board.

PDES (product data exchange specification) A specification for broadening IGES to incorporate both design and manufacturing data. PDES bridges the gap between CAD and CAM by encouraging data exchange among different CAD/CAM systems.

People practices Term to describe how the employees of an organization interact with each other for the benefit of the company and their own job satisfaction.

Peripheral devices Devices such as a printer or mouse that are connected to the computer to improve its effectiveness.

Personal APT PC version of APT.

Personal computer A desktop-sized computer for individual use. Also called microcomputer.

Phase modulation One of the modulation methods in which the phase angle of the carrier signal is varied, with reference to a carrier phase, in proportion to the instantaneous amplitude of the signal to be transmitted.

Physical layer The bottom layer in the ISO's OSI model. It involves electrical connections for data transmission.

PLC See programmable logic controller.

Platform The hardware and operating system required to run an application program.

PM See phase modulation.

Point-to-point The direct connection between two communicating devices. Tool movement between two points with no consideration of the path locus.

Polling A technique of controlling access to a multidrop communications line by asking the connected stations whether they need to transmit.

Portables Another term for personal (or desktop) computers.

Postprocessor A software program that converts the source code into the language of the machine control unit.

Potting A work-holding method in which special adhesives are used to hold the part in place during machining. It is based on phase change, since the adhesive polymerizes.

Precision The quality of an instrument or measuring device to repeat the result closely.

Printed circuit board A group of electronic devices such as transistors and capacitors that are assembled on a board as circuits to achieve certain computing or control functions.

Process capability A measurement of the spread in data collected during normal operation.

Process planning Planning for the steps necessary to manufacture a part or product. The results of process planning are summarized on a route sheet that lists the sequence of operations, machines, and other pertinent information.

Product life cycle An expression to describe the length of time a product lasts in the market. It comprises the periods of conception, design, production, use, and, finally, termination when not needed by the market. Flexible technologies allow shorter product life cycles, which means more frequent design and model changes.

Production control Actions taken to remove the bottlenecks when actual production fails to follow the planned schedule. Production control also encompasses supervisory tasks necessary to ensure that the products are ready for shipment on due dates.

Productivity A measure of the effectiveness of various resources. Productivity is often compared with a baseline (past data) to assess how (relatively) well the resources are used. Alternatively, it can be quantified by the ratio between the value of outputs to that of the inputs.

Profile-21 A term for the recent survey entitled *Countdown to the Future: The Manufacturing Engineer in the 21st Century*, which attempted to predict job descriptions for manufacturing engineers during the next century.

Programmable logic controller A modern controller based on PC technology for controlling a process; its logic is programmable. It monitors parameter levels and initiates programmed actions whenever the levels rise or fall beyond the set limits.

PROLOG (Programming Logic) A programming language, used primarily in developing knowledge-based systems.

Protocol A set of rules for communication.

Pseudo-CIM A CIM environment or operation that is less sophisticated than the perfect one.

PTP See point-to-point.

Public domain Information such as software that is available to anyone who accesses it.

QDC See quick die change.

Quick die change The concept and technology that allows faster die changes. This increases production rates by saving setup time.

RAM See random access memory.

Random access memory Memory space for storing the operating system, application programs, and pertinent data for processing by the CPU.

Rasterization   One of the techniques to create images on a computer screen. In rasterization, the data-specific dots are activated as a beam scans through the screen. The display is basically a grid pattern of dots lighted selectively to create the picture. Television screens work this way.

Rational B-spline   One of the techniques for curve or surface generation in CAD. It is more sophisticated than the B-spline technique, since another degree of control is provided by varying the force of pull on the points under manipulation.

RDBMS   See relational database management system.

Read-only memory   Memory space where permanent information is stored; users cannot store in this space.

Record   Data about an entity or object. Files contain records of individual data items.

Redundancy   Duplicated storage of the same data; obviously undesirable.

Relation   Table of data about an entity type; the *join* operation connects data about different entity types.

Relational database management system   A system based on a relational model in which data are stored in tables, called relations, that consist of rows (tuples) and columns (domains). Relationships among data items are explicitly specified as equally accessible.

Rendering   The final step in sketching in which better sketches are selected and neatly drawn, showing details.

Repeater   A device used at the physical level in the OSI model of communication protocols. It amplifies and conditions the signals received and passes them forward without altering the data.

Reverse engineering   A term describing the process of generating engineering data from existing components or models. The data may be physical dimensions, coordinate values, surfaces, orthographic drawings, and so on. Reverse engineering has been made possible by the digital measurement capabilities of CMMs and the graphical communication features of CAD systems.

RFID   (radio frequency identification) An identification technique in which the object is tracked by transmitting radio signals and capturing the returning signal as obstructed by the object.

Ring   A LAN topology in which each node is connected to two other nodes to form a loop. Data are transmitted from node to node around the loop in the same direction.

RISC   (reduced instruction set computer or computing) A class of computer architecture that improves its performance by minimizing the number and complexity of operations required in the computer's instruction set.

RJE   (remote job entry) Submission of jobs through an input unit remotely located, but having access to the destination computer or device through a data link.

R&D   (research and development) Efforts involved in conceiving, researching, and developing a new product or major modifications in an existing one. It also includes efforts to keep the facilities up to date.

ROI   (return on investment) A percentage value as the basis of selecting the best from possible alternatives. ROI is commonly practiced in industry for making economic decisions.

ROM   See read-only memory.

Route sheet   A list of the processes, machines, and other pertinent information necessary to manufacture a part or product.

Router   A LAN device that receives physical-level signals from a network, carries out required (data-link-level and network-level) protocol processing, and then sends the data via appropriate level protocols onto another network. The primary purpose of a router is to determine how to forward a message packet to its address.

RS-232C   An EIA specification of a modern interface for serial data communication.

RS-408   An EIA specification for a CNC/NC interface.

RS-422   An EIA specification for point-to-point communication, with differential drives, at a high data rate.

RS-423   An EIA standard intended to replace RS-232C and encourage the use of RS-449 connectors.

RS-449   A specification for connectors used in RS-422 and RS-423.

Rule-based system   Another term for knowledge-based system.

Run   (a program) Command or function that makes the computer process a given program.

Scalability   A desirable feature of computer systems or CIM implementations to accommodate growth or divergency. It denotes the capability that enables use of the same software environment on many classes of computers, from PCs to supercomputers, without having to rewrite the code or sacrifice functionality.

Scanner   A device that examines a given pattern and generates corresponding signals.

Scanning   The process of examining signals or data point-by-point in a well-defined sequence.

Schema   A definition of the database. These are of three types: *Conceptual schema* defines the logical structure of the database; *internal schema* specifies how data are physically stored and accessed; and *external schema* describes how applications or users would see the data. Communication among the three schema is the task of the DBMS.

SE   See simultaneous engineering.

Segment   The basic quantum of data handled by application programs under the control of database management software. A segment is made up of several data items.

Semantics   Relates to protocol in communications. It is like the rules that determine when and how people say certain things. In a telephone conversation, for example, the answering party's initial response is Hello or Hi—not Goodbye, which is reserved to signal the end of conversation.

Serial interface   A simple mechanism for converting the parallel bits coming out of computers or from the sender into serial form (one bit after the other) for transmission through a single data line, or vice versa. (See parallel interface.)

Serial transmission   A method of data transmission that uses only one line or wire. The bits forming the character or the packet are transmitted one following the other. (See parallel transmission.)

Server   The computer network's node that provides service. For example, a mainframe computer with large storage capacity may play the role of database server for interactive terminals. The server is a program that provides service such as sharing of files or control of printers to LAN users.

SFC   See shop-floor control.

Shareware   Software procured directly from its authors at low cost. Authors allow the programs to be copied and distributed with few restrictions. Shareware users are encouraged to examine a program, copy it, and pass it on to other users.

Shell   A software package with an inference engine program but without a knowledge base.

Shop-floor control   Control of the various shop floor activities such as materials handling, resource utilization, and efficient routing by keeping track of the containers, tools, WIP, and so on.

SI   (System International) An international system of units.

SIC   An acronym for Standard Industrial Classification. SIC codes are used to classify industrial activities and products; manufacturing falls in the major group 20 through 39. Every type of manufactured product is allocated a four-digit code, of which the first two are in the 20–39 range.

Signal conditioning   Amplification or any other modification or conditioning of the electrical signal to render it suitable for transmission.

Signature analysis   A diagnostic technique in which the "signature" of the normally operating machine is used as a baseline to find any problem in its operation or breakdown. The signature is usually a vibration or noise signal in the time or frequency domain.

Simplex   A system in which signal transmission takes place in one direction only.

Simultaneous engineering   Same as concurrent engineering.

Single tasking   Equipment or a device that can carry out only one task at a time.

Slot reader   A contact-type scanner that physically touches the bar-code symbol or the protective covering over the symbol. Also called wand.

SME   See Society of Manufacturing Engineers.

SNA   (system network architecture) A set of layered communications protocols for networking. IBM's proprietary network.

Society of Manufacturing Engineers   A nonprofit professional organization dedicated to dissemination of manufacturing knowledge through awareness programs, research, and publications. Founded in 1932 and previously known as ASTE (American Society of Tool Engineers) and ASTME (American Society of Tool and Manufacturing Engineers), it became SME in 1970. SME has international recognition and serves the interests of manufacturing professionals worldwide, either directly or through its senior and student chapters. SME membership is almost a necessity for anyone interested in manufacturing. Its address is: One SME Drive, P. O. Box 930, Dearborn, MI 48121, USA.

Soft automation   Same as flexible automation.

Soft gauge   A term to describe inspection on computer monitors, as opposed to that with physical gauges. Software compares the on-line inspection data of the machined part on screen with the geometric model.

Software   Computer programs to make the hardware work or to carry out an application task.

Solid model   The geometry of a part or end product, based on solid primitives, as defined in the CAD system.

Solid primitives   Small, solid elements in the form of spheres, cylinders, boxes, and so on, used to graphically represent the part or complete assembly using the CSG (constructive solid geometry) approach.

Source program   An original program or code that is translated in machine language by an interpreter or compiler prior to running it. The compiled version is usually saved as an object program or code to avoid the need of retranslation.

SPC   (statistical process control) The use of statistical techniques in measuring, analyzing, and controlling a process. SPC helps improve the process to its ultimate capability, thereby improving product quality.

Spline technique   A technique for generating curves and surfaces using CAD systems. A computer equivalent of drafters' French curve. In the spline technique, the control points used to control the shape of the curve are located on the curve itself.

SQC   (statistical quality control) The use of probability and statistics in achieving and maintaining product quality.

SQL   (structured query language) A query language developed by IBM and accepted as a standard language for relational database systems.

SRA   See stored record address.

Station   Equipment such as computers or machining centers attached to a LAN for providing service to users.

Stereo vision   The type of vision technology in which two cameras are used, enabling the measurement of the third dimension such as depth.

STEP   A standard for the exchange of product model data. As ISO 10303, STEP is an international standard dealing with product data representation and exchange. A successor of IGES, it is the basis for PDES.

Stored record address   A specific location in a database whose address is computed as some function (hash function) of a value that appears in the stored record occurrence—usually the primary key value.

Structured light   A light source illuminated through fiber optics to a fixed head mounted on a positioning table that moves the light beam.

Subschema   Refers to an application programmer's view of the data being used, while schema means an overall chart of all the data item and record types stored in the database.

Supercomputer   The most expensive and powerful computer used in solving sophisticated problems.

Synchronous transmission   Data transmission technique in which the time interval between transmitted characters is synchronized between sender and receiver. (See asynchronous transmission.)

Syntax   A set of rules that describe the structure of statements allowed in computer communications. The commands and routines must be in the same syntax.

System   An assemblage of interacting elements to carry out a specific task. Manufacturing is in essence a system. A system approach to manufacturing improves its overall performance.

System commands   A set of commands to activate a CAD/CAM or any other system. As an example, typing RUN runs the current program.

System integrator   A person or device whose role is to integrate the various modules of the system.

System software   A set of software consisting of operating systems, language processors, and utility programs.

Taguchi method   Approach to improving product quality by considering the concept of loss function and using the statistical techniques of experimental design.

Tap   An electrical connection in LANs that permits signal transmission onto, or from, a bus.

Taylorism   Concept of dividing work into well-defined tasks. The practice of this concept has broadened over the years in the name of specialization. Since CIM attempts integration, it may be looked upon as the antithesis of Taylorism.

TDM   See time division multiplexing.

Teach pendant   The control box operators use to program and guide the robot motions, which can be recorded by the control memory and replayed to repeat the cycle.

Terminal   Any input/output device for entering information into a computer or receiving information from it.

Thumbnail sketches   Simple, quick drawings designers generate by collecting ideas from colleagues and putting the ideas on paper to see how they look.

Time division multiplexing   A technique of sharing a transmission line by allocating short periods of time for each pair of transmitting and receiving stations.

Time-shared computer   A computer whose RAM and processing capabilities are shared by more than one user or device.

Token   A unique combination of bits whose receipt by a station or node permits it to transmit on the LAN medium.

Token bus (or ring)   The use of a token to control access to a bus (or ring).

Tool (or tooling) management   The efforts of managing tool inventory to keep the tooling cost low. In modern plants consisting of FMCs and FMSs, tool inventory may become high, especially if the tools are dedicated to individual machines, and not shared.

TOP   (technical office protocol) Initiated by Boeing Co. in 1985, TOP is a specification for facilitating the integration of systems in the technical office so the business side of manufacturing can be integrated with the plant through TOP/MAP.

Touch probes   Electromechanical sensors that can locate the position of tools, workpieces, or other objects by touching them.

TQC   (total quality control) A concept in which quality control efforts are considered as a plantwide system.

TQM   (total quality management) The practice whereby the entire organization, from the president down, is committed to, and involved in, continuous improvement of product quality.

Trailer   A bit pattern that follows the data to mark the end of transmission.

Transfer line   A production facility, usually for automobiles, in which the product is assembled while being moved on a conveyor system from the start end of the line to the finish end.

Transponder   A device that receives signals at one frequency and transmits them at another to avoid interference with the incoming signals.

tto   (tape tryout) The process of verifying that the part program on the (paper) tape is error-free. Tape tryout is usually the last step prior to the program's release to the production department.

tty   (tape tryout) Same as tto.

Tuple   Another name for the row of a table (relation) in the context of RDBMS.

Turning center   A machining center whose primary operation is turning.

UART   (universal asynchronous receiver/transmitter) An interface device for serial-parallel conversion and the associated tasks such as buffering and check-bits manipulation.

UNIX   A multiuser, multitasking interactive operating system developed by AT&T Bell Laboratories. It has been widely used and significantly improved by user universities. UNIX is becoming increasingly popular in a wide range of commercial applications including manufacturing.

Universality   The characteristic of being generic as opposed to being specific.

USART   (universal synchronous/asynchronous receiver/transmitter) An interface that allows both asynchronous and synchronous (variable time intervals between blocks of data, rather than between characters) communication. Compared to customized interfaces, USARTs are relatively inexpensive.

Utility   A product's ability to satisfy a buyer's need.

Utility programs (or utilities)   Standard programs that assist the user in operating data processing or similar systems.

VAL   (Victor's assembly language) A proprietary language developed by Unimation Inc. for robot programming.

Variable data   Data whose value is not fixed.

Vendibility   The characteristics of a product that make it saleable.

VLSI   (very large-scale integration) As a major step in the development of sophisticated electronic circuitry, it denotes the technology in which a huge number of circuit components are integrated in one chip.

Volume   Secondary storage devices connected to the computer via a channel—such as tape and disk units, or drums—and devices on which data are stored in demountable cells or cartridges.

Volumetric accuracy   A term to quantify the quality of CMMs; determined by the ball bar test, which uses a metal rod with precision balls at each end. The rod is measured with the CMM under test, the difference between the minimum and maximum observed measurements being the error.

WAN   (wide area network) A network that allows communication over long distances of large amounts of data. In CIM, WANs would connect two or more plants physically remote from each other.

Wand   Contact-type scanners that physically touch the symbol or the protective covering over the symbol.

Wealth creation   All the activities of individuals, families, and communities that create wealth for the nation. Manufacturing is a major contributor to wealth creation.

WIP   (work in process) Products in various stages of processing; a plant's work inventory awaiting processing.

Word   A set of coded characters that specifies some operation on the machine tool. One or more words make a block of data. The term *word* also means one or more bytes as a group.

Work envelope   The largest space that can be reached by a robot or similar equipment.

WORM   (write-once-read-many) The type of memory that can be written onto once and read as often as needed.

Workstation   A generic term for a place, equipment, process, or device where work is accomplished.

X.25   A commercial packet network access protocol that specifies three levels of connections. Its physical, link, and packet levels correspond to the first three layers of the ISO/OSI model.

XENIX   PC version of UNIX, developed by Microsoft.

ZIP   (Zoning Improvement Plan) code. A five-digit code that identifies a geographic location within the United States.

Bibliography

This bibliography contains prominent books, monographs, reports, and magazines relating to CIM technology and management. References and articles limited to specific issues of CIM are listed at the end of appropriate chapters.

Books

Bray, O. H. (1988). *Computer integrated manufacturing: The data management strategy*. Boston: Digital Press.

Chiantella, N. A. (1986). *Management guide for CIM*. Dearborn, MI: CASA/SME.

Foston, A. L., Smith, C. L., & Au, T. (1991). *Fundamentals of computer-integrated manufacturing*. Englewood Cliffs, NJ: Prentice-Hall.

Goetsch, D. L. (1988). *Fundamentals of CIM technology*. New York: Delmar.

Greenwood, N. R. (1988). *Implementing flexible manufacturing systems*. New York: Wiley.

Harrington, J., Jr. (1985). *Computer integrated manufacturing*. Malabar, FL: Robert E. Krieger Publishing.

Maistre, C. L., & Ahmed, E. (1987). *Computer integrated manufacturing—A systems approach*. UNIPUB/Kraus International Publications.

Mitchell, F. H., Jr. (1991). *CIM systems: An introduction to computer-integrated manufacturing*. Englewood Cliffs, NJ: Prentice-Hall.

Ranky, P. G. (1986). *Computer integrated manufacturing*. Englewood Cliffs, NJ: Prentice-Hall.

Rembold, U., Blume, C., & Dillmann, R. (1985). *Computer-integrated manufacturing technology and systems*. New York: Marcel Dekker.

Rembold, U., Nnaji, B. O., & Storr, A. (1993). *Computer-integrated manufacturing and engineering*. New York: Addison-Wesley.

Scheer, A. W. (1988). *CIM—Computer integrated manufacturing*. New York: Springer-Verlag.

Teicholz, E., & Orr, J. N. (Eds.). (1987). *Computer integrated manufacturing handbook*. New York: McGraw-Hill.

Vail, P. S. (1988). *Computer integrated manufacturing*. Boston: PWS-KENT.

Monographs and Reports

The Automation News factory management glossary. (1984). New York: Grant.

Ecktein, O., et al. (1984). *The DRI report on U.S. manufacturing industries*. Manufacturing Technology Press.

Gettelman, K. M., Marshall, H., Nordquist, W., & Herring, G. (Eds.). (1989). *Modern machine shop 1989 NC/CIM guidebook*. Cincinnati: Gardner. (This is published annually.)

CASA/SME Technical Council, *SME blue book series*. Dearborn, MI: CASA/SME.

Martinez, M. R., & Leu, M. C. (Eds.). (1983). *Computer integrated manufacturing: Vol. 8. PED*. New York: ASME.

National Academy of Engineering. (1988). *The technological dimensions of international competitiveness*. Washington, D.C.: Author.

National Research Council. (1986). *Toward a new era in U.S. manufacturing: The need for a national vision*. Manufacturing Studies Board, Commission on Engineering and Technical Systems. Washington, D.C.: National Academy Press.

Nazemetz, J. W., Hammer, W. E., Jr., & Sadowski, R. P. (Eds.). (1985). *Computer integrated manufacturing: Selected readings*. Norcross, GA: Industrial Engineering and Management Press.

Savage, C. M. (1987). *Fifth generation management*. SME blue book series. Dearborn, MI: SME, pp. 3-5.

Journals and Periodicals

Advanced Manufacturing Engineering. Guildford, Surrey, UK: Butterworth-Heinemann Ltd.

Computer-aided Design. Guildford (UK): Butterworth Scientific Ltd. (Published 10 times a year, it is an international journal that contains articles on the state-of-art in CAD research, development, and applications.)

Computer-integrated Manufacturing Systems. Guilford (UK): Butterworth Scientific Ltd. (A quarterly journal that publishes articles covering all aspects of CIM.)

Industrial Engineering. Norcross, GA: Industrial Engineering and Management Press. (A series of 24 articles on CIM appeared in this magazine from January 1984 through February 1986 under the editorship of Randall P. Sadowski.)

International Journal of Computer Integrated Manufacturing. London: Taylor & Francis. (Started in 1988, the journal is published six times a year. It reports on research and applications of CIM, underlining the possibilities and limitations. Also included are papers on how new technology can be developed and applied in particular manufacturing situations.)

International Journal of Production Research. A British journal.

International Journal of Machine Tool Design and Research. (U.K.).

Journal of Manufacturing Systems. Dearborn, MI: SME.

Manufacturing Engineering. Dearborn, MI: SME.

Mechanical Engineering. New York: ASME.

Modern Machine Shop. Cincinnati: Gardner.

Production. Cincinnati: Gardner. (This monthly magazine aims articles at production managers of metalworking plants. The topics covered are current and the presentation is lucid.)

Transactions of the ASME, Journal of Engineering for Industry

Transactions of the Institute of Industrial Engineers

Conference Proceedings

ASEE Annual Conference Proceedings (The American Society for Engineering Education meets every year in the summer. The annual proceedings is a collection of the articles that have been refereed and presented at the meeting.).

Advances in Machine Tool Design and Research (England).

Annals of the CIRP (International Institution for Production Research, France).

Proceedings of the Canadian CAD/CAM and Robotics Conference.

Proceedings of the Manufacturing International (Organized by the ASME, Manufacturing International is held every two years.).

Proceedings of the North American Manufacturing Research Conference (NAMRC). (This is a conference of researchers for researchers. Most of the research findings reported in this meeting are funded by NSF and similar government agencies.)

Abbreviations

ABET   Accreditation Board for Engineering and Technology

AC   Adaptive control

ADC   Analog-to-digital converter

AGV   Automated guided vehicle

AI   Artificial intelligence

AIAG   Automotive Industry Action Group

AIM Tech   Association for Integrated Manufacturing Technology

ALU   Arithmetic logic unit

AM   Amplitude modulation/modulated

AML   A manufacturing language

AMRF   Advanced Manufacturing Research Facility

AMT   The Association of Manufacturing Technology

AMTDA   American Machine Tool Distributors' Association

ANSI   American National Standards Institute

APICS   American Production and Inventory Control Society

APT   Automatically programmed tools

ARPA   Advanced Research Projects Agency

ASCII   American Standard Code for Information Interchange

ASEE   American Society for Engineering Education

ASM   American Society for Metals (now ASM International)

ASME   American Society of Mechanical Engineers

ASQC   American Society for Quality Control

AS/RS   Automated storage and retrieval system

ATC   Automatic tool changer

AT&T   American Telegraph and Telephone Company

AUTOFACT   Automatic factory

BASIC   Beginner's all-purpose symbolic instruction code

BBS   Bulletin board systems

BCD   Binary coded decimal

BCL   Binary cutter location

bit   binary digit

BOM   Bill of materials

BOMP   Bill-of-materials processor

bps   Bits per second

B-rep   Boundary representation

BTR   Behind-tape reader

CACE   Computer-aided cost estimating

CAD   Computer-aided design

CADD   Computer-aided design and drafting

CAD/CAM   Computer-aided design and manufacturing

CADCAM   Same as CAD/CAM

CAE   Computer-aided engineering

CALCOMP   California computer products

CAM   Computer-aided manufacturing

CAMAC   Computer-assisted measurement and control

CAM-1   Computer Aided Manufacturing—International, Inc.

CAP   Computer-aided planning

CAPP   Computer-aided process planning

CASA/SME   Computer and Automated Systems Association of SME

CC   Computer control

CCD   Charge-coupled device
CCPD   Charge-coupled photo-diode
CD-I   Computer disc—interactive
CD-ROM   Computer disc read-only memory
CEO   Chief executive officer
CGA   Color graphics array
CIE   Computer-Integrated engineering;
    Computer-Integrated Enterprise
CIM   Computer-Integrated manufacturing
CIM I   Computer-Interfaced manufacturing
CIM II   Same as CIM
CIME-ISE   Computers in Mechanical
    Engineering—Information and Software
    Exchange
CIRP   International Institution for Production
    Engineering Research
CL   Cutter location (data/file)
CMM   Coordinate measuring machine
CMOS   Complementary metal-oxide
    semiconductor
CNC   Computer numerical control
COBOL   Common business oriented language
COPICS   Communications-oriented production
    inventory and control system
COSMIC   Computer Software Management and
    Information Center
CPU   Central processing unit
CRC   Cyclic redundancy check
CRP   Capacity requirement/resource planning
CRT   Cathode-ray tube
CSG   Constructive solid geometry
CSMA/CD   Carrier sense multiple access with
    collision detection
DARPA   Defense Advanced Research Projects
    Agency
DASD   Direct access storage device
dB   Decibel
DBA   Database administrator or administration
DBMS   Database management system
DCE   Data communication equipment
DEC   Digital Equipment Corporation
DFA   Design for assemble
DFM   Design for manufacturing/manufacturabil-
    ity/manufacture
DFMA   Design for manufacturing and assembly
DMIS   Dimensional measurement interface
    specification
DNA   Digital Network Architecture
DNC   Direct/distributed numerical control
DoD   (U.S.) Department of Defense

DOS   Disk operating system
DRO   Digital readout
DSS   Decision support system
DTE   Data terminal equipment
DXF   Data exchange format
EAROM   Electrically alterable ROM
EBCDIC   Extended binary coded decimal
    interchange code
Eds.   Editors
EEPROM   Electrically erasable PROM
EGA   Enhanced graphics array
EIA   Electronics Industries Association
EIX   Enterprise information exchange
EMI   Electromagnetic interference
EMO   European Machinery Organization
EOB   End of block
EOT   End of tape
EPROM   Erasable programmable read-only
    memory
EROM   Erasable read-only memory
FAS   Flexible assembly system
FDC   Factory data collection
FDDI   Fiber-optic distributed data interface
FDM   Frequency division multiplexing
FEA   Finite element analysis
FEM   Finite element method
FEP   Front-end processor
FM   Frequency modulation/modulated
FMC   Flexible machining/manufacturing cell
FMS   Flexible machining/manufacturing system
FORTRAN   Formula translator
FTAM   File transfer access management
GDP   Gross domestic product
GM   General Motors Corporation
GNP   Gross national product
GT   Group technology
HCMOS   High-speed complementary metal oxide
    semiconductor
Hex   Hexadecimal system
HLL   High-level language
HP   Hewlett Packard (Company)
Hz   Hertz
IBM   International Business Machines
    Corporation
IC   Integrated circuit
ICAM   Integrated Computer-Aided
    Manufacturing
ID   Identification (tag)
IEEE   Institute of Electrical and Electronic
    Engineers

IGES   Initial graphics exchange specification

IIE   Institution of Industrial Engineers

IITRI   Illinois Institute of Technology Research Institute

IMTS   International Manufacturing Technology Show

INTIME   Information on Technology in Manufacturing Engineering

Inc.   Incorporated

I/O   Input and/or output

IP   Internet protocol

IRMA   Information resource management aid

ISO   International Standards Organization

ITI   Industrial Technology Institute

JIMTOF   Japan International Machine Tool Fair

JIT   Just-in-time

KBS   Knowledge-based system

Kbps   Kilobits per second

LAN   Local area network

Laser   Light amplification by stimulated emission of radiation

LCL   Lower control limit

LEAD   Leadership and excellence in the application and development of CIM

LED   Light emitting diode

LISP   List processing

LOGMARS   Logistics applications of automated marking and reading symbols

LSB   Least significant bit

LSI   Large scale integration

Ltd.   Limited

MANTECH   Manufacturing Technology (program of the U.S. Air Force)

MAP   Manufacturing Automation Protocol

MCD   Machine control data

MCU   Machine control unit

MDI   Manual data input

Mfg.   Manufacturing

MFLOPS   Million floating point operations per second

Micro   Microcomputer

Mini   Minicomputer

MIPS   Million instructions per second

MIS   Management information system

MIT   Massachusetts Institute of Technology

MITI   (Japan's) Ministry of International Trade and Import

MMS   Manufacturing message specification

Modem   Modulator-demodulator

MRP   Materials requirement planning

MRP II   Manufacturing resources planning

MSB   Most significant bit

MTBF   Mean time between failures

MTR   Maximum transmission rate

mW   milliwatt

NAIT   National Association of Industrial Technology

NAMRI   North American Manufacturing Research Institution

NASA   National Aeronautics and Space Administration

NBS   National Bureau of Standards (now NIST)

NC   Numerical control

NCMS   National Center of Manufacturing Sciences

NEMA   National Electrical Manufacturers Association

NGC   Next generation controller

NIST   National Institute of Standards and Technology

NML   Neutral manufacturing language

NMTBA   National Machine Tool Builders' Association (now AMT)

NSF   National Science Foundation

NUBS   Nonuniform B-splines

NURBS   Nonuniform rational B-splines

OCR   Optical character recognition

OIR   Organization for Industrial Research

OSHA   Occupational Safety and Health Administration

OSI   Open systems interconnection

OS/2   Operating system/2

PC   Personal computer

PCB   Printed circuit board

PCI   Process capability index

PDES   Product data exchange specification

PFA   Process/production flow analysis

PFM&C   Plant-floor monitoring and control system

PIC   Production and inventory control

PLC   Programmable logic controller

PM   Phase modulation

POC   Price of conformance

PONC   Price of nonconformance

PP   Payback period

PROLOG   Programming logic

PROM   Programmable read-only memory

PTP   Point-to-point

QA   Quality assurance

QC   Quality control

QDC   Quick die change
RAM   Random access memory
RDBMS   Relational database management system
RFI   Radio frequency interference
RFID   Radio frequency identification
RIA   Robotic Industries Association
RISC   Reduced instruction set computing/computer
RJE   Remote job entry
R&D   Research and development
ROI   Return on investment
ROM   Read-only memory
SE   Simultaneous engineering
SFC   Shopfloor control (system)
SI   System International (of units)
SIC   Standard industrial classification
SME   Society of Manufacturing Engineers
SNA   System Network Architecture
S/N   Signal-to-noise (ratio)
SPC   Statistical process control
SQC   Statistical quality control
SQL   Structured query language
SRA   Stored record address
STEP   Standard for the exchange of product model data (ISO 10303)
SVGA   Super video graphics array
TCP   Transmission control protocol
TDM   Time division multiplexing

TOP   Technical and Office Protocol
TQC   Total quality control
TQM   Total quality management
TTO   Tape tryout
TTY   Teletype; tape tryout
UART   Universal asynchronous receiver/transmitter
UCL   Upper control limit
UPC   Universal product code
USART   Universal synchronous/asynchronous receiver/transmitter
UV   Ultraviolet
VAL   Victor's assembly language
VGA   Video graphics array
VHLL   Very high level language
VLSI   Very large scale integration
WAN   Wide area network
WESTEC   Western Metal and Tool Exposition and Conference
WIP   Work-in-process
WORM   Write-once-read-many
ZIP   Zoning Improvement Plan (code)
μP   Microprocessor
2D   Two-dimensional
3D   Three-dimensional
4GL   Fourth generation computer language

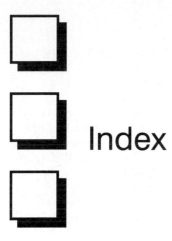

Index